Encyclopedia of Microbiology

Volume I

Encyclopedia of Microbiology
Volume I

Edited by **Lucy Phillip**

R **C**ALLISTO
REFERENCE

New York

Published by Callisto Reference,
106 Park Avenue, Suite 200,
New York, NY 10016, USA
www.callistoreference.com

Encyclopedia of Microbiology: Volume I
Edited by Lucy Phillip

International Standard Book Number: 978-1-63239-269-5 (Hardback)

Printed in the United States of America.

Contents

Preface

Derived from the Greek word 'mikros' or 'small', Microbiology is the study of organisms at the microscopic level and their interactions with humans, plants, animals and the environment. It is focused on unicellular, multicellular and a-cellular organisms as well as bacteria, viruses, protozoa and fungi. The name obviously refers to the small size of these organisms and such organisms are usually known as bacteria, viruses, archaea as well as eukaryotes. Microbiology is an area of study that is multidisciplinary and needs the skills and knowledge form various fields of science like life science, engineering and environmental science. It also has a base in sub-disciplines that include virology, mycology, parasitology, immunology, public health, vaccinology and bacteriology. Medical microbiology is also often termed as an application of microbiology and can be introduced with medical principles of immunology. Microbiologists can be found to be working in environmental protection, basic and applied research, manufacturing of food and other goods, public health, clinical settings and other areas of study and application. Projections for the future suggest that there will be a demand for trained and skilled microbiologists to fill position in all aspects of microbiology.

This book is an attempt to compile and understand new and ongoing research data in the field of microbiology. I am thankful to those who put their effort and hard work into this field. I am also grateful to my family and friends who supported me in this endeavour. I especially wish to acknowledge the efforts of my publisher and the publishing team who has worked tirelessly to make this book a success.

Editor

Influence of Sanitizers on the Lipopolysaccharide Toxicity of *Escherichia coli* Strains Cultivated in the Presence of *Zygosaccharomyces bailii*

Lerato Mogotsi,[1] Olga De Smidt,[1] Pierre Venter,[2] and Willem Groenewald[1]

[1] Unit for Applied Food Science and Biotechnology, Central University of Technology, Free State, P.O. Box 20539, Bloemfontein 9300, South Africa

[2] Fonterra Co-Operative Group Limited, Private Bag 11029, Palmerston North 4442, Dairy Farm Road 1, Palmerston North, New Zealand

Correspondence should be addressed to Willem Groenewald; whgroene@gmail.com

Academic Editor: Fernando Rodrigues

The influence of sublethal concentrations of two sanitizers, liquid iodophor and liquid hypochlorite (LH), on the growth rates and toxicity of food-borne pathogenic *Escherichia coli* strains grown in the presence of spoilage yeast *Zygosaccharomyces bailii* was assessed. When grown in combination with *Z. bailii* both *E. coli* O113 and *E. coli* O26 exhibited slower growth rates, except when *E. coli* O113 was grown in combination with *Z. bailii* at 0.2% LH. The growth rate of *Z. bailii* was not impacted by the addition of the sanitizers or by communal growth with *E. coli*. LAL and IL-6 results indicated a decrease in toxicity of pure *E. coli* cultures with comparable profiles for control and sanitizer exposed samples, although the LAL assay proved to be more sensitive. Interestingly, pure cultures of *Z. bailii* showed increased toxicity measured by LAL and decreased toxicity measured by IL-6. LAL analysis showed a decrease in toxicity of both *E. coli* strains grown in combination with *Z. bailii*, while IL-6 analysis of the mixed cultures showed an increase in toxicity. The use of LAL for toxicity determination in a mixed culture overlooks the contribution made by spoilage yeast, thus demonstrating the importance of using the appropriate method for toxicity testing in mixed microbe environments.

1. Introduction

Microorganisms associated with food spoilage and food-borne diseases pose a considerable threat to human welfare and economy, especially that of developing countries [1]. Food is a complex material, more than often nutrient rich, and can generally support a diverse number of microorganisms that interact with each other and that can attach to a variety of surfaces to form biofilm in the food processing environment [2, 3]. Microorganisms associated with food spoilage and food-borne diseases are problematic in various sectors of the food industry and support continuous investigation. *Escherichia coli*, a Gram-negative bacterium, can be associated with food-borne diseases such as hemorrhagic colitis [4]. The last few decades have seen an increase in awareness of yeasts as food spoilage agents [5] and *Zygosaccharomyces bailii* has been described as the most important of all food spoilage yeasts [6]. This yeast originates from fruit trees exude and results in the spoilage of sweetened wine during processing. In addition, *Z. bailii* may lead to explosion of canned food as a result of gas production due to vigorous alcoholic fermentation [7].

To control food spoilage organisms and food-borne pathogens food processors have relied on physical and chemical methods [2]. However, if cleaning chemicals are not properly rinsed from food processing equipment they may end up in the product in sublethal concentrations. This could influence the response of the microorganisms to residual chemicals. Previous studies focused on the effect of sanitation procedures and sublethal doses of preservatives on the toxicity of specific microorganism [8]; however, these were conducted on pure cultures. There exists a need to investigate this phenomenon in conditions that more accurately reflect the situation found in industries where microorganisms exist not as pure cultures, but in communities.

In the food processing environment, a parallel exits between the resistance of microorganism and the efficiency of sanitizers. Resistance to sanitizers and preservatives may be attributed to cellular barriers of a microorganism such as the lipopolysaccharide (LPS) layer in Gram-negative bacteria that limits diffusion of molecules into the cell [9] and exopolysaccharide (EPS) from yeast cells, viral pathogens, and fungal spores [10]. The LPS layer has been shown to interfere with many host mediation systems, leading to hypotensive shock, disseminated intravascular coagulation, and metabolic abnormalities [11, 12]. Exopolysaccharides produced by attached cells assist the colonization of other organisms to surfaces leading to biofilm formation which are usually highly resistant to antibiotics [13]. The food industry relies on viable counts for microbial testing which do not take into account toxicity contributed by the remaining debris (LPS) of dead cells or the effect that membrane adaptation could have on toxicity of living cells. The limulus amoebocyte lysate (LAL), commonly used for endotoxin determination, can detect 3 pg mL^{-1} (0.03 EU mL^{-1}) of LPS [14]. It does not detect other pyrogenic molecules, and numerous substances can interfere with the assay, for instance, the presence of inhibitory proteins in plasma [15]. The IL-6 ELISA method is used to determine the concentration of porcine specific IL-6 present within a sample. This assay can be performed on plasma, serum, and cell culture supernates. Thus, the aim of this study was to assess the effect of sublethal concentrations of sanitizers on communal growth of E. coli O113, E. coli O26, and Z. bailii and to monitor changes in toxicity as a result of sublethal concentrations of sanitizer exposure on pure and mixed cultures of E. coli and Z. bailii by means of LAL and IL-6 ELISA method.

2. Materials and Methods

2.1. Strains Used. Escherichia coli O113 (smooth strain) and Escherichia coli O26 (rough strain) were isolated from food samples and a spoilage yeast strain, while Zygosaccharomyces bailii Y-1535 was obtained from the University of Free State culture collection. The rough and smooth status of the E. coli were confirmed by the salt aggregation test method, which entailed the "salting out" of rough strains with 4 M ammonium sulphate [16].

2.2. Determining Sublethal Doses of Sanitizers. Liquid iodophor (LI) and liquid hypochlorite (LH) were purchased from a leading supplier of sanitizers for industrial use. A preliminary study was performed to determine sublethal concentrations by using the use-dilution method [17]. A series of dilutions of the sanitizers from 0.05%–0.6% LI and 0.075%–0.6% LH were inoculated with a loopful of either bacteria or yeast. The tubes were shaken at 200 rpm at 30°C and optical density was measured at different time intervals.

Inadequate mixing and aeration in the tubes resulted in unusable data. Volumes were changed to 100 mL flasks and standardized inocula of initial OD$_{690 nm}$ values of 0.1 were used. Flasks were incubated at 30°C with shaking. Thus, the selected working concentrations in this study were as follows: 0.05% LH, 0.2% LH, and 0.075% LI. Two LH concentrations were used, in order to allow both E. coli strains to grow. E. coli O26 (rough strain) grew in the presence of 0.05% LH as it was more susceptible to 0.2% LH, while the growth of E. coli O113 (smooth) was less inhibited at this concentration.

2.3. Growth Conditions. At midexponential phase preinocula of Escherichia coli O113, E. coli O26 and Z. bailii Y-1535 were inoculated in 200 mL yeast malt (YM) media (Biolab, Midrand, South Africa) and incubated at 30°C with shaking at 200 rpm. Strains were cultivated as pure cultures and in combinations of E. coli and Z. bailii. Growth was monitored by optical density (OD$_{690 nm}$) and viable plate counts on Violet Red Bile (Biolab) and YM media with 5% tartaric acid for E. coli and Z. bailii, respectively. Maximum specific growth rates (μ_{max}) were calculated by fitting a straight line to the data points that appeared to best represent the exponential growth phase using (1), where N_1 and N_2 are colony forming units at time 1 (t_1) and time 2 (t_2). Differences in growth of communal cultures from their corresponding control at two selected time intervals were calculated using (2) where [T_{C_1}] represents concentration (%) of pure culture control (with sanitizer) at time 1 and [T_{G_1}] represents the concentration (%) of communal growth (with sanitizer) at time 1. A value of zero was considered to indicate no difference in growth when the strains were grown in combination. When a difference between the two concentrations was below zero, the strain indicated better growth in combination and the opposite when the two concentrations were greater than zero:

$$\mathrm{Ln}\frac{N_2/N_1}{t_2 - t_1},\qquad (1)$$

$$\left[T_{C_1}\right] - \left[T_{G_1}\right] = 0.\qquad (2)$$

For toxicity analysis 1 mL of culture was sampled during the exponential growth phase (10 h for E. coli and 12 h for Z. bailii). The influence of the selected sanitizers was evaluated by growing the strains under the same conditions as described for the controls except for those with added sanitizers at predetermined concentrations of 0.2% or 0.05% liquid hypochlorite (LH) and 0.075% liquid iodophor (LI).

2.4. Limulus Amoebocyte Lysate Assay (LAL). LPS toxicity was determined using the QCL-1000 chromogenic LAL assay (Lonza, Bloemfontein, South Africa) for all samples for time zero and at the late exponential phase of growth (10–12 h). This quantitative test for Gram-negative bacterial endotoxin was performed by the microplate method as prescribed by the manufacturer. The test kits included E. coli endotoxin standards with approximately 50–648 endotoxin units (EU) lyophilized endotoxins. Toxicity was calculated from a standard curve and change in toxicity was expressed as ΔEU$\cdot$ mL$^{-1}\cdot$OD$_{690 nm}^{-1}$ which was converted to

Influence of Sanitizers on the Lipopolysaccharide Toxicity of Escherichia coli Strains Cultivated in the Presence
of Zygosaccharomyces bailii

3

TABLE 1: Growth parameters of pure and mixed cultures of *E. coli* O113, *E. coli* O26, and *Z. bailii* Y-1535 grown in YM medium in the presence of the following sanitizer concentrations: 0%, 0.05% LH, 0.2% LH, and 0.075% LI.

			E. coli O113	*E. coli* O26	*Z. bailii* Y-1535	O113 (ZB + O113)	ZB (ZB + O113)	O26 (ZB + O26)	ZB (ZB + O26)
		μ_{max} (h^{-1})	0.496	0.636	0.333	0.103	0.300	0.36	0.294
		r^2	0.99	0.99	0.99	0.99	0.99	0.99	0.96
LI	0.075%	μ_{max} (h^{-1})	0.488	0.629	0.306	0.073	0.259	0.099	0.249
		r^2	0.99	0.98	0.98	0.99	0.99	0.99	0.96
LH	0.2%	μ_{max} (h^{-1})	0.176	—	0.381	0.23	0.352		
		r^2	0.99	—	0.98	0.76	0.99		
LH	0.05%	μ_{max} (h^{-1})	—	0.589	0.259			0.0032	0.355
		r^2	—	0.99	0.99			0.71	0.98

ZB: *Z. bailii* Y-1535.
(ZB + O113): *Z. bailii* cultivated together with *E. coli* O113.
(ZB + O26): *Z. bailii* cultivated together with *E. coli* O26.
μ_{max}: maximum specific growth rate.

Δpg$\cdot$ mL$^{-1}\cdot$ OD$_{690\,nm^{-1}}$ (1 EU = 100 pg endotoxin). Therefore, a positive value indicates an increase and a negative value a decrease in toxicity, while the Δtoxicity $\cdot$ OD$_{690\,nm^{-1}}$ value represents the magnitude of change. The absorbance of released p-nitroaniline from the synthetic substrate resulting in a yellow colour was read at 410 nm with a microplate reader (Bio-Rad, Johannesburg, South Africa). All equipment used was pyrogen free.

2.5. Interleukin-6 (IL-6). LPS/EPS toxicity on all samples from time zero and the late exponential phase of growth (10–12 h) was determined using the in vitro enzyme-linked immunosorbent assay for the quantitative measurement of porcine IL-6 (RayBiotech, Pretoria, South Africa). Blood was randomly collected from 10 pigs, pooled and diluted to 1 : 3 with sterile saline, and allowed to separate for 12 h at 37°C in a CO_2 incubator. The supernatant was used for IL-6 determination measuring bound tetramethylbenzidine at 540 nm. The toxicity values were calculated from a standard curve. The change in toxicity values was expressed as Δpg$\cdot$ mL$^{-1}\cdot$ OD$_{690\,nm^{-1}}$. Pyrogen free equipment, reagents, and consumables were used and all experiments were performed as independent duplicates.

3. Results and Discussion

3.1. Influence of Sanitizers on Maximum Specific Growth Rate of Pure and Mixed Cultures. The addition of sublethal concentrations of 0.075% LI resulted in no change in growth rates for *E. coli* O113 and O26 pure cultures compared to their unchallenged controls. However, the growth rate of mixed cultures of *E. coli* O113 and *Z. bailii and E. coli* O26 and *Z. bailii* showed a decrease with the addition of 0.075% LI as compared to the control which can be attributed to the influence of communal growth (Table 1).

Growth rate of *E. coli* O113 pure culture was noticeably impaired by exposure to 0.2% LH. However, the addition of a lower LH concentration (0.05%) did not influence the growth rate of *E. coli* O113 pure culture.

The growth rate of *E. coli* O26 pure culture was influenced by the addition of 0.05%. However, the addition of liquid iodophor resulted in no change in growth rate. Communal growth with *Z. bailii*, however, resulted in markedly slower growth rates. No negative impact was observed for *Z. bailii* growth rates by either communal growth with *E. coli* strains or the exposure to sublethal concentration of sanitizers. Interaction is considered to have taken place when the growth rate of the target microorganism in a mixed culture is decreased by 10% [11]. Generally, communal growth had a marked impact on the growth rates of the *E. coli* strains compared to their controls (pure cultures), where growth rates decreased by 79.2% and 78.6% for O113 and O26, respectively. The growth rate of *Z. bailii* was also influenced when grown in combination with O113 (9.9%) and O26 (11.7%), but clearly this interaction had the greater effect on the bacterial strains. These results are not entirely unexpected since the production of organic acids [18] or ethanol by yeasts in restraining the growth of some microorganisms is not uncommon [19]. However, it was necessary to determine changes in toxicity in mixed cultures in order to investigate the survival adaptation of the bacterial strains in terms of LPS. According to Giotis et al. [20] extended exposure to chemicals in the media might have an effect on growth, which results in the organisms competing with each other for nutrients, leading to poor growth of target organisms.

3.2. Influence of Sanitizers on LPS and EPS Toxicity. For both *E. coli* strains (O113 and O26) growth without sanitizer or in the presence of sanitizers resulted in no increase in toxicity (Figure 1).

When *E. coli* O113 was grown without sanitizers (control) a pronounced change in toxicity was observed as a decrease. The same trend was evident in the presence of both sanitizers indicating that the addition of sanitizers to the medium did not influence the toxicity of O113 in pure culture. Growth of *E. coli* O26 also resulted in a decrease in toxicity with the change over time. Exposure to sanitizers still resulted in a decrease in toxicity, in both sanitizers. However, there is a difference in

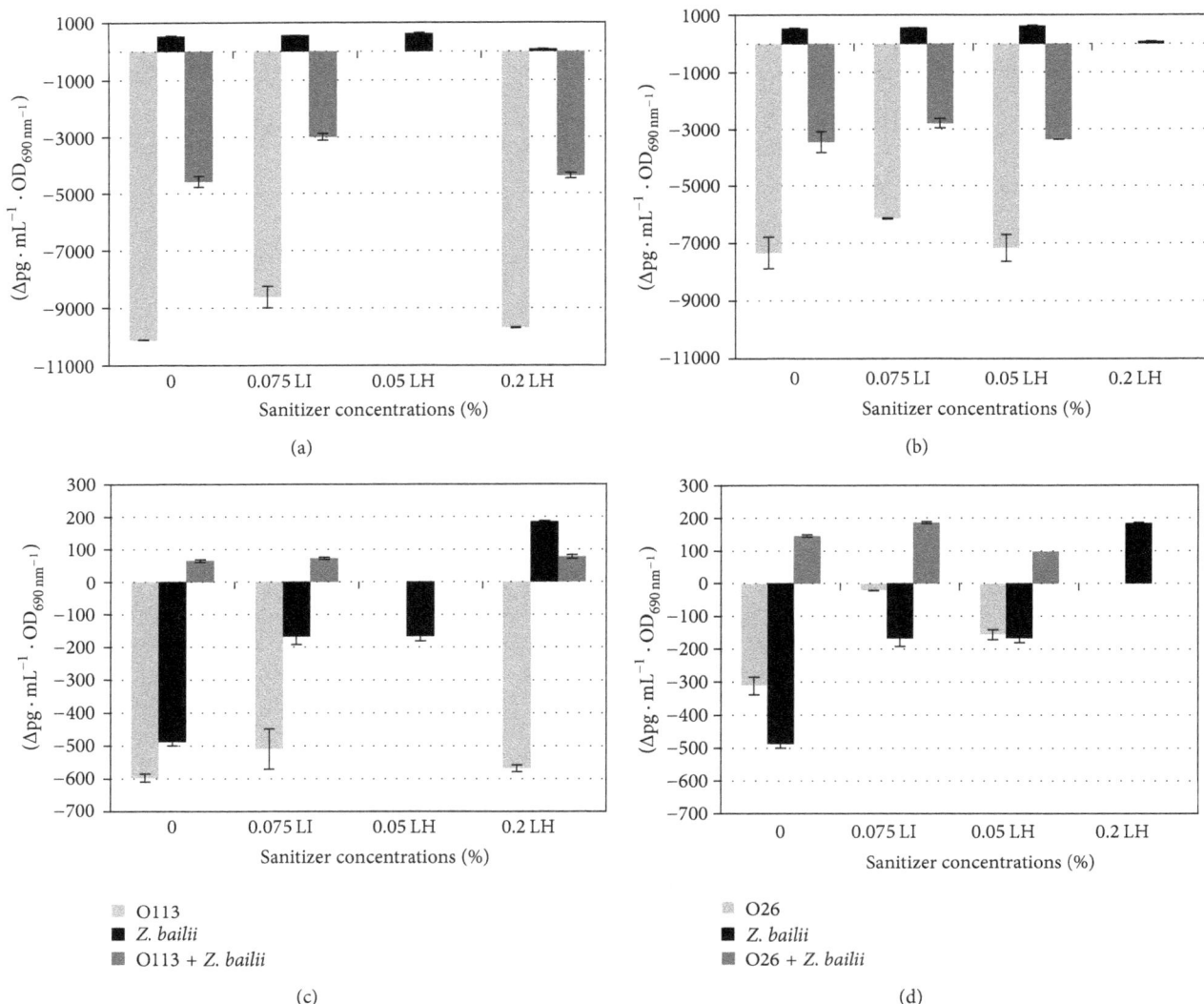

FIGURE 1: Changes in EPS/LPS toxicity of pure and mixed cultures of *E. coli* and *Z. bailii* following exposure to different concentrations of liquid iodophor and liquid hypochlorite.

the magnitude of change in the toxicity of O26 in the presence of 0.075% LI (Figure 1(d)). This can be explained by the different modes of action displayed by the sanitizers. Liquid hypochlorite is a highly active oxidizing agent and thereby destroys the cellular activity of proteins. However, their penetration is maximal when they are in a unionized state [21]. Liquid iodophor rapidly penetrates into microorganisms and attacks proteins, nucleotides, and fatty acids; by attacking fatty acids, it already interferes with the lipid A structure measured by LAL method. *Zygosaccharomyces bailii* grown in pure cultures revealed higher toxicity levels when compared to both *E. coli* strains using LAL assay. Limulus amoebocyte lysate interacts with the lipid component of Gram-negative LPS. Although the toxicity values are considerably lower than that detected for the Gram-negative bacteria, this is the first evidence of toxicity detected from eukaryal EPS using the LAL assay. It is tempting to speculate on the similarities that might exist between the targeted sections of the LPS and EPS

as has previously been described for Gram-positive EPS [22]. Notably the communal growth of the *E. coli* strains and *Z. bailii* produced different toxicity profiles when using the IL-6 method. In both cases where *E. coli* O113 (Figure 1(c)) or O26 (Figure 1(d)) were cultured in the presence of *Z. bailii* the toxicity of the mixture increased. This occurrence was evident in the presence of both sanitizers indicating that it may be attributed to communal growth. The largest increase in toxicity was brought about by the communal growth of *E. coli* O26 and *Z. bailii* in the presence of sublethal concentration of sanitizer LI. Limulus amoebocyte lysate method on the other hand detected a decrease in toxicity for both O113 (Figure 1(a)) and O26 (Figure 1(b)). The increase in toxicity in communal growth could be a result of cultivation conditions. Communal growth might have influenced the liberation of 3-hydroxy fatty acids resulting in a significant change in saccharide composition of EPS, which affects immune stimulation [14, 23].

Influence of Sanitizers on the Lipopolysaccharide Toxicity of Escherichia coli Strains Cultivated in the Presence of Zygosaccharomyces bailii

5

Interleukin-6 compared to the LAL assay may be influenced by several parameters which have to be standardized. For example, the whole procedure of LAL assay takes 1 to 3 h, while it takes 4 to 5 h for the IL-6 rendering it time consuming and laborious. Differences between the LAL and IL-6 methods may also depend on the samples used, for example, communal growth samples, showing low values of LAL assay in spite of high toxicity values in IL-6 because the porcine IL-6 is able to detect other pyrogens (e.g., EPS in yeast). In contrast the LAL assay is highly sensitive to endotoxin activity [24].

4. Conclusions

In light of these results care should be taken in the food industry where contamination with sanitizers is a risk factor. Incorrect dosage can increase growth of spoiler/pathogenic yeasts. To eliminate such contaminants, food processing equipment needs to be thoroughly rinsed. Factors responsible for the different responses of *E. coli* and *Z. bailii* to the presence of sanitizers, including maximum specific growth rates and toxicity profiles, should be further investigated.

Escherichia coli and *Z. bailii* mixed cultures showed increase in toxicity using IL-6 analysis indicating that the use of LAL for the detection of toxicity in a mixed microorganism environment overlooks the contribution made by spoilage yeast. The importance of using the appropriate method for toxicity testing in mixed microbe environments was clearly demonstrated. Although the LAL assay is regarded as sensitive, reproducible, and simplistic, it did not give a representative account of yeast EPS toxicity in mixed cultures, as is relevant to the food processing environment.

Conflict of Interests

The authors declare that there is no conflict of interests regarding the publication of this paper.

Acknowledgments

The authors gratefully acknowledge the National Research Foundation (RSA) and the Central University of Technology for their financial support.

References

[1] M. Jevšnik, V. Hlebec, and P. Raspor, "Consumers' awareness of food safety from shopping to eating," *Food Control*, vol. 19, no. 8, pp. 737–745, 2008.

[2] S. K. Hood and E. A. Zottola, "Isolation and identification of adherent gram-negative microorganisms from four meat-processing facilities," *Journal of Food Protection*, vol. 60, no. 9, pp. 1135–1138, 1997.

[3] I. W. Sutherland, "Biofilm exopolysaccharides: a strong and sticky framework," *Microbiology*, vol. 147, no. 1, pp. 3–9, 2001.

[4] S. Viazis and F. Diez-Gonzalez, "Enterohemorrhagic Escherichia coli. The twentieth century's emerging foodborne pathogen: a review," *Advances in Agronomy*, vol. 111, pp. 1–50, 2011.

[5] V. Loureiro and A. Querol, "The prevalence and control of spoilage yeasts in foods and beverages," *Trends in Food Science and Technology*, vol. 10, no. 11, pp. 356–365, 1999.

[6] P. Martorell, M. Stratford, H. Steels, M. T. Fernández-Espinar, and A. Querol, "Physiological characterization of spoilage strains of *Zygosaccharomyces bailii* and *Zygosaccharomyces rouxii* isolated from high sugar environments," *International Journal of Food Microbiology*, vol. 114, no. 2, pp. 234–242, 2007.

[7] F. Rodrigues, M. Côrte-Real, C. Leao, J. P. Van Dijken, and J. T. Pronk, "Oxygen requirements of the food spoilage yeast *Zygosaccharomyces bailii* in synthetic and complex media," *Applied and Environmental Microbiology*, vol. 67, no. 5, pp. 2123–2128, 2001.

[8] T. Gutsmann, A. B. Schromm, and K. Brandenburg, "The physicochemistry of endotoxins in relation to bioactivity," *International Journal of Medical Microbiology*, vol. 297, no. 5, pp. 341–352, 2007.

[9] P. M. Davidson and M. A. Harrison, "Resistance and adaptation to food antimicrobials, sanitizers, and other process controls," *Food Technology*, vol. 56, no. 11, pp. 69–78, 2002.

[10] M. Daneshian, A. Wendel, T. Hartung, and S. von Aulock, "High sensitivity pyrogen testing in water and dialysis solutions," *Journal of Immunological Methods*, vol. 336, no. 1, pp. 64–70, 2008.

[11] D. C. Morrison and R. J. Ulevitch, "The effects of bacterial endotoxins on host mediation systems: a review," *American Journal of Pathology*, vol. 93, no. 2, pp. 526–617, 1978.

[12] D. C. Morrison and J. L. Ryan, "Bacterial endotoxins and host immune responses," *Advances in Immunology*, vol. 28, no. C, pp. 293–450, 1980.

[13] C. G. Kumar and S. K. Anand, "Significance of microbial biofilms in food industry: a review," *International Journal of Food Microbiology*, vol. 42, no. 1-2, pp. 9–27, 1998.

[14] N. Binding, S. Jaschinski, S. Werlich, S. Bletz, and U. Witting, "Quantification of bacterial lipopolysaccharides (endotoxin) by GC-MS determination of 3-hydroxy fatty acids," *Journal of Environmental Monitoring*, vol. 6, no. 1, pp. 65–70, 2004.

[15] S. Laitinen, J. Kangas, K. Husman, and P. Susitaival, "Evaluation of exposure to airborne bacterial endotoxins and peptidoglycans in selected work environments," *Annals of Agricultural and Environmental Medicine*, vol. 8, no. 2, pp. 213–219, 2001.

[16] M. L. Sorongon, R. A. Bloodgood, and R. P. Burchard, "Hydrophobicity, adhesion, and surface-exposed proteins of gliding bacteria," *Applied and Environmental Microbiology*, vol. 57, no. 3, pp. 3193–3199, 1991.

[17] J. C. Du Preez, *Microbial Growth and Death Practical Course*, Department of Microbiology and Biochemistry University of the Free State, Bloemfontein, South Africa, 2004.

[18] P. K. Malakar, D. E. Martens, M. H. Zwietering, C. Béal, and K. Van 'T Riet, "Modelling the interactions between Lactobacillus curvatus and Enterobacter cloacae: II. Mixed cultures and shelf life predictions," *International Journal of Food Microbiology*, vol. 51, no. 1, pp. 67–79, 1999.

[19] F. Liu, Y.-Z. Guo, and Y.-F. Li, "Interactions of microorganisms during natural spoilage of pork at 5°C," *Journal of Food Engineering*, vol. 72, no. 1, pp. 24–29, 2006.

[20] E. S. Giotis, I. S. Blair, and D. A. McDowell, "Morphological changes in Listeria monocytogenes subjected to sublethal alkaline stress," *International Journal of Food Microbiology*, vol. 120, no. 3, pp. 250–258, 2007.

[21] G. Mcdonnell and A. D. Russell, "Antiseptics and disinfectants: activity, action, and resistance," *Clinical Microbiology Reviews*, vol. 12, no. 1, pp. 147–179, 1999.

[22] M. Abraham, P. Venter, J. Lues, I. Ivanov, and O. De Smidt, "The exopolysaccharide (EPS) ultra structure of Staphylococcus aureus: changes occurring in EPS resulting from exposure to physical and chemical food preservation practises in South Africa," *Annals of Microbiology*, vol. 59, no. 3, pp. 499–503, 2009.

[23] V. Liebers, H. Stubel, M. Düser, T. Brüning, and M. Raulf-Heimsoth, "Standardization of whole blood assay for determination of pyrogenic activity in organic dust samples," *International Journal of Hygiene and Environmental Health*, vol. 212, no. 5, pp. 547–556, 2009.

[24] Z. Cadieux, *Biosynthesis of nucleotide sugar monomers for exopolysaccharide production in Myxococcus Xanthus [Master of Science in Biological Science]*, 2002.

2

Adaptation of Mycoplasmas to Antimicrobial Agents: *Acholeplasma laidlawii* Extracellular Vesicles Mediate the Export of Ciprofloxacin and a Mutant Gene Related to the Antibiotic Target

Elena S. Medvedeva,[1] **Natalia B. Baranova,**[1] **Alexey A. Mouzykantov,**[1]
Tatiana Yu. Grigorieva,[1] **Marina N. Davydova,**[1] **Maxim V. Trushin,**[1,2]
Olga A. Chernova,[1,2] **and Vladislav M. Chernov**[1,2]

[1] *Kazan Institute of Biochemistry and Biophysics, Kazan Scientific Centre of Russian Academy of Sciences, Kazan 420111, Russia*
[2] *Kazan Federal University, Kazan 420008, Russia*

Correspondence should be addressed to Vladislav M. Chernov; chernov@mail.knc.ru

Academic Editors: A. El-Shibiny and D. Zhou

This study demonstrated that extracellular membrane vesicles are involved with the development of resistance to fluoroquinolones by mycoplasmas (class Mollicutes). This study assessed the differences in susceptibility to ciprofloxacin among strains of *Acholeplasma laidlawii* PG8. The mechanisms of mycoplasma resistance to antibiotics may be associated with a mutation in a gene related to the target of quinolones, which could modulate the vesiculation level. *A. laidlawii* extracellular vesicles mediated the export of the nucleotide sequences of the antibiotic target gene as well as the traffic of ciprofloxacin. These results may facilitate the development of effective approaches to control mycoplasma infections, as well as the contamination of cell cultures and vaccine preparations.

1. Introduction

The suppression of mycoplasmas that infect humans, animals, and plants, as well as contaminating cell cultures and vaccines, is a serious problem [1–3]. This is because mycoplasmas rapidly acquire resistance to antibiotics. However, antibiotic therapy remains the primary tool used for the treatment of mycoplasma infections and the decontamination of cell cultures. The most widely used fluoroquinolones, which are synthetic antibacterials, are enrofloxacin, sparfloxacin, ofloxacin, ciprofloxacin, and levofloxacin [3–5]. The mechanisms that facilitate the rapid development of resistance to fluoroquinolones in mycoplasmas remain unclear [6, 7]. The development of resistance to quinolones is associated with mutations in genes that encode antibiotic-targeted proteins, and a limitation is that the antibiotics used to treat microbial cells are not considered to be effective against mycoplasmas

[4, 8]. Thus, it would be beneficial to elucidate the mechanisms that facilitate the rapid development of antibiotic resistance to allow the treatment of mycoplasma infections, which appear to be associated with the adaptation of mycoplasmas to stress conditions [9, 10]. The successful implementation of genome projects for a number of mycoplasmas has opened up the possibility of using postgenomic technologies to study their antibiotic resistance processes [11, 12].

A unique species of mycoplasma with adaptive properties is *Acholeplasma laidlawii*, which is a causative agent of phytomycoplasmoses and the main contaminant of cell cultures and vaccines [2, 13, 14]. In our study [9, 10], transcriptome-proteome analysis and nanoscopy were identified as the stress-reactive proteins and genes of *Acholeplasma laidlawii*, which showed that the adaptation of this mycoplasma to stressful factors was related to the production of extracellular vesicles (EVs). The EVs of bacteria are spherical nanostructures surrounded by a membrane (20–200 nm in diameter), which mediate the traffic of a wide variety of compounds that participate in signaling, intercellular interactions, and pathogenesis [15, 16]. Recent studies suggest the possible involvement of EVs in the development of resistance to antibiotics in bacteria [17, 18]. However, there are no previous studies of the roles of EVs in antibiotic resistance in mycoplasmas. Thus, the present study demonstrated the participation of *A. laidlawii* EVs in the development of resistance to fluoroquinolones (ciprofloxacin).

2. Materials and Methods

2.1. Bacterial Strain and Plasmids. *Acholeplasma laidlawii* PG8 (from the N.F. Gamalei Research Institute of Epidemiology and Microbiology, Moscow, Russia), clinical isolate of *Staphylococcus aureus*, *E. coli* NovaBlue strain, and plasmid vector pGEM-T Easy Vector Systems (Promega) were used in this work.

2.2. Cultivation of A. laidlawii. *A. laidlawii* PG8 cells were cultivated for 1 day at 37°C in Edward's medium (tryptose, 2% (w/v); NaCl, 0.5% (w/v); KCl, 0.13% (w/v); Tris base, 0.3% (w/v); serum of horse blood, 10% (w/v); fresh yeast extract, 5% (w/v); glucose solution, 1% (w/v); benzylpenicillin (500,000 IE mL^{-1}), 0.2% (w/v)) to obtain the control cells. To obtain ciprofloxacin-resistant clones, a mycoplasma culture was grown from a single colony of the laboratory *A. laidlawii* PG8 strain and sequentially inoculated in a broth medium that contained increasing concentrations of the antibiotic. To determine the minimal inhibitory concentration (MIC) of cells, *A. laidlawii* PG8R was subcultured into Edward's medium containing an appropriate concentration of ciprofloxacin. Resistant cultures of *A. laidlawii* PG8R were plated onto solid agar that contained appropriate concentrations of ciprofloxacin and individual colonies were analyzed.

2.3. Determination of the MIC. To determine the MIC, the original mycoplasma cell culture was passaged in broth medium with different concentrations of ciprofloxacin: 0.1, 0.2, 0.3, 0.4, 0.5, 0.6, 0.7, 0.8, 0.9, 1.0, and 1.5 μg mL^{-1}. The MIC values were determined based on three independent replicates.

2.4. Transmission Electron Microscopy and Atomic Force Microscopy. Transmission electron microscopy was performed according to the method of Cole [19]. Samples were fixed with 2.5% glutaraldehyde ("Fluka," Germany) in 0.1 M phosphate-buffered saline (PBS) (pH 7.2) for 2 h. The fixed samples were then dehydrated using an acetone, ethanol, and propylene series, before postfixing in 0.1% OsO$_4$ with 25 mg mL^{-1} of saccharose. After treatment with epoxy resin ("Serva," Switzerland), ultrathin sections were cut using an LKB-III ultramicrotome (Sweden), which were stained with uranyl acetate for 10 min and lead citrate for 10 min. The stained samples were examined using a JEM-1200EX transmission electron microscope ("Joel," Japan).

To prepare samples for atomic force microscopy (AFM) analysis, 1 mL aliquots of the *A. laidlawii* PG8 cells and EVs were centrifuged at 12000 rpm for 20 min at room temperature. The pellets were resuspended in 1 mL of PBS × 1 (pH 7.2). The cells were centrifuged once more at 12000 rpm for 15 min at room temperature and repeatedly resuspended in 0.5 mL of the same buffer. The prepared cells were placed onto mica (Advanced Technologies Center, Moscow, Russia) where the upper layer was removed. The cells were air-dried and washed twice with redistilled water. The samples were air-dried after each wash.

AFM imaging was performed using a Solver P47H atomic force microscope (NT-MDT, Moscow, Russia), which operated in the tapping mode with fpN11S cantilevers ($r \leq 10$ nm, Advanced Technologies Center, Moscow, Russia). The height, Mag (signal from lock-in amplifier), RMS (signal from RMS detector), and phase (signal from the phase detector) were set using the Nova 1.0.26 RC1 software (NT-MDT). The scan rate was 1 Hz and the image resolution was 512 × 512 pixels. Numerical data were expressed as the mean ± SE.

2.5. Accumulation of Ciprofloxacin. The presence of antibiotics in the EVs was determined using a fluorimetric method [20, 21]. The fluorescence was measured with a Fluorolog 3 spectrofluorometer (Horiba Jobin Yvon SAS, France) at an excitation wavelength of 282 nm and an emission wavelength of 442 nm. A blank sample that contained an equivalent amount of cells without the addition of ciprofloxacin was used as a control. The results were expressed as nanograms of ciprofloxacin per milligram of protein. All of the experiments were performed at least three times to ensure reproducibility. The mean and standard error were calculated.

2.6. Antimicrobial Activity. The Kirby-Bauer disc diffusion method with a ciprofloxacin-sensitive test strain of *Staphylococcus aureus* was used to evaluate the bacteriostatic effects of *A. laidlawii* EVs [22].

2.7. Isolation of EVs. EVs were isolated from *A. laidlawii* cultures (logarithmic growth phase) according to the method of Kolling and Matthews, with some modifications [9, 14, 23]. The cells were precipitated by centrifugation at 5000 g for 20 min. The supernatant was filtered through a 0.10 μm filter (Sartorius Minisart, France), and the filtrate was concentrated 20-fold by ultrafiltration (Vivacell 100, 100000 MWCO, "Sartorius Stedim Biotech GmbH", Germany). Vesicles were collected by ultracentrifugation (100000 g, 1 h, 8°C) with a MLA-80 rotor (Beckman Coulter Optima MAX-E), twice washed, resuspended in buffer (50 mM Tris-HCl, pH = 7, 4; 150 mM NaCl; 2 mM MgCl$_2$) and filtered through a sterile filter (Sartorius Minisart, France) with a pore size of 200 nm.

2.8. Isolation of DNA from Cells and EVs. DNA was isolated from mycoplasma cells according to the method of Maniatis [24]. DNA was isolated from EVs using a commercial DNA-express kit ("Litekh," Moscow). Before extracting the nucleic acids, the EV samples were treated with DNAse I and RNase (at 37°C for 30 min).

2.9. Sequencing. The PCR primers were constructed by NSF Litekh (Moscow, Russia) using the nucleotide sequences

Adaptation of Mycoplasmas to Antimicrobial Agents: Acholeplasma laidlawii Extracellular Vesicles Mediate the Export
of Ciprofloxacin and a Mutant Gene Related to the Antibiotic Target

9

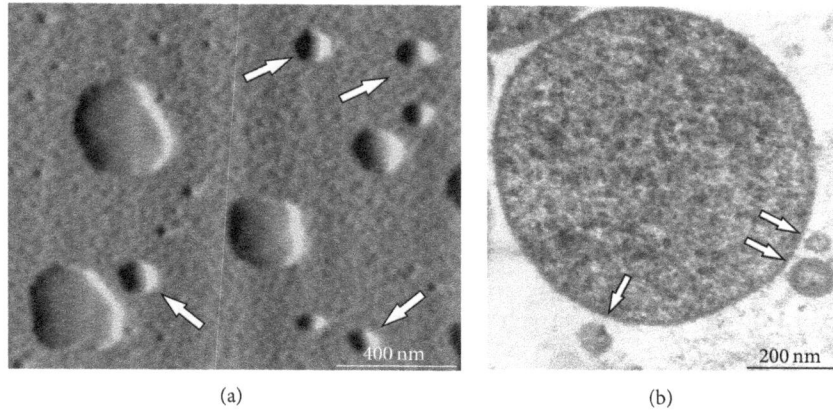

FIGURE 1: Atomic force microscopy (a) and transmission electron microscopy (b) images of *Acholeplasma laidlawii* PG8 cells and extracellular vesicles (indicated by arrows).

of *A. laidlawii* PG8A genes (GenBank accession number NC_010163): *ftsZ* (Ala1F 5′-ggttttttggatttaacgatg-3′ Ala1R 5′-gcttccgcctcttttattt-3′); spacer 16S-23S of ribosome operon (A16LF 5′-ggaggaaggtggggatgacgtcaa-3′ A23LR 5′-ccttag-gagatggtcctcctatcttcaaac-3′); and *parC* (GenBank accession number NC_010163) (AqF15: 5′-ata cgc aat ggg aca aat g-3′; AqR15: 5′-ggt tct tgt tcc tca tca tca-3′). PCR was performed in the following regime: for primers Ala1, 95°C, 3 min (95°C, 20 sec; 52°C, 20 sec; 72°C, 20 sec) (30 cycles); 72°C, 10 min. For primers A23LR, 95°C, 3 min (95°C, 5 sec; 63°C, 5 sec; 72°C, 20 sec) (30 cycles); 72°C, 5 min. For primers Aq15, 95°C, 3 min (95°C, 5 sec; 63°C, 5 sec; 72°C, 5 sec) (35 cycles); 72°C, 5 min.

DNA sequencing was performed using BigDye Terminator v3.1 Cycle Sequencing kits ("Applied Biosystems," USA) and a 3130 Genetic Analyzer ("Applied Biosystems," USA). The nucleotide sequences were analyzed using the Sequencing Analysis 5.3.1 program ("Applied Biosystems," USA) and the NCBI (National Center for Biotechnology Information, http://blast.ncbi.nlm.nih.gov/Blast.cgi) database.

3. Results and Discussion

To determine the role of membrane EVs in the development of resistance to fluoroquinolones in *A. laidlawii*, this study used mycoplasma strains with different levels of susceptibility to ciprofloxacin, that is, PG8 (MIC 0.5 μg/mL) and PG8R (MIC 1 μg/mL). *A. laidlawii* strain PG8R was obtained by stepwise selection from *A. laidlawii* PG8. The analysis of micrographs obtained using variants of transmissive and probe microscopy showed that the mycoplasma cells secreted EVs in normal conditions (Figures 1 and 2(a)). The treatment of the mycoplasma culture with ciprofloxacin elicited a significant increase in the level of vesiculation (sixfold) (Figure 2(b)). The level of vesiculation in *A. laidlawii* PG8R was significantly higher than that in *A. laidlawii* PG8. Recently, similar results were obtained with *Pseudomonas aeruginosa* [25] in another study, which also showed that EVs were involved with the traffic of antibiotics, because the EVs produced by the bacterial cells contained gentamicin [26].

The present study showed that after *A. laidlawii* was cultured in a medium that contained ciprofloxacin, the EVs produced by the mycoplasma contained the antibiotic (0.66 ± 0.15 ng/mg of protein). The EVs that contained ciprofloxacin also exhibited bacteriostatic effects against a clinical isolate of ciprofloxacin-sensitive *S. aureus* (Table 1). These results suggest that the EVs of *A. laidlawii* may be involved with the export of ciprofloxacin and possibly with the mechanism of mycoplasma adaptation to antibiotics.

The main mechanisms that facilitate the development of resistance to quinolones in bacteria are considered to be associated with mutations in specific loci, that is, the quinolone-resistance-determining regions (QRDRs) of genes that encode antibiotic-targeted proteins [7, 27, 28]. The most significant locus is the QRDR in *parC* of DNA topoisomerase IV [7, 29]. To determine whether this locus was involved with the increased resistance to ciprofloxacin in *A. laidlawii* PG8R, the *parC* QRDRs of the mycoplasma strains were amplified, cloned, and sequenced. The DNA was extracted from mycoplasmas cultivated in broth culture, which were obtained from a single colony grown on solid medium. The DNA sequences isolated from the cells of *A. laidlawii* PG8 and *A. laidlawii* PG8R were not identical (Figure 3(a)). The nucleotide sequence of *parC* QRDR from *A. laidlawii* PG8R contained a C → T transition at position 272, which caused a serine to leucine (Ser (91) Leu) replacement in the amino acid sequence of the antibiotic-targeted protein (Figure 3(b)). Previously, this mutation was found in a strain of *A. laidlawii* PG8B, which was characterized by its elevated resistance to quinolones (6–10 μg/mL) [27]. Thus, the nucleotide sequence of the QRDR in the gene encoding the C subunit of DNA topoisomerase IV in *A. laidlawii* PG8R contained a sense mutation.

The development of resistance to quinolones in bacterial populations associated with point mutations in the QRDR of *parC* should lead to the rapid spread of the corresponding nucleotide sequences in bacterial communities. The main means of spreading antibiotic resistance genes in bacteria is horizontal transfer [30]. Recently, it was shown that the EVs

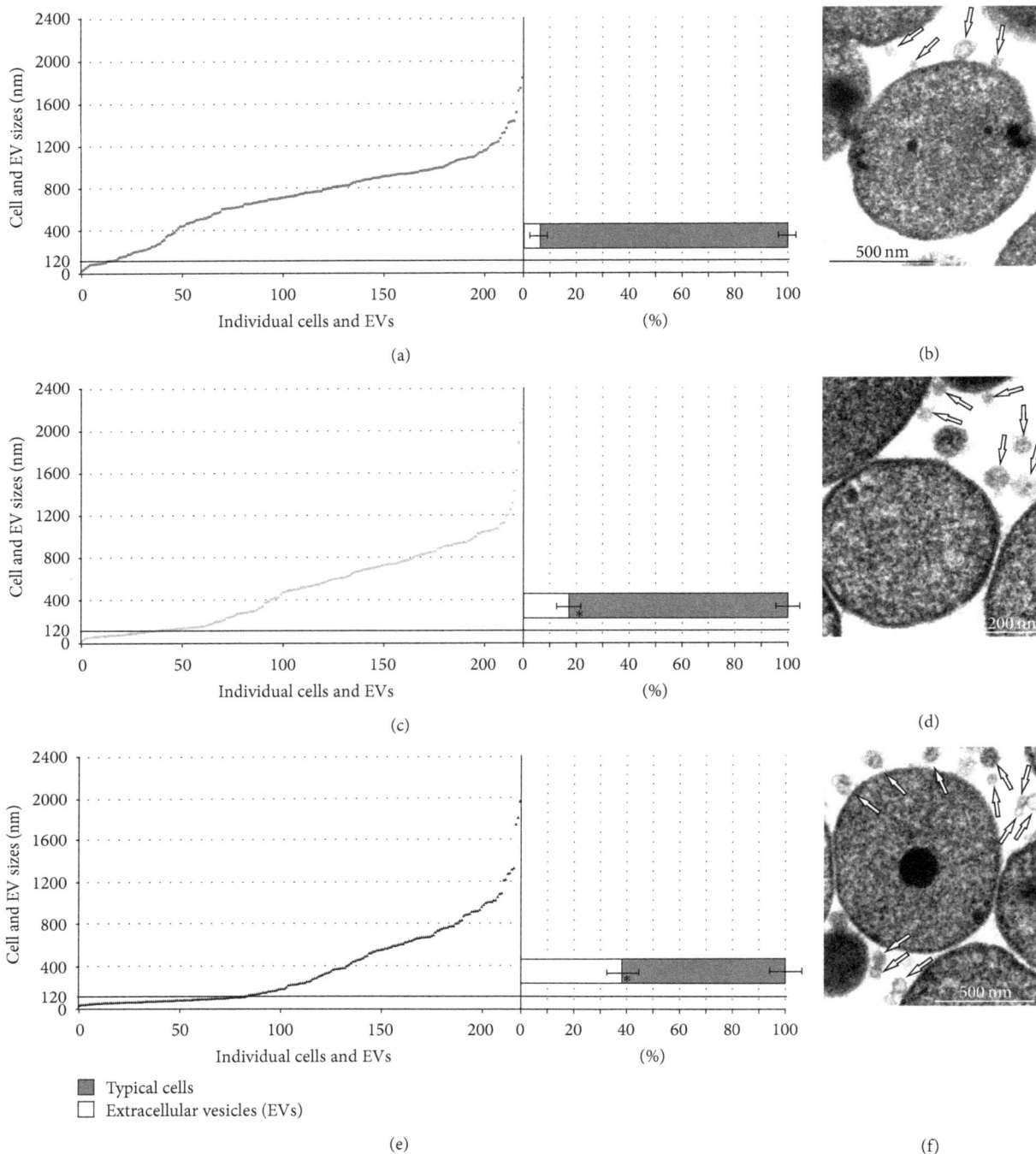

FIGURE 2: Relationships between typical cells and extracellular vesicles (EVs) (a, c, e) and transmission electron microscopy images (b, d, f) (extracellular vesicles (indicated by arrows)) of *Acholeplasma laidlawii* PG8 without ciprofloxacin (a, b), in the presence of antibiotic **(c, d) and *A. laidlawii* PG8R (e, f) (according to transmission electron microscopy). Each point corresponds to the linear size of individual cells or EVs. *$P < 0.05$. **Treatment with ciprofloxacin for 80 min.

of some bacteria may allow the traffic of genes that encode antibiotic-targeted proteins, thereby mediating the horizontal transfer of nucleotide sequences that determine antibiotic resistance in bacterial populations [31, 32].

Thus, to estimate the potential participation of the mycoplasma EVs in the traffic of the mutant gene that encoded the fluoroquinolone-targeted protein, the EVs of *A.*

laidlawii PG8 and *A. laidlawii* PG8R were tested to determine the presence of the nucleotide sequences of *parC* QRDR. Previously [9, 14], we reported that the EVs of *A. laidlawii* PG8 contained the nucleotide sequences of several genes, which can be used as specific markers of the mycoplasma EVs, thereby allowing the detection of EVs and providing a control to detect the absence of bacterial cells in EV

Adaptation of Mycoplasmas to Antimicrobial Agents: Acholeplasma laidlawii Extracellular Vesicles Mediate the Export of Ciprofloxacin and a Mutant Gene Related to the Antibiotic Target

11

```
        1      400   800   1200  1600  2000  2400  2800
        [                                              ]

        QRDR ▼
     AqF
  5' ━━━━━━━━━━━━━━━━━━━━━━━━ parC ━━━━━━━━━━━━━━━▶ 3'   2574 bp
              AqR
```

```
         1                                                                    80
1 (1)   ATACGCAATGGGACAAATGGGTTTAGCCTTTAACAAATCATATAAAAAATCTGCAAGAATTGCAGGAGAAGTTATGGGGA
2 (1)   ATACGCAATGGGACAAATGGGTTTAGCCTTTAACAAATCATATAAAAAATCTGCAAGAATTGCAGGAGAAGTTATGGGGA
3 (1)   ATACGCAATGGGACAAATGGGTTTAGCCTTTAACAAATCATATAAAAAATCTGCAAGAATTGCAGGAGAAGTTATGGGGA
         81                                                                   160
1 (81)  AATATCACCCTCATGGGGATTCATCTATTTATGAAGCAATGGTTCGTATGAGCCAGGACTTTAAAATGAGAGTACCTCTA
2 (81)  AATATCACCCTCATGGGGATTTATCTATTTATGAAGCAATGGTTCGTATGAGCCAGGACTTTAAAATGAGAGTACCTCTA
3 (81)  AATATCACCCTCATGGGGATTTATCTATTTATGAAGCAATGGTTCGTATGAGCCAGGACTTTAAAATGAGAGTACCTCTA
         161                                                                  240
1 (161) GTTGATATGCATGGTAACAATGGTTCCATAGATGGTGACTCTGCTGCGGCAATGCGTTATACCGAGGCTAGACTATCTAA
2 (161) GTTGATATGCATGGTAACAATGGTTCCATAGATGGTGACTCTGCTGCGGCAATGCGTTATACCGAGGCTAGACTATCTAA
3 (161) GTTGATATGCATGGTAACAATGGTTCCATAGATGGTGACTCTGCTGCGGCAATGCGTTATACCGAGGCTAGACTATCTAA
         241                                                            321
1 (241) AGAAGCAGAATTTTTACTTAAAGATATCGATAAAAGAACAGTAAACTTTGTACCTAACTTTGATGATGAGGAACAAGAACC
2 (241) AGAAGCAGAATTTTTACTTAAAGATATCGATAAAAGAACAGTAAACTTTGTACCTAACTTTGATGATGAGGAACAAGAACC
3 (241) AGAAGCAGAATTTTTACTTAAAGATATCGATAAAAGAACAGTAAACTTTGTACCTAACTTTGATGATGAGGAACAAGAACC
```

(a)

```
        Quinolone resistance-determining region (QRDR)  ▼
     N-━━━━━━━━━━━━━ TOP4c ━━━━━━━━━━━━━━━━ -C   857 aa
```

```
         1                                                                    80
1 (1)   YAMGQMGLAFNKSYKKSARIAGEVMGKYHPHGDSSIYEAMVRMSQDFKMRVPLVDMHGNNGSIDGDSAAAMRYTEARLSK
2 (1)   YAMGQMGLAFNKSYKKSARIAGEVMGKYHPHGDLSIYEAMVRMSQDFKMRVPLVDMHGNNGSIDGDSAAAMRYTEARLSK
3 (1)   YAMGQMGLAFNKSYKKSARIAGEVMGKYHPHGDLSIYEAMVRMSQDFKMRVPLVDMHGNNGSIDGDSAAAMRYTEARLSK
         81          106
1 (81)  EAEFLLKDIDKRTVNFVPNFDDEEQE
2 (81)  EAEFLLKDIDKRTVNFVPNFDDEEQE
3 (81)  EAEFLLKDIDKRTVNFVPNFDDEEQE
```

▭ C-terminal domain of DNA gyrase
▼ Active center

(b)

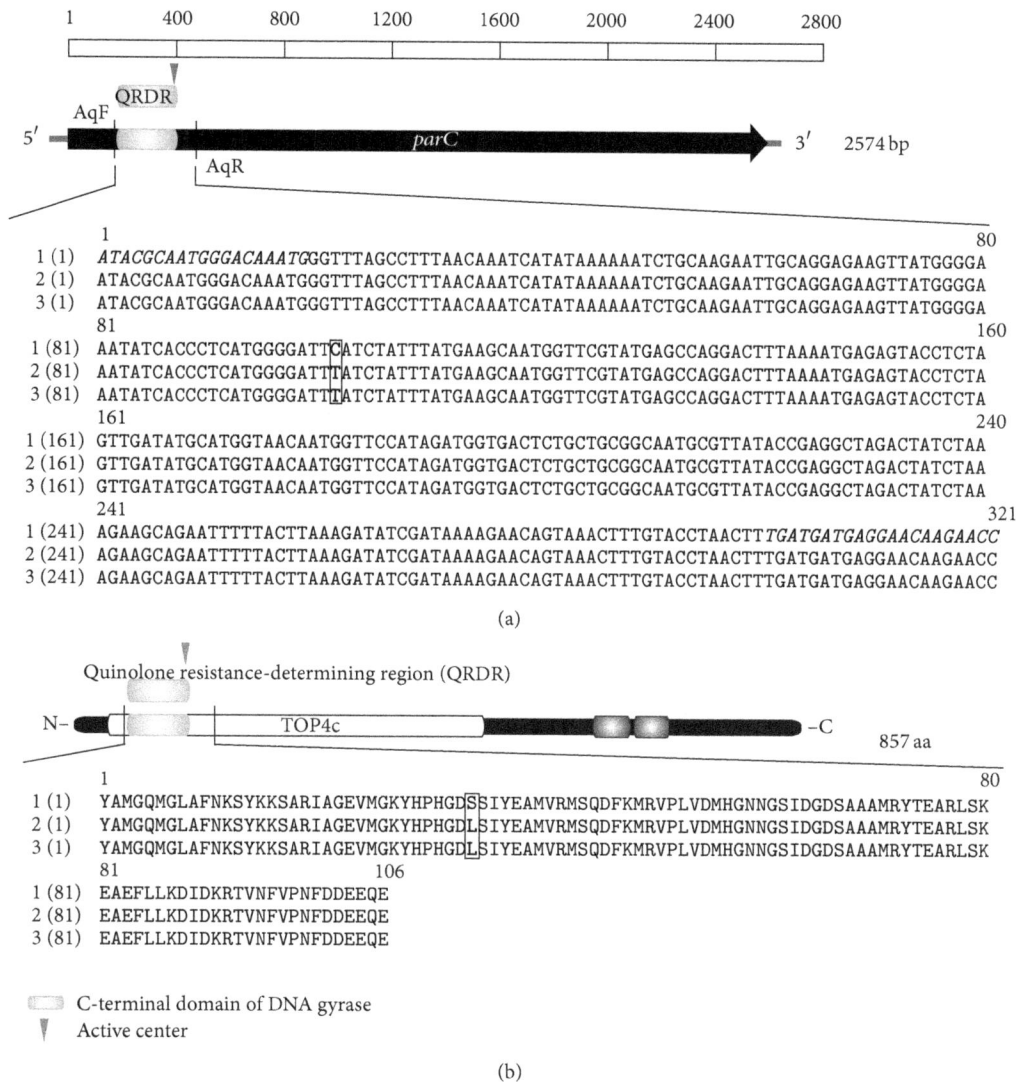

FIGURE 3: Results of the alignments of the nucleotide (a) and amino acid (b) sequences of the *parC* gene of *Acholeplasma laidlawii* PG8 cells (1), cells of *A. laidlawii* PG8R (2), and the extracellular vesicles of *A. laidlawii* PG8R (3). Italics indicate the sequences of the forward and reverse primers. Nucleotide substitutions and amino acid replacements are enclosed in rectangles, the active site.

TABLE 1: Ciprofloxacin content and bacteriostatic activities of the extracellular vesicles of *Acholeplasma laidlawii* PG8R.

Test samples	Ciprofloxacin (5 g/mL)	Extracellular vesicles of *A. laidlawii* PG8R[*]	
		Native	Destroyed
Ciprofloxacin content, ng/mg of protein	—	—	0.66 ± 0.15
Lysis area (mm)[**]	21.2 ± 0.3	3.4 ± 0.21 (P < 0.05)	8.8 ± 0.16 (P < 0.05)

[*] Mean ± standard deviation.
[**] Native extracellular vesicles of *A. laidlawii* PG8R were used as a control to estimate the bacteriostatic effects of the vesicle components.

preparations. These data were considered when conducting this study.

PCR using primers that amplified the nucleotide sequences of marker genes in the EVs of *A. laidlawii* PG8 and *parC* QRDR from the mycoplasmas detected the *parC* QRDR-specific PCR signal, where DNA obtained from the EVs of *A. laidlawii* PG8 and *A. laidlawii* PG8R were used as templates (Figure 4). The sequencing data indicated that the PCR signals corresponded to *parC* QRDR from the mycoplasmas. The C → T transition at position 272 was also detected with *A. laidlawii* PG8R (Figure 3). Thus, this study demonstrated that *A. laidlawii* PG8 and *A. laidlawii* PG8R produced EVs that contained *parC* QRDR nucleotide sequences, which were copies of the gene sequences from the respective mycoplasma strains.

A. laidlawii PG8R EVs mediated the export of a *parC* QRDR sequence with a C → T transition at position 272,

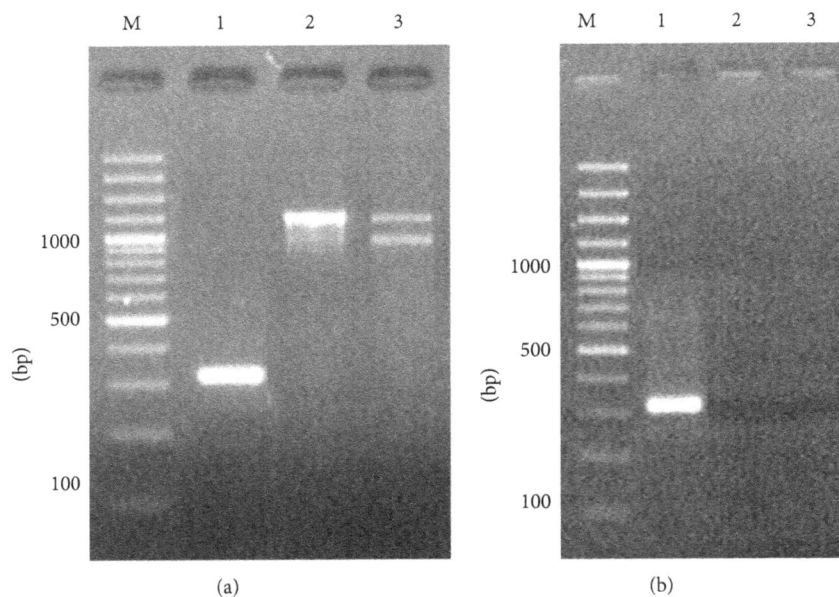

FIGURE 4: Electrophoregrams of the amplification products of the nucleotide sequences of *parC* (1), *ftsZ* (2), and A23LR (3) of *Acholeplasma laidlawii* PG8, which were obtained by PCR using the total DNA (as a template) isolated from the cells (a) and extracellular vesicles (b) of *A. laidlawii* PG8R. M: molecular weight marker.

which produced a substitution in the amino acid sequence of the antibiotic-targeted protein. This suggests the possibility that EVs can spread the mutant gene in bacterial communities via horizontal transfer. This capacity was demonstrated recently in *E. coli* and *P. aeruginosa* model systems [33]. A study of similar processes in *A. laidlawii* model systems has yet to be made.

4. Conclusions

This study showed that the EVs secreted by *A. laidlawii* cells are involved with the development of the resistance to ciprofloxacin. The results also indicate that the mechanism of antibiotic resistance in this mycoplasma was associated with a mutation in a gene that encoded a quinolone-targeted protein, while the vesiculation level was also modulated. Furthermore, the EVs of *A. laidlawii* mediated the export of the nucleotide sequences of the gene that encoded the ciprofloxacin-targeted protein, as well as the antibiotic. These results demonstrate the necessity of applying appropriate approaches to the development of effective methods for controlling mycoplasma infections, as well as contamination of cell cultures and vaccines.

Conflict of Interests

The authors declare that there is no conflict of interests regarding the publication of this paper.

Acknowledgments

This work was supported by the grant of the Ministry of Education and Science of the Russian Federation (the 8048 Agreement), Russian Foundation for Basic Research 11-04-01406a, and 12-04-01052a, 12-04-31396, the Grant of President of Russian Federation (MK-3823.2023.4), and the Grant for state support of leading scientific schools of the Russian Federation (no. NSH-825.2012.4). The authors are grateful to G. F. Shaimardanova (PhD) for her help in the transmissive electron microscopy.

References

[1] M. Folmsbee, G. Howard, and M. Mcalister, "Nutritional effects of culture media on mycoplasma cell size and removal by filtration," *Biologicals*, vol. 38, no. 2, pp. 214–217, 2010.

[2] H. M. Windsor, G. D. Windsor, and J. H. Noordergraaf, "The growth and long term survival of *Acholeplasma laidlawii* in media products used in biopharmaceutical manufacturing," *Biologicals*, vol. 38, no. 2, pp. 204–210, 2010.

[3] C. C. Uphoff and H. G. Drexler, "Detection of mycoplasma contaminations," *Methods in Molecular Biology*, vol. 946, pp. 1–13, 2013.

[4] C. Bébéar, S. Pereyre, and O. Peuchant, "*Mycoplasma pneumoniae* susceptibility and resistance to antibiotics," *Future Microbiology*, vol. 6, no. 4, pp. 423–431, 2011.

[5] E. Mariotti, F. D'alessio, P. Mirabelli, R. di Noto, G. Fortunato, and L. del Vecchio, "*Mollicutes* contamination: a new strategy for an effective rescue of cancer cell lines," *Biologicals*, vol. 40, no. 1, pp. 88–91, 2012.

[6] S. Raherison, P. Gonzalez, H. Renaudin, A. Charron, C. Bébéar, and C. M. Bébéar, "Evidence of active efflux in resistance to ciprofloxacin and to ethidium bromide by *Mycoplasma hominis*," *Antimicrobial Agents and Chemotherapy*, vol. 46, no. 3, pp. 672–679, 2002.

[7] Y. Yamaguchi, M. Takei, R. Kishii, M. Yasuda, and T. Deguchi, "Contribution of topoisomerase IV mutation to

Adaptation of Mycoplasmas to Antimicrobial Agents: Acholeplasma laidlawii Extracellular Vesicles Mediate the Export
of Ciprofloxacin and a Mutant Gene Related to the Antibiotic Target

13

quinolone resistance in *Mycoplasma genitalium*," *Antimicrob Agents Chemother*, vol. 57, no. 4, pp. 1772–1776, 2013.

[8] D. L. Couldwell, K. A. Tagg, N. J. Jeoffreys, and G. L. Gilbert, "Failure of moxifloxacin treatment in *Mycoplasma genitalium* infections due to macrolide and fluoroquinolone resistance," *International Journal of STD & AIDS*, vol. 24, no. 10, pp. 822–828, 2013.

[9] V. M. Chernov, O. A. Chernova, A. A. Mouzykantov et al., "Extracellular vesicles derived from *Acholeplasma laidlawii* PG8," *TheScientificWorldJournal*, vol. 11, pp. 1120–1130, 2011.

[10] V. M. Chernov, O. A. Chernova, E. S. Medvedeva et al., "Unadapted and adapted to starvation *Acholeplasma laidlawii* cells induce different responses of *Oryza sativa*, as determined by proteome analysis," *Journal of Proteomics*, vol. 74, no. 12, pp. 2920–2936, 2011.

[11] M. F. Nicolás, F. G. Barcellos, P. N. Hess, and M. Hungria, "ABC transporters in *Mycoplasma hyopneumoniae* and *Mycoplasma synoviae*: insights into evolution and pathogenicity," *Genetics and Molecular Biology*, vol. 30, no. 1, pp. 202–211, 2007.

[12] S. Razin and L. Hayflick, "Highlights of mycoplasma research-an historical perspective," *Biologicals*, vol. 38, no. 2, pp. 183–190, 2010.

[13] I. G. Scripal, *Microorganisms—Agents of the Plant Diseases*, Edited by V. I. Bilay, Naukova Dumka, Kiev, Ukraine, 1989.

[14] V. M. Chernov, O. A. Chernova, A. A. Mouzykantov et al., "Extracellular membrane vesicles and phytopathogenicity of *Acholeplasma laidlawii* PG8," *The Scientific World Journal*, vol. 2012, Article ID 315474, 6 pages, 2012.

[15] A. Kulp and M. J. Kuehn, "Biological functions and biogenesis of secreted bacterial outer membrane vesicles," *Annual Review of Microbiology*, vol. 64, pp. 163–184, 2010.

[16] J. W. Schertzer and M. Whiteley, "Bacterial outer membrane vesicles in trafficking, communication and the host-pathogen interaction," *Journal of Molecular Microbiology and Biotechnology*, vol. 23, pp. 118–130, 2013.

[17] O. Ciofu, T. J. Beveridge, J. Kadurugamuwa, J. Walther-Rasmussen, and N. Hoiby, "Chromosomal β-lactamase is packaged into membrane vesicles and secreted from *Pseudomonas aeruginosa*," *Journal of Antimicrobial Chemotherapy*, vol. 45, no. 1, pp. 9–13, 2000.

[18] J. Lee, E. Y. Lee, S. H. Kim et al., "*Staphylococcus aureus* extracellular vesicles carry biologically active β-lactamase," *Antimicrobial Agents and Chemotherapy*, vol. 57, pp. 2589–2595, 2013.

[19] R. M. Cole, *Methods in Mycoplasmology*, Edited by S. H. Razin, J. G. Tully, Academic Press, New York, NY, USA, 1983.

[20] J. S. Chapman and N. H. Georgopapadakou, "Fluorometric assay for fleroxacin uptake by bacterial cells," *Antimicrobial Agents and Chemotherapy*, vol. 33, no. 1, pp. 27–29, 1989.

[21] Y. Sun, M. Dai, H. Hao et al., "The role of RamA on the development of ciprofloxacin resistance in *Salmonella enterica* serovar Typhimurium," *PLoS ONE*, vol. 6, no. 8, Article ID e23471, 2011.

[22] "Performance standards for antimicrobial susceptibility testing," CLSI Approved Standard M100-S15, Clinical and Laboratory Standards Institute, Wayne, Pa, USA, 2005.

[23] G. L. Kolling and K. R. Matthews, "Export of virulence genes and Shiga toxin by membrane vesicles of *Escherichia coli* O157:H7," *Applied and Environmental Microbiology*, vol. 65, no. 5, pp. 1843–1848, 1999.

[24] T. Maniatis, E. E. Fritsch, and J. Sabrook, *Molecular Cloning. A Laboratory Manual*, Cold Spring Harbor, NY, USA, 1982.

[25] R. Maredia, N. Devineni, P. Lentz et al., "Vesiculation from *Pseudomonas aeruginosa* under SOS," *The Scientific World Journal*, vol. 2012, Article ID 402919, 8 pages, 2012.

[26] J. L. Kadurugamuwa and T. J. Beveridge, "Bacteriolytic effect of membrane vesicles from *Pseudomonas aeruginosa* on other bacteria including pathogens: conceptually new antibiotics," *Journal of Bacteriology*, vol. 178, no. 10, pp. 2767–2774, 1996.

[27] K. D. Taganov, A. E. Gushchin, N. I. Abramycheva, and V. M. Govorun, "Role of mutations in the a-subunit of *Acholeplasma laidlawii* PG-8B topoisomerase IV in formation of resistance to fluoroquinolones," *Molekuliarnaia Genetika, Mikrobiologiia I Virusologiia*, no. 2, pp. 30–33, 2000 (Russian).

[28] J. Ruiz, "Mechanisms of resistance to quinolones: target alterations, decreased accumulation and DNA gyrase protection," *Journal of Antimicrobial Chemotherapy*, vol. 51, no. 5, pp. 1109–1117, 2003.

[29] C. Bébéar, H. Renaudin, A. Charron, J. M. Bové, C. Bébéar, and J. Renaudin, "Alterations in topoisomerase IV and DNA gyrase in quinolone-resistant mutants of *Mycoplasma hominis* obtained in vitro," *Antimicrobial Agents and Chemotherapy*, vol. 42, no. 9, pp. 2304–2311, 1998.

[30] M. Ip, S. S. L. Chau, F. Chi, J. Tang, and P. K. Chan, "Fluoroquinolone resistance in atypical pneumococci and oral streptococci: evidence of horizontal gene transfer of fluoroquinolone resistance determinants from *Streptococcus pneumoniae*," *Antimicrobial Agents and Chemotherapy*, vol. 51, no. 8, pp. 2690–2700, 2007.

[31] S. Yaron, G. L. Kolling, L. Simon, and K. R. Matthews, "Vesicle-mediated transfer of virulence genes from *Escherichia coli* O157:H7 to other enteric bacteria," *Applied and Environmental Microbiology*, vol. 66, no. 10, pp. 4414–4420, 2000.

[32] C. Rumbo, E. Fernández-Moreira, M. Merino et al., "Horizontal transfer of the OXA-24 carbapenemase gene via outer membrane vesicles: a new mechanism of dissemination of carbapenem resistance genes in *Acinetobacter baumannii*," *Antimicrobial Agents and Chemotherapy*, vol. 55, no. 7, pp. 3084–3090, 2011.

[33] A. J. Manning and M. J. Kuehn, "Contribution of bacterial outer membrane vesicles to innate bacterial defense," *BMC Microbiology*, vol. 11, article 258, 2011.

3

Strategic Feeding of Ammonium and Metal Ions for Enhanced GLA-Rich Lipid Accumulation in *Cunninghamella bainieri* 2A1

Shuwahida Shuib,[1] Wan Nazatul Naziah Wan Nawi,[1] Ekhlass M. Taha,[1] Othman Omar,[1] Abdul Jalil Abdul Kader,[2] Mohd Sahaid Kalil,[3] and Aidil Abdul Hamid[1]

[1] School of Biosciences and Biotechnology, Faculty of Science and Technology, Universiti Kebangsaan Malaysia (UKM), 43600 Bangi, Selangor, Malaysia
[2] Faculty of Science and Technology, Universiti Sains Islam Malaysia, Bandar Baru Nilai, 71800 Nilai, Negeri Sembilan, Malaysia
[3] Department of Chemical and Process Engineering, Faculty of Engineering, Universiti Kebangsaan Malaysia (UKM), 43600 Bangi, Selangor, Malaysia

Correspondence should be addressed to Aidil Abdul Hamid; aidilmikrob@gmail.com

Academic Editor: João B. T. Rocha

Strategic feeding of ammonium and metal ions (Mg^{2+}, Mn^{2+}, Fe^{3+}, Cu^{2+}, Ca^{2+}, Co^{2+}, and Zn^{2+}) for enhanced GLA-rich lipid accumulation in *C. bainieri* 2A1 was established. When cultivated in nitrogen-limited medium, the fungus produced up to 30% lipid (g/g biomass) with 12.9% (g/g lipid) GLA. However, the accumulation of lipid stopped at 48 hours of cultivation although glucose was abundant. This event occurred in parallel to the diminishing activity of malic enzyme (ME), fatty acid synthase (FAS), and ATP citrate lyase (ACL) as well as the depletion of metal ions in the medium. Reinstatement of the enzymes activities was achieved by feeding of ammonium tartrate, but no increment in the lipid content was observed. However, increment in lipid content from 32% to 50% (g/g biomass) with 13.2% GLA was achieved when simultaneous feeding of ammonium, glucose, and metal ions was carried out. This showed that the cessation of lipid accumulation was caused by diminishing activities of the enzymes as well as depletion of the metal ions in the medium. Therefore, strategic feeding of ammonium and metal ions successfully reinstated enzymes activities and enhanced GLA-rich lipid accumulation in *C. bainieri* 2A1.

1. Introduction

C. bainieri 2A1 is an oleaginous fungus isolated from Malaysian soil. This fungus has the capability of producing up to 30% lipid (g/g biomass) containing 10–15% GLA [1]. Lipid produced in some oleaginous fungi contains large amount of high-valued essential polyunsaturated fatty acids (PUFAs) such as GLA and arachidonic acid (AA) [2]. GLA is an important intermediate in the biosynthesis of biologically active prostaglandin that is derived from linolenic acid. However, GLA cannot be synthesized by human and therefore must be consumed in the daily diet. In Europe, GLA is known as "King's Cure-All" because of its nutritional effects to cure diseases such as decreasing blood cholesterol,

suppressing acute and chronic inflammations, and improving atopic eczema [3].

Previously, GLA is commercially produced from the seeds of Evening Primrose (*Oenothera biennis*), Borage (*Borago officinalis*), and Blackcurrant (*Ribes nigrum*) [4]. However, there are a number of problems associated with mass production of GLA from these seeds such as low productivity, long duration of crop cultivation, and requirement of huge area for harvesting [5]. Therefore, to overcome these problems, zygomycetes such as *Cunninghamella*, *Mucor*, and *Mortierella* are often mentioned in the literature as alternatives of GLA producers due to their desirable characteristics such as high productivity of the GLA, short process cycle, and easier scalability. Currently, sugar-based low-cost materials and several types of residues (e.g., glycerol, lignocellulosic sugars, cheese whey, molasses, tomato waste hydrolysates, etc.) have been successfully considered as substrates for the production of SCO by zygomycetes. Glycerol is a major by-product discharged after the biodiesel manufacturing process has been used as a substrate for SCO production in *C. echinulata* and *M. isabellina* [6]. Besides, Vamvakaki et al. [7] reported that *Mortierella isabellina* had an outstanding performance in biomass, lipid, and GLA production when grown on cheese whey. Moreover, *Cunninghamella echinulata* and *M. isabellina* showed capability of using molasses for SCO production due to its high sugar content [8]. Tomato waste hydrolysate also has been applied as a substrate for SCO production in *C. echinulata* [9]. Among the above mentioned low-cost waste materials, lignocellulose is of great importance because of its continuous supply as a result of land cultivation. As xylose is the second most abundant sugar of lignocellulosic biomass, growth of *C. echinulata* and *M. isabellina* on xylose containing nitrogen-limited media resulted in the accumulation of significant quantities of lipid as well as GLA [6].

Lipid accumulation in oleaginous microorganisms is triggered by a nutrient imbalance in the culture medium. When nitrogen sources are depleted, excess carbon in the medium continues to be assimilated by the cells and converted into storage lipid [10]. Several studies have shown that media with varying composition of trace elements affect growth and lipid accumulation in various fungal species. In relation to lipid and PUFAs production, Mg^{2+}, Mn^{2+}, Fe^{2+}, Ca^{2+}, Cu^{2+} and Zn^{2+} have been shown to influence lipid and AA as well as GLA production by *Mortierella rammanniana var rammaniana* [11].

Oleaginous microorganisms are capable of accumulating large amount of lipid because of their ability to produce a sufficient supply of NADPH and a continuous supply of acetyl-CoA which is a necessary precursor for lipid biosynthesis [12]. The lipid biosynthesis pathway in oleaginous microbes occurs due to the concerted action of a few key lipogenic enzymes, such as ME, FAS, and ACL [10]. It has been suggested that ME specifically functions as the sole provider of NADPH, which is required for FAS activity in *Mucor circinelloides* and *Mortierella alpina*. Thus, ME is implicated as the key enzyme in regulating the extent of lipid accumulation [13]. ACL is a cytosolic enzyme that catalyzes the cleavage of citrate to generate acetyl CoA, which can be used as a precursor for lipid biosynthesis [14, 15]. On the other hand, FAS catalyzes the synthesis of palmitate, in which acetyl-CoA and malonyl-CoA are substrates and NADPH is the reducing agent. Other than these key lipogenic enzymes, NAD : ICDH was reported as being an important enzyme involved in the initiation of lipid accumulation. It was proposed that the activity of this enzyme would be inhibited after nitrogen depletion, as a result of significant decrease in AMP concentration via the activation of AMP deaminase; its activity is dependent on AMP concentration. This would then result in the accumulation of citrate which serves as the precursor for ACL which generates acetyl-CoA, a precursor for lipid synthesis [13].

Our previous observations showed that *C. bainieri* 2A1 is capable of accumulating up to 30% lipid (g/g biomass) containing 10–15% GLA [1]. However, cessation of lipid accumulation occurred after 48 hours of growth, although glucose was still present in the medium [16]. Further analysis showed that cessation of lipid accumulation coincides with a significant decrease in the specific activities of ME, FAS, and ACL which suggest that diminishing enzymes activities is the probable cause which leads to the cessation of lipid accumulation. Although this has also been reported to occur in the other oleaginous fungi such as *M. circinelloides* and *M. alpina* [13], no further evidence is currently reported. Our preliminary studies have also shown that an initial concentration of metal ions in the medium affects lipid accumulation as well as lipogenic enzyme activities [17]. As they are important in lipogenesis, the metal ions probable involvement in contributing to the enhancement of lipid accumulation should be carried out. Therefore, in this study, the effect of reinstating the activities of ME, FAS, and ACL as well as the possible involvement of metal ions in increasing lipid accumulation in *C. bainieri* 2A1 was investigated.

2. Materials and Methods

2.1. Microorganism. *C. bainieri* 2A1, a local soil isolate, was obtained as a stock culture from the School of Biosciences and Biotechnology, Faculty of Science and Technology, Universiti Kebangsaan Malaysia, Bangi, Selangor. The cultures were maintained at 4°C on Potato Dextrose agar (PDA), and subculture was performed every 2 months.

2.2. Preparation of Media and Culture Conditions. A nitrogen-limited medium [18] containing the following constituents (g/L): $(NH_4)_2C_4H_4O_6$ 1.0, KH_2PO_4 7.0, Na_2HPO_4 2.0, $MgSO_4 \cdot 7H_2O$ 1.5, yeast extract 1.5, $CaCl_2$ 0.1, $FeCl_3 \cdot 6H_2O$ 0.008, $Co(NO_3)_2 \cdot 6H_2O$ 0.0001, $ZnSO_4 \cdot 7H_2O$ 0.0001, $CuSO_4 \cdot 5H_2O$ 0.0001, and $MnSO_4 \cdot 5H_2O$ 0.0001 was sterilized at 121°C for 40 min. Glucose (30 g/L) was added separately after sterilization. Seed culture was prepared by transferring spore suspension into 500 mL shake flasks containing 200 mL of nitrogen-limited medium to a final concentration of 1×10^5 spore/mL. The cultures were incubated at 30°C and agitated at 200 rpm for 48 h. Ten percent (v/v) of the culture was then used for subsequent inoculations. All experiments were carried out using 500 mL conical flasks containing 200 mL of the nitrogen-limited medium as described above. Cultivation was carried out at 30°C, with agitation at 200 rpm for 120 h. For fed-batch experiments, simultaneous feeding of concentrated ammonium tartrate, glucose, and each of the metal ions (Mg^{2+}, Mn^{2+}, Fe^{3+}, Cu^{2+}, Co^{2+}, Ca^{2+}, and Zn^{2+}) was carried out at 72 h to reach their initial concentrations. Controls that consisted of cultures fed with glucose and either the metal ions or ammonium tartrate were also conducted. Cultures were sampled every 24 h and assayed for enzyme activities (ME, FAS, and ACL), glucose, ammonium, biomass concentrations, lipid content, and concentration of metal ions in the biomass and culture broths. Fed-batch experiments were also conducted in 5 L fermentor (Minifors, Switzerland) equipped with an online data acquisition and control system. Culture pH and dissolved O_2 were monitored with a pH meter (Mettler-Toledo, Switzerland), and oxygen probe (Mettler-Toledo, Switzerland). The cultivation conditions were as follows: 10% (v/v) seed culture, temperature of 30°C, pH 6 (by automating control using 5 M NaOH), agitation rate at 600 rpm, aeration rate at 0.8 v/v/m, and dissolved oxygen at 40%–50% saturation. The initial culture contained 4 L nitrogen-limited medium.

2.3. Analytical Methods. Ammonium concentration was determined using the indophenol method [19], while glucose concentration was determined using a glucose oxidase GOD-PERID test kit (Boehringer Mannheim) according to the manufacture's instruction. A 1 mL of culture medium was filtered using Whatman nylon membrane filter with pore size of 0.45 μm and was diluted. Then, quantification of metal ions concentration in the medium was performed using inductively coupled plasma mass spectrometry (ICP-MS) (Perkin-Elmer Elan 5000) [20]. Fungal biomass was harvested by filtration of 200 mL culture through pre-weight Whatman No. 1 filter paper followed by washing with 400 mL distilled water. The filtered mycelia were then freeze-dried overnight to a constant weight. Dried mycelia were then ground into a powder using pestle and mortar, followed by lipid extraction using chloroform/methanol 2:1 (v/v) [21]. The mixture was filtered through Whatman No. 1 filter paper to remove cell debris and the organic fractions, pooled in separating funnel, and washed with 1% sodium chloride then with distilled water using separating funnel. The chloroform extract was rotary evaporated. The remaining lipid was dissolved in 3 mL diethyl ether and transferred to pre-weight vial and the diethyl ether evaporated at room temperature under the hood. The sample was dried in desiccator for 24 h and weighted. Lipid content was expressed as % (g/g of biomass). Fatty acid methyl esters (FAMES) were analyzed using a Shimadzu GC-2010 FID. Separation was achieved by using a packed column (DB-23) and flame ionization detector (FID). The column was maintained at 250°C. The fatty acids composition presence in the sample was calculated based on the peak area of corresponding methyl ester against reference standard of FAME mixture. All data presented were the mean values of triplicates and the standard deviation was not more than 20%.

2.4. Production of Cell-Free Extract. Harvested mycelia were suspended in 20% (w/v) extraction buffer containing 100 mM KH_2PO_4/KOH (pH 7.5), 20% (v/v) glycerol, 1 mM benza-midine, 1 mM mercaptoethanol, and 1 mM EDTA and were disrupted using pestle and mortar. The cell suspension was centrifuged at 10 000 g for 15 minutes at 4°C, and the resulting supernatant, termed the cell-free extract (cfe), was used for enzyme assays. Protein concentration was determined using the method of Bradford [22] with BSA as a standard.

2.5. Enzyme Assays. The activities of ME, FAS, and ACL were determined using continuous assays following the oxidation or reduction of NAD(P)(H) at A_{340nm} and carried out at 30°C [23–25]. The change in absorbance was followed continuously for 15 min using software (UV PROBE 2.31). Specific activity is expressed as nmol/min·mg protein.

3. Results and Discussion

3.1. Relationship between Lipid Accumulation and ME, FAS, and ACL Specific Activity in C. bainieri 2A1. C. bainieri 2A1 showed similar lipid accumulation and ME, FAS, and ACL profiles as reported in other filamentous fungi in batch cultivation [13]. Lipid accumulation was initiated after total depletion of ammonium at 12 hours of cultivation with lipid content showing an increase from 12% to 32% (g/g biomass) at 48 h with 10.1% GLA (g/g lipid) (Figure 1). However, lipid accumulation ceased after 48 hours of cultivation although excess glucose was still present in the medium. This observation was similar to various reports involving filamentous fungi such as *M. circinelloides*, *M. alpina* [13], and *M. isabellina* [26, 27]. Conversely, in *C. echinulata* lipid accumulation was significantly delayed and was reported to probably be due to delayed decrease in the activity of NAD- and NADP isocitrate dehydrogenase [28] though the concentration of intracellular AMP was not mentioned to relate with the activity of the enzyme present. Diminishing activity of NAD:ICDH through repression or significant decrease in intracellular AMP concentration via the activation of AMP deaminase would result in the accumulation of citrate which serves as the precursor for ATP citrate lyase (ACL). In *C. bainieri*, NAD:ICDH activity decreased after N limitation but remained detectable until 96 h (results not shown), whereas NADP:ICDH remained active during lipogenesis until the end of cultivation [16]. This was similar to what was reported in *M. circinelloides* and *M. alpina*, where the activity of NAD:ICDH was detectable until 120 hours of cultivation. However, the activities in *M. alpina* decreased more rapidly which corresponds to its higher lipid level achieved [13].

When the specific activities of ME, FAS, and ACL of the culture throughout the cultivation period were investigated, results showed that the highest specific activities of ME, FAS, and ACL were detected at 24 hours of cultivation (12.3, 21.1, and 24.7 nmol/min·mg protein, resp.). However, the activities of ME, FAS, and ACL markedly decreased at 48 h and this coincided with the cessation of lipid production with percentage of reduction of the activity of 73%, 52%, and 64%, respectively, and diminished at 120 h (Figure 2). This suggests the existence of vital relationship between the specific enzymes activities (ME, ACL, and FAS) and lipid accumulation. At least two other oleaginous zygomycetes (*M. circinelloides* and *M. alpina*) were reported to show corresponding decline in lipid production with decreasing ME, FAS, and ACL activity [13]. As the reintroduction of limited concentration of ammonium has been shown to reinstate the activity of ME, FAS, and ACL in *M. circinelloides* and *M. alpina* [13], effects of feeding of ammonium tartrate at 72 h on the activities of the enzymes were investigated.

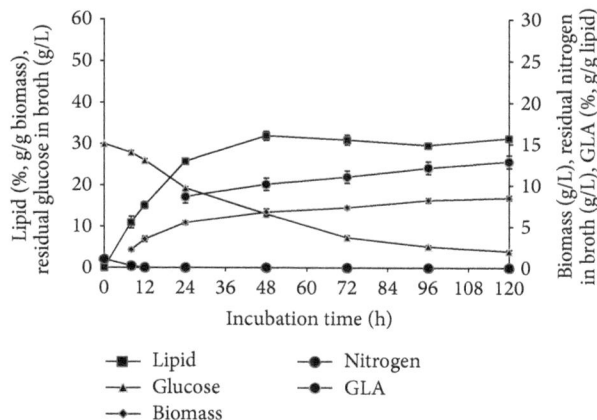

FIGURE 1: Profile of growth and lipid accumulation of *C. bainieri* 2A1. The culture was cultivated in 500 mL shake flask containing 200 mL of nitrogen-limited medium at 30°C, 200 rpm, and starting inoculums of 10% (v/v). The data represent means ± SDs, $n = 3$.

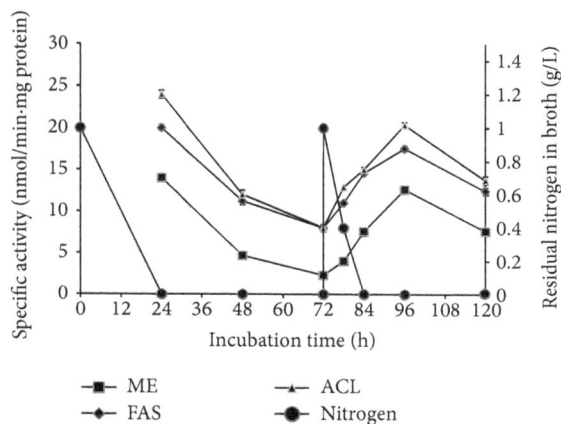

FIGURE 3: Effect of feeding ammonium tartrate at 72 hours of cultivation on ME, FAS, and ACL specific activities of *C. bainieri* 2A1. The culture was cultivated in 500 mL shake flask containing 200 mL of nitrogen-limited medium at 30°C, 200 rpm, and starting inoculums of 10% (v/v). The data represent means ± SDs, $n = 3$.

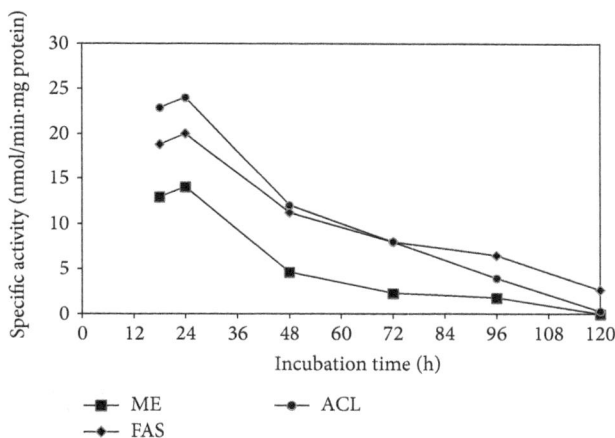

FIGURE 2: Specific activity of ME, FAS, and ACL of *C. bainieri* 2A1. The culture was cultivated in 500 mL shake flask containing 200 mL of nitrogen-limited medium at 30°C, 200 rpm, and starting inoculums of 10% (v/v). The data represent means ± SDs, $n = 3$.

3.2. Effects of Feeding of Ammonium Tartrate on ME, FAS, and ACL Specific Activity and Lipid Accumulation in C. bainieri 2A1. Results showed that the specific activities of ME, FAS, and ACL increased significantly from 2.3 to 12.6, from 8.0 to 17.3, and from 7.9 to 20.3 nmol/min·mg protein, respectively, within 24 h after feeding of ammonium tartrate and decreased at 120 hours of cultivation (Figure 3). Ammonium is not a positive effector for these enzymes in this fungus as no significant increase in the specific activities of ME, FAS, and ACL was observed when ammonium tartrate was incorporated into the assay mixtures. This suggests that the increase in specific activities after the feeding is more likely a result of an increase in the synthesis of the enzymes due to the availability of ammonium as the nitrogen source. Although reinstatement of ME, FAS, and ACL activities was achieved, no increment in lipid content was observed, where the lipid content remained at 31% (g/g biomass).

As glucose concentration at the point of feeding was low, the experiment was repeated with simultaneous feeding of concentrated ammonium tartrate and glucose to reach their initial concentrations (1 and 30 g/L, resp.) at 72 hours of cultivation. However, similar results were observed where the specific activities of the three enzymes increased but with no increase in lipid content (data not shown). These results indicate that lipid accumulation stopped although the cultures were in the most optimal condition for lipid accumulation, that is, limited N, excess C, and in the presence of ME, FAS, and ACL activities.

As the limitation of lipid accumulation occurred at the latter stage of the cultivation, it was thought that the limitation could be due to the depletion of components of the medium such as metal ions. According to Dyal et al. [11], Mg^{2+}, Mn^{2+}, Fe^{2+}, Ca^{2+}, Cu^{2+}, and Zn^{2+} have been shown to influence lipid accumulation in *Mortierella rammanniana var rammaniana*. Previous work also showed that different initial concentrations of Mg^{2+}, Fe^{3+}, and Zn^{2+} have significant effects on lipid accumulation in *C. bainieri* 2A1 [17]. It was thought that the possible limiting concentration of these metal ions in the medium contributed to the limitation of lipid production despite the fact that the successful reinstatement of the enzymes was achieved. When analysis of the medium throughout the cultivation using ICP-MS was carried out, a pronounced decrease in the concentrations of each of the metal ions (Mg^{2+}, Mn^{2+}, Ca^{2+}, Cu^{2+}, Fe^{3+}, Co^{2+}, and Zn^{2+}) was observed, though at varying rates, thus supporting the suggestion (Figure 4). Most of the metal ions (Fe^{3+}, Mg^{2+}, Mn^{2+}, Cu^{2+}, Co^{2+}, and Zn^{2+}) diminished within 24 h soon after nitrogen limitation, whereas Ca^{2+} was still present at approximately 50% of its initial concentration at 120 h. To the best of the authors' knowledge, the trend and relationship of utilization of metal ions during growth and lipid accumulation phase have not been reported. Therefore, to further establish the significance of the activities of ME,

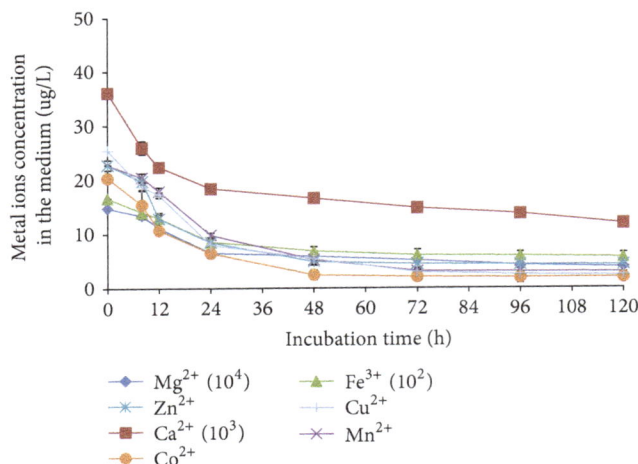

FIGURE 4: Concentration of metal ions left in culture medium throughout the cultivation of *C. bainieri* 2A1. The culture was cultivated in 500 mL shake flask containing 200 mL of nitrogen-limited medium at 30°C, 200 rpm, and starting inoculums of 10% (v/v). The data represent means ± SDs, $n = 3$.

FAS, and ACL as well as the depletion of metal ions in determining the extent of lipid production in *C. bainieiri* 2A1, the effects of simultaneous reinstatement of the activities of the enzymes and replenishment of metal ions in the medium on lipid production were carried out.

3.3. Possible Involvement of Metal Ions in the Limitation of Lipid Accumulation. When simultaneous feeding of ammonium tartrate, glucose, and all metal ions (Mg^{2+}, Mn^{2+}, Fe^{3+}, Cu^{2+}, Ca^{2+}, Co^{2+}, and Zn^{2+}) was carried out at 72 h, reinstatement of the enzyme activities was followed by an increase in lipid content from 32% to 50% (g/g biomass) with 13.2% (g/g lipid) of GLA content at 120 h (Table 1). Total of dry biomass and lipid yields which were 0.34 and 0.17 g gram of glucose consumed, respectively, were similar to *M. isabellina* grown on high sugar content media [26, 27]. On the other hand, in *C. echinulata* ATHUM 4411 that was grown on glucose and tomato waste hydrolysate, the concentration of GLA produced per liter of culture medium was reported to be 0.8 and 0.78 g/L, respectively [9, 26, 27]. Interestingly, in our present works, a much higher concentration of GLA was obtained, 0.97 g GLA per liter of culture medium. To the best of the authors' knowledge, this amount is amongst the highest achieved compared to other zygomycetes such as *M. circinelloides*, *M. rammanniana*, *M. isabellina*, and *C. echinulata* which commonly produced between 0.2 to 0.8 g GLA per liter of culture medium [26, 27]. On the contrary, no increment in lipid content was observed when the enzymes activities were not reinstated (by feeding of metal ions or glucose only) or when the reinstatement of the enzymes activities was performed with the omission of metal ions during the feeding. These results showed that the cessation of lipid accumulation observed previously was as a result of

diminishing activities of the enzymes as well as the depletion of metal ions.

When further experiments were carried out by simultaneous feeding of ammonium tartrate, glucose with individual ions, that is, either Fe^{3+}, Mg^{2+}, Mn^{2+}, or Zn^{2+}, similar increment of lipid content (from 32% to up to 48%, g/g biomass) was observed with the GLA content between 10 and 15% (g/g lipid) (Table 1). The highest lipid and GLA yields were achieved when Fe^{3+} was fed compared to other metal ions (0.15 and 0.02 g, resp., per gram of glucose consumed) (Figure 5). The metal ions such as Cu^{2+}, Co^{2+}, and Ca^{2+} also have been tested alone for the lipid and GLA production. However, no increment in lipid content was observed when these metal ions were fed into the culture and also in controls, which were fed with the omission of either the metal ions or ammonium tartrate. Metal ions play important roles in the biological function of many enzymes as the presence of the metal ions is crucial for the activity of some enzymes. These enzymes are known as metalloenzymes. Therefore, lipid accumulation is more pronounced in the presence of metal ions such as Fe^{3+} because the ions may serve as a cofactor for the key lipogenic enzymes such as ME and ACL. In addition, cultivation of *C. bainieri* 2A1 in 5 L fermentor with the same feeding strategy showed that lipid accumulation was enhanced from 48% (g/g biomass) (in shake flaks) to 55% (g/g biomass) (Figure 6). This result was similar to *M. isabellina* cultivated in 3 L fermentor with glucose as a substrate where increment in lipid content was observed following scale up of the process from 9.9 g/L of lipid in shake flask to 12.7 g/L of lipid in 3 L fermentation setup [8]. Conversely, in *Thamnidium elegans*, better results were reported in shake flasks compared to larger fermenter setup [29]. It is noteworthy to point out that the increment of lipid content achieved after feeding at 72 h was not followed by any decrease in GLA content, the effect commonly observed in oleaginous fungi when increment in lipid accumulation is achieved through increased glucose concentration [9, 26, 27, 30, 31].

In oleaginous microorganisms, lipid turnover typically occurred when carbon source was exhausted in the culture medium. Papanikolaou et al. [28] reported that repression of lipid turnover could be due to multiple limitation factors such as the nitrogen and several microelements like Mg^{2+}, Fe^{3+} or elements derived from the yeast extract, whilst lipid turnover was activated when metal ions were added. However, no proof of lipid turnover was observed in *C. bainieri* 2A1 as the result showed constant lipid content from 48 h onwards, in the presence of glucose (Figure 1). This result was similar to at least two other zygomycetes: *M. circinelloides* and *M. alpina* [15]. In our experiments, glucose was always kept over 10 g/L and lipid turnover was not observed in this fungus, although metal ions were added (Figure 4). As proved by our data, lipid synthesis was reinitiated when metal ions were added in the culture whereby multiple limitations resulted in the cessation of lipid accumulation in *C. bainieri* 2A1. Therefore, these results showed that, as reported by Papanikolaou et al. [28], presence of metal ions is vital for lipid synthesis as well as lipid turnover for zygomycetes. In addition, the cessation

TABLE 1: Effect of feeding on lipid and GLA production. Cultures were fed with several medium components at 72 hours of cultivation. Lipid was extracted and analyzed for GLA content at 120 hours of cultivation. The data represent means $\pm$ SDs, $n = 3$.

Feeding	Biomass, X (g/L)	Lipid, L % (g/g biomass)	GLA % (g/g lipid)	Yield of biomass per glucose consumed, $Y_{X/G}$	Yield of lipid per glucose consumed, $Y_{L/G}$	Yield of GLA per glucose consumed $Y_{GLA/G}$
Ammonium tartrate, glucose, and all metal ions (Mg^{2+}, Mn^{2+}, Fe^{3+}, Cu^{2+}, Ca^{2+}, Co^{2+} and Zn^{2+})	14.77 ± 0.45	50.95 ± 1.07	13.24 ± 0.13	0.34	0.17	0.02
Ammonium tartrate, glucose, and Fe^{3+}	13.95 ± 0.60	48.75 ± 0.67	12.66 ± 0.05	0.31	0.15	0.02
Ammonium tartrate, glucose, and Mn^{2+}	12.05 ± 0.42	44.22 ± 0.80	11.86 ± 0.32	0.22	0.13	0.01
Ammonium tartrate, glucose, and Mg^{2+}	13.28 ± 0.40	39.24 ± 0.71	12.13 ± 0.39	0.32	0.13	0.01
Ammonium tartrate, glucose and Zn^{2+}	11.04 ± 0.55	37.73 ± 0.54	11.45 ± 0.27	0.26	0.09	0.01
Ammonium tartrate and glucose	12.57 ± 0.54	31.57 ± 0.79	12.19 ± 0.49	0.34	0.11	0.01
Without feeding	8.42 ± 0.58	31.64 ± 0.88	12.95 ± 0.14	0.33	0.31	0.01

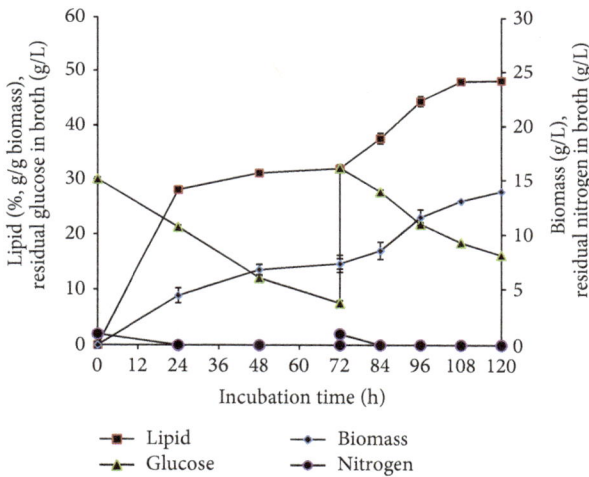

FIGURE 5: Effect of feeding ammonium tartrate, glucose, and Fe^{3+} at 72 h in lipid accumulation of *C. bainieri* 2A1. The culture was cultivated in 500 mL shake flask containing 200 mL of nitrogen-limited medium at 30°C, 200 rpm, and starting inoculums of 10% (v/v). The data represent means $\pm$ SDs, $n = 3$.

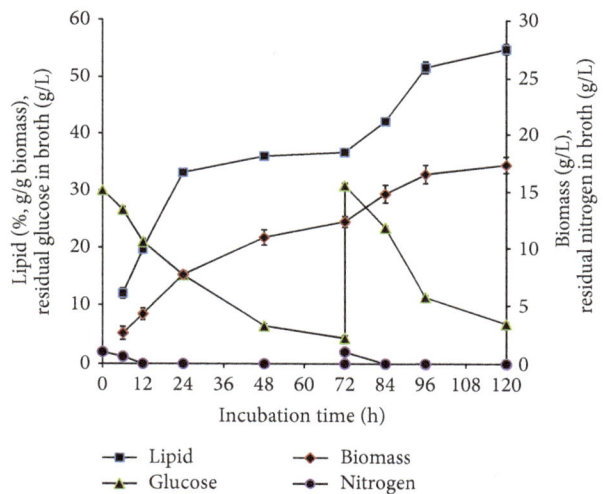

FIGURE 6: Effect of feeding ammonium tartrate, glucose, and Fe^{3+} at 72 h in lipid accumulation of *C. bainieri* 2A1. The culture was cultivated in a 5 L fermenter with 4 L of nitrogen-limited medium with 30°C, pH 6.00, aeration 0.8 v/v/m, agitation 600 rpm, and starting inoculums of 10% (v/v). The data represent means $\pm$ SDs, $n = 3$.

of lipid accumulation at 48 h in *C. bainieri* 2A1 was not caused by saturation of lipid accumulated inside this fungus as reported to occur in *C. echinulata* and *M. isabellina* [28]. This is because the increment of lipid content in *C. bainieri* 2A1 was observed only when feeding of ammonium, glucose, and metal ions was carried out at 72 h.

In addition, fatty acid compositions such as C16:0 (palmitic acid), C18:0 (stearic acid), C18:1 (oleic acid), C18:2 (linoleic acid), and C18:3n-6 (γ-linolenic acid) were analyzed at the end of the cultivation period (120 h) by gas

chromatography (Table 2). The results showed that C18:1 was the predominant fatty acid accumulated by this fungus in all experiments performed. It was followed by the amount of C16:0, whereas low content of C18:0 and C18:2 (12.03% and 8.78%, g/g lipid, resp.) was observed. The GLA composition was found in significant amount (12%, g/g lipid) in all experiments performed. These profiles were different in comparison to lipid production by several oleaginous fungi. Previous reports showed that C18:1 was the main fatty acid

TABLE 2: Fatty acid composition of the cellular lipids of *C. bainieri* 2A1 at 120 hours of cultivation.

Feeding	Fatty acid composition (%)				
	C16:0	C18:0	C18:1	C18:2	C18:3 n-6
Ammonium tartrate, glucose, and all metal ions (Mg^{2+}, Mn^{2+}, Fe^{3+}, Cu^{2+}, Ca^{2+}, Co^{2+}, and Zn^{2+})	19.95	12.03	30.66	8.78	13.24
Ammonium tartrate, glucose, and Fe^{3+}	21.35	8.95	35.47	9.77	12.66
Ammonium tartrate, glucose, and Mn^{2+}	19.21	9.75	37.17	9.69	11.86
Ammonium tartrate, glucose, and Mg^{2+}	27.87	12.99	32.03	4.29	12.13
Ammonium tartrate, glucose, and Zn^{2+}	21.47	12.71	42.75	7.83	11.04
Without feeding	18.62	10.27	40.35	15.72	12.95

in *M. isabellina* when grown on high sugar content medium. However significant amount of C16:0 and C18:2 (22% and 18%, g/g lipid, resp.) but with low amount of C18:0 and C18:3n-6 was reported [26, 27]. In *C. echinulata* ATHUM 4411 that was grown on tomato waste hydrolysate with the feeding of glucose, after 12 days of cultivation, the concentration of GLA fell dramatically from 25% (before the feeding) to 12% (g/g lipid) with the increment of the concentrations of C18:1, C16:0, and C18:0 being reported [9]. Therefore, generally, the differences in fatty acid composition could be related to strains, fermentation time, initial sugar concentration, types of carbon and nitrogen sources used, and initial molar of C/N ratio.

4. Conclusions

Cessation of lipid accumulation in *C. bainieri* 2A1 is caused by diminishing activities of ME, ACL, and FAS as well as depletion of metal ions in the medium. Simultaneous feeding of ammonium (which resulted in the reinstatement of the activities of the enzymes) and metal ions at 72 h successfully reinstated the enzymes activities and enhanced lipid accumulation, followed by consistent percentage of GLA content throughout the cultivation of *C. bainieri* 2A1.

Abbreviations

PUFAs:	Polyunsaturated fatty acids
SCO:	Single cell oil
GLA:	γ-linolenic acid
AA:	Arachidonic acid
NADPH:	Nicotinamide adenine dinucleotide phosphate
AMP:	Adenosine monophosphate
Acetyl Co-A:	Acetyl co-enzyme A
ME:	Malic enzyme
FAS:	Fatty acid synthase
ACL:	ATP citrate-lyase
NAD : ICDH:	Nicotinamide adenine dinucleotide : isocitrate dehydrogenase
N:	Nitrogen
C:	Carbon
X:	Biomass (g/L)
L:	Lipid.

Conflict of Interests

The authors declare that there is no conflict of interests regarding the publication of this paper.

Acknowledgments

The authors acknowledge the financial supports from the Ministry of Higher Education, Malaysia, under the Grant FRGS/1/2011/ST/UKM/02/2.

References

[1] A. A. Hamid, W. M. W. Yusof, R. M. Illias, and K. Nadarajah, "Pengasingan Strain Baru Kulat dari Tanah di Malaysia untuk Penghasilan Asid γ-linolenik (GLA)," *Jurnal Teknologi*, vol. 34, no. 3, pp. 1–8, 2001.

[2] S. S. Mamatha, R. Ravi, and G. Venkateswaran, "Medium optimization of gamma linolenic acid production in *Mucor rouxii* CFR-G15 using RSM," *Food and Bioprocess Technology*, vol. 1, no. 4, pp. 405–409, 2008.

[3] D. F. Horrobin, "Nutritional and medical importance of gamma-linolenic acid," *Progress in Lipid Research*, vol. 31, no. 2, pp. 163–194, 1992.

[4] C. Ratledge, "Yeasts, moulds, algae and bacteria as sources of lipid," in *Technological Advances in Improved and Alternative Sources of Lipid*, B. S. Kamel and Y. Kakuda, Eds., pp. 235–291, Blackie Academic and Professional, London, UK, 1994.

[5] F. D. Gunstone, "Gamma linolenic acid—occurrence and physical and chemical properties," *Progress in Lipid Research*, vol. 31, no. 2, pp. 145–161, 1992.

[6] S. Fakas, S. Papanikolaou, A. Batsos, M. Galiotou-Panayotou, A. Mallouchos, and G. Aggelis, "Evaluating renewable carbon sources as substrates for single cell oil production by *Cunninghamella echinulata* and *Mortierella isabellina*," *Biomass and Bioenergy*, vol. 33, no. 4, pp. 573–580, 2009.

[7] A.-N. Vamvakaki, I. Kandarakis, S. Kaminarides, M. Komaitis, and S. Papanikolaou, "Cheese whey as a renewable substrate for microbial lipid and biomass production by Zygomycetes," *Engineering in Life Sciences*, vol. 10, no. 4, pp. 348–360, 2010.

[8] A. Chatzifragkou, S. Fakas, M. Galiotou-Panayotou, M. Komaitis, G. Aggelis, and S. Papanikolaou, "Commercial sugars as substrates for lipid accumulation in *Cunninghamella echinulata* and *Mortierella isabellina* fungi," *European Journal of Lipid Science and Technology*, vol. 112, no. 9, pp. 1048–1057, 2010.

[9] S. Fakas, M. Čertik, S. Papanikolaou, G. Aggelis, M. Komaitis, and M. Galiotou-Panayotou, "γ-Linolenic acid production by *Cunninghamella echinulata* growing on complex organic nitrogen sources," *Bioresource Technology*, vol. 99, no. 13, pp. 5986–5990, 2008.

[10] C. Ratledge, "Regulation of lipid accumulation in oleaginous micro-organisms," *Biochemical Society Transactions*, vol. 30, no. 6, pp. 1047–1050, 2002.

[11] S. D. Dyal, L. Bouzidi, and S. S. Narine, "Maximizing the production of γ-linolenic acid in *Mortierella ramanniana var. ramanniana* as a function of pH, temperature and carbon source, nitrogen source, metal ions and oil supplementation," *Food Research International*, vol. 38, no. 7, pp. 815–829, 2005.

[12] Z. Ruan, M. Zanotti, X. Wang, C. Ducey, and Y. Liu, "Evaluation of lipid accumulation from lignocellulosic sugars by *Mortierella isabellina* for biodiesel production," *Bioresource Technology*, vol. 110, pp. 198–205, 2012.

[13] J. P. Wynn, A. B. A. Hamid, and C. Ratledge, "The role of malic enzyme in the regulation of lipid accumulation in filamentous fungi," *Microbiology*, vol. 145, no. 8, pp. 1911–1917, 1999.

[14] C. Ratledge, "Fatty acid biosynthesis in microorganisms being used for single cell oil production," *Biochimie*, vol. 86, no. 11, pp. 807–815, 2004.

[15] J. P. Wynn, A. Kendrick, and C. Ratledge, "Sesamol as an inhibitor of growth and lipid metabolism in *Mucor circinelloides* via its action on malic enzyme," *Lipid*, vol. 32, pp. 605–610, 1997.

[16] A. Abdul Hamid, N. F. Mokhtar, E. M. Taha, O. Omar, and W. M. W. Yusoff, "The role of ATP citrate lyase, malic enzyme and fatty acid synthase in the regulation of lipid accumulation in *Cunninghamella* sp. 2A1," *Annals of Microbiology*, vol. 61, no. 3, pp. 463–468, 2011.

[17] F. Muhid, W. N. N. W. Nawi, A. J. A. Kader, W. M. W. Yusoff, and A. A. Hamid, "Effects of metal ion concentrations on lipid and gamma linolenic acid production by *Cunninghamella* sp. 2A1," *Online Journal of Biological Sciences*, vol. 8, no. 3, pp. 62–67, 2008.

[18] A. Kendrick and C. Ratledge, "Lipids of selected molds grown for production of n-3 and n-6 polyunsaturated fatty acids," *Lipids*, vol. 27, no. 1, pp. 15–20, 1992.

[19] A. L. Chaney and E. P. Marbach, "Modified reagents for determination of urea and ammonia," *Clinical Chemistry*, vol. 8, pp. 130–132, 1962.

[20] S. Iglesia-Turiño, A. Febrero, O. Jauregui, C. Caldelas, J. L. Araus, and J. Bort, "Detection and quantification of unbound phytochelatin 2 in plant extracts of *Brassica napus* grown with different levels of mercury," *Plant Physiology*, vol. 142, no. 2, pp. 742–749, 2006.

[21] J. Folch, M. Lees, and G. H. Sloane-Stanley, "A simple method for the isolation and purification of total lipides from animal tissues," *The Journal of Biological Chemistry*, vol. 226, no. 1, pp. 497–509, 1957.

[22] M. M. Bradford, "A rapid and sensitive method for the quantitation of microgram quantities of protein utilizing the principle of protein dye binding," *Analytical Biochemistry*, vol. 72, no. 1-2, pp. 248–254, 1976.

[23] R. Y. Hsu and H. A. Lardy, "Malic enzyme," *Methods in Enzymology*, vol. 13, pp. 230–235, 1969.

[24] F. Lynen, "Yeast fatty acid synthase," *Methods in Enzymology*, vol. 14, pp. 17–33, 1969.

[25] P. A. Srere, "Citrate-cleavage enzyme," *Methods in Enzymology*, vol. 5, pp. 641–644, 1962.

[26] S. Papanikolaou, M. Komaitis, and G. Aggelis, "Single cell oil (SCO) production by *Mortierella isabellina* grown on high-sugar content media," *Bioresource Technology*, vol. 95, no. 3, pp. 287–291, 2004.

[27] V. K. Eroshin, A. D. Satroudinov, E. G. Dedyukhina, and T. I. Chistyakova, "Arachidonic acid production by *Mortierella alpina* growth-coupled lipid synthesis," *Process Biochemistry*, vol. 35, no. 10, pp. 1171–1175, 2000.

[28] S. Papanikolaou, S. Sarantou, M. Komaitis, and G. Aggelis, "Repression of reserve lipid turnover in *Cunninghamella echinulata* and *Mortierella isabellina* cultivated in multiple-limited media," *Journal of Applied Microbiology*, vol. 97, no. 4, pp. 867–875, 2004.

[29] E. Zikou, A. Chatzifragkou, A. A. Koutinas, and S. Papanikolaou, "Evaluating glucose and xylose as cosubstrates for lipid accumulation and gamma-linolenic acid biosynthesis of *Thamnidium elegans*," *Journal of Applied Microbiology*, vol. 114, no. 4, pp. 1020–1032, 2013.

[30] H. Gema, A. Kavadia, D. Dimou, V. Tsagou, M. Komaitis, and G. Aggelis, "Production of γ-linolenic acid by *Cunninghamella echinulata* cultivated on glucose and orange peel," *Applied Microbiology and Biotechnology*, vol. 58, no. 3, pp. 303–307, 2002.

[31] A. Kavadia, M. Komaitis, I. Chevalot, F. Blanchard, I. Marc, and G. Aggelis, "Lipid and γ-linolenic acid accumulation in strains of zygomycetes growing on glucose," *Journal of the American Oil Chemists' Society*, vol. 78, no. 4, pp. 341–346, 2001.

4

Microbiological Quality of Fresh Produce from Open Air Markets and Supermarkets in the Philippines

Pierangeli G. Vital,[1,2] **Kris Genelyn B. Dimasuay,**[1,2]
Kenneth W. Widmer,[3] **and Windell L. Rivera**[1,2]

[1] *Institute of Biology, College of Science, University of the Philippines, Diliman, 1101 Quezon City, Philippines*
[2] *Natural Sciences Research Institute, University of the Philippines, Diliman, 1101 Quezon City, Philippines*
[3] *International Environmental Analysis and Education Center, Gwangju Institute of Science and Technology, 261 Cheomdan-Gwagiro, Buk-gu, Gwangju 500-712, Republic of Korea*

Correspondence should be addressed to Windell L. Rivera; wlrivera@science.upd.edu.ph

Academic Editor: Heléne Norder

This study is the first in the Philippines to conduct a comprehensive assessment of the prevalence of bacterial pathogens and somatic phages in retailed fresh produce used in salad preparation, namely, bell pepper, cabbage, carrot, lettuce, and tomato, using culture and molecular methods. Out of 300 samples from open air and supermarkets, 16.7% tested positive for thermotolerant *Escherichia coli*, 24.7% for *Salmonella* spp., and 47% for somatic phages. Results show that counts range from 0.30 to 4.03 $\log_{10}$ CFU/g for *E. coli*, 0.66 to $\geq$2.34 $\log_{10}$ MPN/g for *Salmonella* spp., and 1.30 to $\geq$3.00 $\log_{10}$ PFU/g for somatic phages. Statistical analyses show that there was no significant difference in the microbial counts between open air and supermarkets ($\alpha = 0.05$). TaqMan and AccuPower Plus DualStar real-time polymerase chain reaction (RT-PCR) was used to confirm the presence of these organisms. The relatively high prevalence of microorganisms observed in produce surveyed signifies reduction in shelf-life and a potential hazard to food safety. This information may benefit farmers, consumers, merchants, and policy makers for foodborne disease detection and prevention.

1. Introduction

Food-borne pathogens take a serious toll on public health. It is estimated in the United States alone that approximately 14 million incidents of food related illness occur [1]. A recognized source for food-borne pathogens is fecal contamination of water used for irrigation, or for processing, of fresh produce [2]. While many agricultural products are cooked prior to eating, many Southeast Asian cultures also consume uncooked produce either directly or as fresh condiments to other dishes, such as soups.

Surveys of agricultural produce, meats, and shellfish have been conducted finding relatively high microbial loads in Southeast Asia [3, 4] indicating that contamination of water for agriculture and aquaculture, compounded by poor food handling during distribution, can have a negative impact on public health. Some have also investigated risk assessment models based on consumption of fresh produce [5]. While these survey studies have been conducted, actual sampling data is lacking for many developing countries making attempts to accurately develop risk assessment studies problematic. Some approaches to assess microbial risk associated with drinking water based on theoretical values for developing countries have been attempted [6]. Yet, the additional variables of the transfer of waterborne pathogens to produce through irrigation and washing of produce can make such theoretical values difficult to calculate. Other approaches involving direct sampling of produce for pathogens to evaluate microbial risk assessment have been employed [7]; however, sufficient comprehensive survey data of produce in Southeast Asia, particularly in the Philippines, is lacking.

Additionally, studies for the impact of enteric viruses on produce in the Philippines are also limited. As microbial contamination of agricultural products is of concern, additional survey data would be a key to assessing their impact on food safety [1]. A comprehensive survey for the presence and enumeration of bacterial pathogens, in addition to establishing the quantification of somatic bacteriophages as an indicator for viral pathogens, would be ideal for determining microbial contaminant loads on fresh produce used in salad preparation through culture and molecular methods. This study investigates the prevalence of thermotolerant *Escherichia coli*, *Salmonella* spp., and somatic bacteriophages in fresh produce consumed uncooked, namely, bell pepper, cabbage, carrot, lettuce, and tomato, found in both open air markets and supermarkets in the Philippines. This information would be a cornerstone to determining the risk associated with microbial contamination of freshly consumed foods for populations in the Philippines and safer vending practices. This study is the first in the country to address microbial contamination of fresh produce that may directly or indirectly benefit farmers, consumers, merchants, vendors, and policy makers towards food quality and safety.

2. Materials and Methods

2.1. Produce Sampling. A total of 300 fresh produce samples typical for raw consumption were surveyed. Sampling areas included five large local and open air markets where most residents and retailers purchase produce from National Capital Region, Laguna (province in South Luzon), and Pampanga (province in Central Luzon). Five common and known supermarkets were also surveyed to compare the microbial quality of produce from that of open air markets. Five types of vegetable produce that are consumed uncooked, namely, bell pepper (*Capsicum frutescens*, family Solanaceae), cabbage (*Brassica oleracea*, family Brassicaceae), carrot (*Daucus carota*, family Apiaceae), lettuce (*Lactuca sativa*, family Asteraceae), and tomato (*Lycopersicon esculentum*, family Solanaceae), were collected from a variety of sources to encompass the different produce handling and distribution practices. A total of 50 vegetables were collected per market, ten of each kind, and another 50 from supermarkets. All samples purchased were placed in individual polyethylene bags. Vegetable produce samples were transported to the Natural Sciences Research Institute, University of the Philippines, in an improvised ice box (kept under 10°C) and processed within 6–8 h after collection.

2.2. Produce Analysis: Culture Method. Using sterile scalpel blades, 5 g portion was excised from the samples, weighed in sterile plastic boats, and transferred to sterile whirl pack bags (Nasco Whirl-Pak, USA). Thirty millilitres of sterile 0.1% buffered peptone water (BPW) (MB Cell, Korea) was added; then bags were sealed and rotated on a shaking platform for 10 min. The produce wash was then processed for microbial analysis.

2.2.1. Detection and Enumeration of Escherichia coli. After undergoing the washing step, the buffer was examined for the presence of thermotolerant *E. coli*, an opportunistic pathogen and classic indicator organism for fecal contamination, using a membrane filtration technique [8]. Briefly, tenfold serial dilutions were performed from the 10 mL wash solution in BPW. They were filtered through 0.45 μm filters (PALL Corporation, USA), plated on membrane-fecal coliform (m-FC) agar plates (MB Cell, Korea), and incubated at 44.5 ± 0.2°C for 24 h where CFU/g of fresh produce was obtained. Blue to deep blue colonies were considered presumptive for *E. coli* and at least four isolated colonies were streak plated on MacConkey agar (MCA) plates (MB Cell, Korea) and incubated at 35 ± 0.5°C for 24 h. Streaked colonies were considered thermotolerant *E. coli* indicated by the presence of light pink or red colonies.

2.2.2. Detection and Enumeration of Salmonella Strains. The wash buffer solution was also examined for the presence and enumeration of *Salmonella* spp. using MPN method in Rappaport-Vassiliadis Enrichment Broth (RV) (MB Cell, Korea). Briefly, 5 mL volume of the wash was mixed with 5 mL double strength BPW, and 3 further tenfold serial dilutions were performed in BPW as an enrichment step. All tubes were incubated at 35 ± 0.5°C for 24 h. Afterwards, 0.2 mL per dilution series was added to 1.8 mL RV in three respective wells (12 wells total) using a sterile multiwell plate and incubated at 42 ± 0.5°C for 24 h. Results were recorded as positive (+) or negative (−) and compared to an MPN table to obtain MPN/g of fresh produce. Positive wells were streak plated in triplicate onto xylose lysine deoxycholate (XLD) agar plates (Difco, Becton, Dickinson and Company, USA) and incubated at 35 ± 0.5°C for 18–24 h. *Salmonella* spp. isolates were indicated by red colonies with a dark center, confirming results from the RV broth MPN.

2.2.3. Detection and Enumeration of Somatic Phages. The wash buffer was also tested for the presence of somatic bacteriophages. Somatic phages are classic indicator organisms of fecal contamination and have been demonstrated as efficient model organisms for viral pathogens [9–11]. A stock culture of *E. coli* CN-13, a nalidixic acid resistant strain, was maintained in tryptic soy agar (TSA) (MB Cell, Korea) with nalidixic acid and used to serve as somatic bacteriophage host for this assay. Portions of the sample wash were processed for somatic phage enumeration using a double agar layer method. Briefly, 0.7% tryptic soy broth (TSB) (MB Cell, Korea) soft agar tubes were prepared with nalidixic acid, and log host culture was added, followed by either 100 μL or 10 μL wash solution. Soft TSB agar tubes were then poured on TSA plates supplemented with nalidixic acid. Formation of plaques after incubation at 35 ± 0.5°C for 18–24 h was counted to obtain CFU·g^{-1} counts.

Additional controls were prepared using both a high titer of Phi X-174 (DSM-4497) and a negative control (without any phage) to confirm specificity of *E. coli* CN-13 and to ensure no phage contamination occurred when propagating the host culture. Controls were incubated in conditions similar to the wash samples.

TABLE 1: Microbiological quality of fresh produce from open air and supermarkets in the Philippines[a].

Markets and fresh produce	E. coli ($\log_{10}$CFU/g)	Salmonella spp. ($\log_{10}$MPN/g)	Coliphages ($\log_{10}$PFU/g)
Open air markets			
Bell pepper	1.62–3.95	0.66–2.34≤	2.15–3.00≤
Cabbage	0.30–2.58	0.66–2.34≤	1.41–3.00≤
Carrot	2.30–4.03	0.66–2.34≤	2.41–3.00≤
Lettuce	1.20–3.92	0.66–2.34≤	1.60–3.00≤
Tomato	0.88–3.66	0.66–2.34≤	3.00
Supermarkets			
Bell pepper	2.26–4.15	0.66–2.34≤	1.30–3.00≤
Cabbage	1.00–2.88	0.66–1.68≤	1.30–2.15
Carrot	1.75–2.78	0.66–2.34≤	1.60–2.15
Lettuce	3.09–3.15	0.66–2.34≤	1.30–3.00≤
Tomato	1.94–3.12	0.00	0.00

[a]Values are ranges of 50 fresh produce samples per type from all open air markets; 10 fresh produce samples per type from all supermarkets; a total of 300 fresh produce samples.

2.3. Produce Analysis: Molecular Method. Isolates were obtained from the confirmed positive samples in the culture method (atypical colonies on MCA plates for *E. coli* or XLD agar plates for *Salmonella* spp.). For DNA extraction of these isolates, the culture stocks underwent boil lysis method following the protocol of de Medici et al. [12] with modifications. Briefly, 0.05 N NaOH was added and boiled at 95°C for 15 min. The DNA extracts were then kept at −20°C and transferred to International Environmental Analysis and Education Center, Gwangju Institute of Science and Technology, Korea, in an ice box for molecular analysis.

Confirmation of bacterial isolates was done using Taq-Man real-time PCR (RT-PCR) with amplification of DNA performed using a Rotor-Gene 3000 (Corbett Research) real-time PCR instrument. TaqMan Environmental Master Mix (Applied Biosystems, USA) was used for the RT-PCR assay of *E. coli*, with primers specific for *E. coli* adapted from Takahashi et al. [13]. The forward primer ECN1254F (5′-GCA AGG TGC ACG GGAATA TT-3′) and reverse primer ECN1328R (5′-CAG GTG ATC GGA CGC GT-3′) employed in this study amplify the *uidA* gene (β-glucuronidase). The dual labelled probe, ECL1277p (5′-CGCCACTGGCGG-AAGCAACG-3′), was added to the reaction.

Real-time PCR assay of *Salmonella* spp. was conducted using AccuPower Plus DualStar qPCR Premix (Bioneer, Korea), and primers for this assay were adapted from Elizaquível and Aznar [14]. The forward primer OriP1 (5′-TTA TTA GGA TCG CGC CAG GC-3′) and reverse primer OriP3 (5′-GGA CCA CGA TCA CCG ATC A-3′) amplify the *OriC* gene (replication origin sequence). The dual labelled probe, OriP214 (5′-TCA ATG CGT TGG AAA GGA TCA CTA GCT GT-3′) was included to the reaction.

E. coli ATCC strain 15597 and *Salmonella* KCTC strain 2421 were used as standards in this part of the experiment. They were grown in TSB as a simple enrichment medium, serially diluted in 0.1% BPW, and spread plated on TSA plates to enumerate the sample as CFU/mL. DNA was extracted from these standards and run in RT-PCR instrument.

The 20 µL reaction consisted of 2x RT-PCR Mastermix, 0.5 µM of forward and reverse primers, 0.25 µM of probe, 0.1 µg/µL bovine serum albumin, and 2 µL of DNA template. Each set of samples assayed included a nontemplate control (NTC) and DNA standards in duplicate. RT-PCR for *E. coli* was performed with an initial denaturation at 95°C for 15 min, followed by 40 cycles at 95°C for 15 sec, and 63°C extension step for 60 sec where the fluorescent signal was acquired. For reactions with *Salmonella* isolates, the conditions had an initial denaturation step at 95°C for 5 min, followed by 40 cycles at 95°C for 15 sec, and 63°C extension step for 60 sec. All samples with fluorescent signals at the end of 40 cycles compared with no signal from the negative controls were scored as positive.

2.4. Statistical Analysis. Data gathered from the microbiological assessment of fresh produce were subjected to single-factor analyses of variance (ANOVA) using the General Linear Model Procedure (PROC GLM) of the SAS statistical package version 9.1.3 [15] with Duncan Multiple Range Test (DMRT) for post hoc determinations of significant differences between the five open air markets and between open air markets and supermarkets ($\alpha = 0.05$).

3. Results

A total of 300 fresh produce samples eaten raw, namely, bell pepper, cabbage, carrot, lettuce, and tomato from five open air markets and five supermarkets, were surveyed for their microbial quality in the Philippines using culture technique and confirmed by molecular analysis. *E. coli* isolates were observed in 50 of the 300 fresh produce samples surveyed. Further, of the 300, 74 produce samples were found contaminated with *Salmonella* spp., and 141 samples recovered somatic phage from produce wash solutions. Table 1 shows the typical ranges of results of microbial contamination that were observed in fresh produce.

TABLE 2: Prevalence of microbial contamination in fresh produce from open air markets and supermarkets in the Philippines by culture and real-time PCR techniques[a].

Fresh produce	E. coli		Salmonella spp. (% and no. of positive/total)		Coliphages	
	Open air	Supermarket	Open Air	Supermarket	Open air	Supermarket
Bell pepper	10% (5/50)	20% (2/10)	20% (10/50)	30% (3/10)	62% (31/50)	60% (6/10)
Cabbage	10% (5/50)	20% (2/10)	22% (11/50)	10% (1/10)	52% (26/50)	60% (6/10)
Carrot	22% (11/50)	30% (3/10)	38% (19/50)	50% (5/10)	38% (19/50)	100% (10/10)
Lettuce	24% (12/50)	20% (2/10)	24% (12/50)	30% (3/10)	64% (32/50)	80% (8/10)
Tomato	16% (8/50)	0% (0/10)	18% (9/50)	10% (1/10)	6% (3/50)	0% (0/10)
Total	41/250	9/50	61/250	13/50	111/250	30/50

[a]Values are expressed in percentages of contaminated fresh produce from 60 per type from all open air markets, 10 fresh produce samples per type from all supermarkets. Values in parentheses are number of positive samples for E. coli, Salmonella spp., and somatic phages over the total produce surveyed.

TABLE 3: Statistical analysis of microbial quality of fresh produce collected from open air and supermarkets in the Philippines.

Microorganisms	Difference between open air markets	Difference between open air and supermarkets
E. coli	0.0001[a]	0.9147
Salmonella spp.	0.1280	0.3229
Somatic phage	0.2152	0.8161

[a]Value is statistically significant at $\alpha = 0.05$.

Thermotolerant E. coli had values as high as 4.15 $\log_{10}$ CFU/g, while Salmonella spp. values were observed as high as ≥ 2.34 $\log_{10}$ MPN/g determined when all MPN dilution wells were positive after incubation in RV. All positive isolates were confirmed through RT-PCR using fluorescent probes specific for these microorganisms. For somatic phage, most values are found to be higher than 3.00 $\log_{10}$ PFU/g, with plaques being too numerous to count even with 10 μL sample wash solution being used for the assay (Table 1).

Table 2 shows the prevalence of E. coli, Salmonella, and somatic phages in the fresh produce surveyed. All samples positive for E. coli (50 isolates) and Salmonella (74 isolates) spp. were subjected to RT-PCR for confirmation. RT-PCR revealed fluorescent signals at the end of each cycle for positive isolates. It was observed that carrots had the highest microbial counts. This may be attributed to environmental factors such as the manner of planting and harvesting. Soil can greatly contaminate produce, transferring the microorganisms from environment to produce [16]. Pre- and postharvest processes utilizing contaminated water may also contribute to contamination. On the other hand, tomatoes had the least number of microbial contaminants in both open air and supermarkets. This could be due to its smooth pericarp that may possibly inhibit attachment of microorganisms.

Open air markets were observed to have statistically similar Salmonella spp. and somatic phages counts, except for E. coli (Table 3). Post hoc analysis of open air markets showing E. coli contamination revealed that these markets formed two groups, with markets from the provinces in South and Central Luzon forming one cluster and markets from the National Capital Region forming another group.

Although supermarkets are thought to have more stringent processing prior to vending, all log values were statistically insignificant between open air and supermarkets ($P \leq 0.05$) (Table 3). The data suggest that incidence of microbial contamination among markets and supermarkets are likely similar.

Varying molecular techniques can be utilized for the detection of coliforms. While immunological techniques have shown promise for their detection, there are some problems of low antibody specificity. PCR has shown more promise in the detection of coliform bacteria in both sensitivity and specificity [17–19]. The confirmation by RT-PCR revealed fluorescent signals for all positive isolates compared to a lack of signal in the nontemplate controls, while positive controls also indicated expected amplification (data not shown). Additionally, the specificity of the RT-PCR primers and probes confirms the results of the culture techniques that were employed in this study.

4. Discussion

Consumption of fresh produce from health conscious consumers has resulted in a growing industry with global production for fruits and vegetables increasing from 1979 to 2004 [20]. However, it is facing new challenges that require attention, such as the protection of consumers against food-borne pathogens [21]. Contamination of vegetables may occur during growth, harvest, or processing. Under certain conditions, microorganisms can also become localized within vegetables. Conditions that promote entry of microorganisms include damage to the natural structure such as punctures, stem scars, cuts, and splits [22]. In the past several decades, there has been an increase in the occurrence of food-borne illness linked to fresh fruits and vegetables. Fresh vegetables and herbs including those of the leafy variety have been attributed as vehicles for the transmission of microbial food-borne disease worldwide [23]. For instance, in the USA between 1998 and 2002, vegetables were associated with 2.9% (192/6647) of the total food-borne outbreaks recorded [24].

The target bacteria and phage studied in this survey were bacterial pathogens, potentially opportunistic pathogens, or indicator organisms for fecal contamination. *E. coli* is a Gram-negative, rod-shaped bacterium commonly found in the gastrointestinal tract and is also one of the most implicated pathogens on diarrheal cases worldwide [25, 26]. Leafy vegetables are commonly linked to food-borne infections where *E. coli* serves as the responsible disease agent [27–29]. As shown in this study, a number of lettuces and cabbages were contaminated with *E. coli*. Gilbert et al. [30] reviewed the dose-response estimates for *E. coli* O157:H7 and original estimates of infectious dose were less than a few hundred cells. While *E. coli* isolates obtained in this study have not been identified as pathogenic strains, they demonstrated the ability to be cultured at high temperature of 44.5°C for 24 h in relatively high prevalence (Tables 1 and 2).

Salmonella spp. are the most commonly identified aetiological agent associated with fresh produce-related infection, isolated in 48% of cases between 1973 and 1997 in the USA [31] and in 41% of cases during 1992–2000 in the UK. A range of fresh fruit and vegetable products have been implicated in *Salmonella* infection, most commonly lettuce, sprouted seeds, melon, and tomatoes [32]. Jay et al. [33] included data on the incidence of salmonellae in fruit, vegetables, and spices with the prevalence shown to be below 10%. They noted that numbers of salmonellae on raw vegetables are usually $<1 \log_{10}$ CFU/g. On the other hand, this study shows that *Salmonella* can be as high as $2.34 \log_{10}$ MPN/g (Table 2). The high detection rate of *Salmonella* can be alarming as many food-borne outbreaks have been associated with this microorganism (Table 2).

Additionally, studies for the impact of enteric viruses on produce in Southeast Asian countries are also limited. While direct detection of enteric viruses would be problematic due to the intensive processing needed (molecular methods and cell culture), a possible alternative was utilized in this study, surveying the presence of somatic bacteriophages [34]. Such bacteriophages have served as ideal indicators of enteric viruses for water contamination [35] and could serve as model organisms for virus contamination of agricultural produce. Enteric viruses have a low infective dose and remain active even after exposure to low pH (<3) and temperature extremes [36]. While somatic phages are not pathogens, they are classic indicator organisms for fecal contamination and can be utilized to mimic and, in turn, serve as a predictor for the survival of enteric viral pathogens on produce. This study is the first in the Philippines to present and investigate somatic bacteriophages in a wide array of fresh produce.

Gabriel et al. [37] surveyed the microbiological composition of one produce sample, mung bean sprouts, vended in public markets from the National Capital Region; however, only *Salmonella* spp. were observed in that study. Additionally, the data presented were only qualitative in nature and reported as the number positive (or negative) per 25 g samples [37]. In contrast, this study has shown that there is high prevalence of bacterial pathogens and fecal indicator organisms from various fresh produce samples from both open air and supermarkets (Tables 1 and 2). Furthermore,

these microbes are not only observed for their presence but also quantified and compared.

This study observed that carrots had the highest microbial counts and may be attributed to environmental factors from soil as there is extensive exposure during cultivation of this produce type. Soil can greatly contaminate produce, transferring the microorganisms from environment to produce [16]. Pre- and postharvest processes utilizing contaminated water may also contribute to contamination. Alternately, tomatoes had the least number of microbial contaminants in both open air and supermarkets which could be due to its smooth pericarp that may possibly inhibit attachment of microorganisms.

It is also interesting to note and compare the prevalence of bacterial pathogens and phages in the two types of markets. Although produce samples from supermarkets are thought to be less contaminated due to their processing, all log values were found to be statistically similar between open air and supermarkets ($P \leq 0.05$) (Table 3). The data suggest that microbial contamination from all types of markets and supermarkets has a similar prevalence despite the differences in operation.

It is important to note that confirmation of bacterial isolates was done using TaqMan RT-PCR which is highly specific. This method has several advantages as identification of bacterial isolates can be rapid and accurate. It has been demonstrated that RT-PCR assay can not only be specific, but also have amplification kinetics suitable for the specific detection of *E. coli*, irrespective of strain [13]. This study investigated the presence of the *uidA* gene, which is encoded for β-D glucuronidase and is found in almost all strains including the pathogenic strains, *E. coli* O157:H7 [14], and also demonstrated the specificity of OriP1/P3 primers and OriP214 probe for *Salmonella* spp. [14]. While enumeration of such bacterial organisms in this study used simple culture techniques such as multiple-tube fermentation and membrane filtration, these routine methods have limitations. Potential growth of heterotrophic microbial flora from the samples despite using selective media and lack of specificity are some issues. Hence, the use of PCR incorporating a probe was ideal for rapid, sensitive, and specific confirmation of bacterial isolates.

The usefulness of the microbial indicators as tools for risk assessment can be significantly enhanced by the development of testing methods and analysis techniques that can define specific sources of these organisms. In particular, real-time PCR has several advantages including enhanced speed and the absence of post-PCR processing steps. Rapid and accurate identification of bacterial pathogens from food samples is important, both for food quality assurance and for tracing outbreaks of bacterial pathogens within the food supply. Growing concerns regarding the safety of fresh produce warrant a greater emphasis on the development of more rapid, specific, and highly sensitive detection methods [38]. It was shown in this study that RT-PCR can be used as a rapid method in confirming bacterial isolates and shows potential for using similar techniques for determining microbial contamination in fresh produce.

In the Philippines, policies that can help reduce microbial contamination of foods are strengthened by the Department of Agriculture (DA) due to the country's high dependence on agriculture [39, 40]. The Good Agricultural Practice for Fruits and Vegetable Farming (GAPVF) is a set of consolidated safety and quality standards formulated by the DA for on-farm fruit and vegetable production. These codes of practices are based on concept of Hazard Analysis of Critical Control Points (HACCP) and quality management principles. The basis of GAPVF program is to provide safe food product for the consumers and its focus is to reduce risk of microbial and pesticide contamination. Additional benefits of the program are worker safety and protection of the environment. However, the current technologies employed cannot absolutely eliminate food safety hazards associated with fresh produce which are to be consumed raw, as shown in the data and results from this study.

5. Conclusions

This study is the first comprehensive survey of microbial contamination of fresh produce in the Philippines. The increasing awareness of Filipinos of a healthy diet, that is, consuming fresh produce, may also pose an unintentional risk of increasing the incidence of gastrointestinal illnesses (and other related diseases) by the consumption of contaminated foods. The recent undocumented food outbreaks in the country also make it timely to investigate the prevalence of bacterial pathogens and phages in food. Thus, it is imperative to conduct surveillance on these produce samples for microbial contamination to educate consumers buying either in open air markets or in supermarkets resulting in disease detection and prevention. Further, the results established in the study may be of use to farmers, retailers, food safety educators, and policy makers in improving the microbiological quality and safety of fresh produce in the Philippines and in preventing the occurrence of diseases associated with it.

Conflict of Interests

The authors declare that there is no conflict of interests regarding the publication of this paper.

Acknowledgments

This work was supported by the UNU & GIST Joint Programme on Science and Technology for Sustainability, Gwangju Institute of Science and Technology, Korea; Office of the Chancellor of the University of the Philippines Diliman, through the Office of the Vice Chancellor for Research and Development; and Office of the Vice President for Academic Affairs, University of the Philippines System.

References

[1] P. S. Mead, L. Slutsker, V. Dietz et al., "Food-related illness and death in the United States," *Emerging Infectious Diseases*, vol. 5, no. 5, pp. 607–625, 1999.

[2] H. Shuval, Y. Lampert, and B. Fattal, "Development of a risk assessment approach for evaluating wastewater reuse standards for agriculture," *Water Science and Technology*, vol. 35, pp. 15–20, 1997.

[3] N. T. van Ha, M. Kitajima, N. V. M. Hang et al., "Bacterial contamination of raw vegetables, vegetable-related water and river water in Ho Chi Minh City, Vietnam," *Water Science and Technology*, vol. 58, no. 12, pp. 2403–2411, 2008.

[4] A. Yajima and H. Kurokura, "Microbial risk assessment of livestock-integrated aquaculture and fish handling in Vietnam," *Fisheries Science*, vol. 74, no. 5, pp. 1062–1068, 2008.

[5] M. B. Diallo, A. J. Anceno, B. Tawatsupa, E. R. Houpt, V. Wangsuphachart, and O. V. Shipin, "Infection risk assessment of diarrhea-related pathogens in a tropical canal network," *Science of the Total Environment*, vol. 407, no. 1, pp. 223–232, 2008.

[6] G. Howard, S. Pedley, and S. Tibatemwa, "Quantitative microbial risk assessment to estimate health risks attributable to water supply: can the technique be applied in developing countries with limited data?" *Journal of Water and Health*, vol. 4, no. 1, pp. 49–65, 2006.

[7] K. D. Mena and S. D. Pillai, "An approach for developing quantitative risk-based microbial standards for fresh produce," *Journal of Water and Health*, vol. 6, no. 3, pp. 359–364, 2008.

[8] *EPA-814B-92-002, Manual for the Certification of Laboratories Analyzing Drinking Water*, Office of Ground Water and Technical Support Division, U. S. Environmental Protection Agency, Cincinnati, Ohio, USA, 1992.

[9] J. Lasobras, J. Dellunde, J. Jofre, and F. Lucena, "Occurrence and levels of phages proposed as surrogate indicators of enteric viruses in different types of sludges," *Journal of Applied Microbiology*, vol. 86, no. 4, pp. 723–729, 1999.

[10] H. S. Lee and M. D. Sobsey, "Survival of prototype strains of somatic coliphage families in environmental waters and when exposed to UV low-pressure monochromatic radiation or heat," *Water Research*, vol. 45, no. 12, pp. 3723–3734, 2011.

[11] R. S. Wentsel, P. E. O'Neill, and J. F. Kitchens, "Evaluation of coliphage detection as a rapid indicator of water quality," *Applied and Environmental Microbiology*, vol. 43, no. 2, pp. 430–434, 1982.

[12] D. de Medici, L. Croci, E. Delibato, S. di Pasquale, E. Fileti, and L. Toti, "Evaluation of DNA extraction methods for use in combination with SYBR green I real-time PCR to detect *Salmonella* enterica serotype Enteritidis in poultry," *Applied and Environmental Microbiology*, vol. 69, no. 6, pp. 3456–3461, 2003.

[13] H. Takahashi, B. Kimura, Y. Tanaka, J. Shinozaki, T. Suda, and T. Fujii, "Real-time PCR and enrichment culture for sensitive detection and enumeration of *Escherichia coli*," *Journal of Microbiological Methods*, vol. 79, no. 1, pp. 124–127, 2009.

[14] P. Elizaquível and R. Aznar, "A multiplex RTi-PCR reaction for simultaneous detection of *Escherichia coli* O157:H7, *Salmonella* spp. and *Staphylococcus aureus* on fresh, minimally processed vegetables," *Food Microbiology*, vol. 25, no. 5, pp. 705–713, 2008.

[15] SAS 9.1.3 ETL Studio, SAS Institute Inc, User's Guide, Cary, NC, USA, 2004.

[16] E. H. Garrett, U. R. Gorny, L. R. Beuchat et al., "Microbiological safety of fresh and fresh-cut produce: description of the situation and economic impact," *Comprehensive Reviews in Food Science and Food Safety*, vol. 2, no. 1, pp. 13–37, 2003.

[17] B. Mull and V. R. Hill, "Recovery and detection of *Escherichia coli* O157:H7 in surface water, using ultrafiltration and real-time PCR," *Applied and Environmental Microbiology*, vol. 75, no. 11, pp. 3593–3597, 2009.

[18] P. Rose, J. M. Harkin, and W. J. Hickey, "Competitive touchdown PCR for estimation of *Escherichia coli* DNA recovery in soil DNA extraction," *Journal of Microbiological Methods*, vol. 52, no. 1, pp. 29–38, 2003.

[19] A. Rompré, P. Servais, J. Baudart, M. R. de-Roubin, and P. Laurent, "Detection and enumeration of coliforms in drinking water: current methods and emerging approaches," *Journal of Microbiological Methods*, vol. 49, no. 1, pp. 31–54, 2002.

[20] FAO/WHO, "Microbiological hazards in fresh leafy vegetables and herbs," 2013, http://www.who.int/foodsafety/publications/micro/mra_fruitveges/en/.

[21] L. R. Beuchat, "Pathogenic microorganisms associated with fresh produce," *Journal of Food Protection*, vol. 59, pp. 204–216, 1996.

[22] NSW Food Authority, Food Safety Risk Assessment of NSW Food Safety Schemes Australia, 2013, http://www.foodauthority.nsw.gov.au.

[23] L. R. Beuchat, "Vectors and conditions for preharvest contamination of fruits and vegetables with pathogens capable of causing enteric diseases," *British Food Journal*, vol. 108, no. 1, pp. 38–53, 2006.

[24] M. Lynch, J. Painter, R. Woodruff, and C. Braden, "Surveillance for foodborne-disease outbreaks—United States, 1998–2002," *Morbidity and Mortality Weekly Report*, vol. 10, pp. 1–34, 2006.

[25] A. Asea, P. Kaur, and A. Chakraborti, "Enteroaggregative *Escherichia coli*: an emerging enteric food borne pathogen," *Interdisciplinary Perspectives on Infectious Diseases*, vol. 2010, Article ID 254159, 10 pages, 2010.

[26] C. N. Berger, S. V. Sodha, R. K. Shaw et al., "Fresh fruit and vegetables as vehicles for the transmission of human pathogens," *Environmental Microbiology*, vol. 12, no. 9, pp. 2385–2397, 2010.

[27] D. A. Rasko, D. R. Webster, J. W. Sahl et al., "Origins of the *E. coli* strain causing an outbreak of hemolytic-uremic syndrome in Germany," *The New England Journal of Medicine*, vol. 365, no. 8, pp. 709–717, 2011.

[28] Center for Disease Control, "Multistate outbreak of Shiga toxin-producing *Escherichia coli* O26 infections linked to raw clover sprouts at Jimmy John's Restaurants," 2013, http://www.cdc.gov/ecoli/2012/O26-02-12/index.html.

[29] Center for Disease Control, "Multistate Outbreak of Shiga toxin-producing *Escherichia coli* O157:H7 infections linked to organic spinach and spring mix blend," 2013, http://www.cdc.gov/ecoli/2012/O157H7-11-12/index.html.

[30] S. E. Gilbert, R. Whyte, G. Bayne, R. J. Lake, and P. van der Logt, "Survey of internal temperatures of New Zealand domestic refrigerators," *British Food Journal*, vol. 109, no. 4, pp. 323–329, 2007.

[31] S. Sivapalasingam, C. R. Friedman, L. Cohen, and R. V. Tauxe, "Fresh produce: a growing cause of outbreaks of foodborne illness in the United States, 1973 through 1997," *Journal of Food Protection*, vol. 67, no. 10, pp. 2342–2353, 2004.

[32] G. Doran, F. Sheridan, N. Delappe et al., "*Salmonella* enterica serovar Kedougou contamination of commercially grown mushrooms," *Diagnostic Microbiology and Infectious Disease*, vol. 51, no. 1, pp. 73–76, 2005.

[33] S. Jay, D. Davos, M. Dundas, E. Frankish, and D. Lightfoot, "*Salmonella*," in *Foodborne Microorganisms of Public Health Significance*, A. D. Hocking, Ed., pp. 207–266, Australian Institute of Food Science and Technology, Waterloo, NSW, Australia, 2003.

[34] J. Debartolomeis and V. J. Cabelli, "Evaluation of an *Escherichia coli* host strain for enumeration of F male-specific bacteriophages," *Applied and Environmental Microbiology*, vol. 57, no. 5, pp. 1301–1305, 1991.

[35] A. H. Havelaar, M. van Olphen, and Y. C. Drost, "F-specific RNA bacteriophages are adequate model organisms for enteric viruses in fresh water," *Applied and Environmental Microbiology*, vol. 59, no. 9, pp. 2956–2962, 1993.

[36] I. J. Seymour and H. Appleton, "Foodborne viruses and fresh produce," *Journal of Applied Microbiology*, vol. 91, no. 5, pp. 759–773, 2001.

[37] A. A. Gabriel, M. C. Berja, A. M. P. Estrada, M. G. A. A. Lopez, J. G. B. Nery, and E. J. B. Villaflor, "Microbiology of retail mung bean sprouts vended in public markets of National Capital Region, Philippines," *Food Control*, vol. 18, no. 10, pp. 1307–1313, 2007.

[38] A. A. Bhagwat, "Simultaneous detection of *Escherichia coli* O157:H7, *Listeria monocytogenes* and *Salmonella* strains by real-time PCR," *International Journal of Food Microbiology*, vol. 84, no. 2, pp. 217–224, 2003.

[39] R. E. de la Cruz, "Philippines: challenges, opportunities, and constraints in agricultural biotechnology," in *Agricultural Biotechnology and the Poor, Philippines*, pp. 58–63, 1998.

[40] Code of Good Agricultural Practices (GAP) for Fresh Fruits and Vegetable Farming, Philippines's Certification Scheme on Good Agriculture Practices(GAP) for Fresh Fruits and Vegetable Farming, 2013, http://www.agritech.tnau.ac.in.

5

Antimicrobial Edible Films and Coatings for Meat and Meat Products Preservation

Irais Sánchez-Ortega,[1,2] **Blanca E. García-Almendárez,**[1] **Eva María Santos-López,**[2]
Aldo Amaro-Reyes,[1] **J. Eleazar Barboza-Corona,**[3] **and Carlos Regalado**[1]

[1] DIPA, PROPAC, Facultad de Química, Universidad Autónoma de Querétaro, 76010 Querétaro, QRO, Mexico
[2] Área Académica de Química, Instituto de Ciencias Básicas e Ingeniería, Universidad Autónoma del Estado de Hidalgo,
  Ciudad del Conocimiento, Carr. Pachuca-Tulancingo Km 4.5 Col Carboneras, 42184 Mineral de la Reforma, HGO, Mexico
[3] División Ciencias de la Vida, Universidad de Guanajuato, Campus Irapuato-Salamanca, 36500 Irapuato, GTO, Mexico

Correspondence should be addressed to Carlos Regalado; regcarlos@gmail.com

Academic Editor: Matias S. Attene Ramos

Animal origin foods are widely distributed and consumed around the world due to their high nutrients availability but may also provide a suitable environment for growth of pathogenic and spoilage microorganisms. Nowadays consumers demand high quality food with an extended shelf life without chemical additives. Edible films and coatings (EFC) added with natural antimicrobials are a promising preservation technology for raw and processed meats because they provide good barrier against spoilage and pathogenic microorganisms. This review gathers updated research reported over the last ten years related to antimicrobial EFC applied to meat and meat products. In addition, the films gas barrier properties contribute to extended shelf life because physicochemical changes, such as color, texture, and moisture, may be significantly minimized. The effectiveness showed by different types of antimicrobial EFC depends on meat source, polymer used, film barrier properties, target microorganism, antimicrobial substance properties, and storage conditions. The perspective of this technology includes tailoring of coating procedures to meet industry requirements and shelf life increase of meat and meat products to ensure quality and safety without changes in sensory characteristics.

1. Introduction

Animal origin foods (AOF) constitute a good nutrients source for human diet, where their protein provides high biological value and essential amino acids which complement the quality of cereals and other vegetable proteins [1]. However, AOF are susceptible to chemical deterioration and microbiological spoilage and therefore represent a high risk for consumer health, in addition to producer economic losses. According to the Centers for Disease Control (CDC), every year foodborne illnesses account for about 48 million cases, 3,000 deaths, and 128,000 hospitalizations, reaching US $77.7 billion economic burden in the United States. In addition, reduced consumer confidence, recall losses, or litigation costs should be met by the food industry, whereas public health agencies pay the cost of responding to illnesses and outbreaks [2]. Losses can be greater in countries where less

stringent regulation system and sanitary control is practiced. Outbreaks involving AOF comprise 40% of total US reported cases [3]. The presence of foodborne pathogens in a country food supply not only affects the health of local population but also represents a potential for pathogens spread by tourists and consumers where these food products are exported [4].

Edible coatings are food grade suspensions which may be delivered by spraying, spreading, or dipping, which upon drying form a clear thin layer over the food surface. Coatings are a particular form of films directly applied to the surface of materials and are regarded as part of the final product [5]. On the other hand, edible films are obtained from food grade filmogenic suspensions that are usually cast over an inert surface, which after drying can be placed in contact with food surfaces. Films can form pouches, wraps, capsules, bags, or casings through further processing and one of the main differences between films and coatings is their thickness.

The use of films in foods dates back to the 12th century in China where waxes were used to coat citric fruits to retard water loss, whereas the first edible film used for food preservation was made in the 15th century from soymilk (Yuba) in Japan. In England lard or fats were used as coating to prolong shelf life of meat products in the 16th century and in Europe; this process was known as "larding" [6, 7]. In the nineteenth century, a US patent was issued in relation to preservation of meat products by gelatin coatings [7, 8].

Edible films and coatings (EFC) are an alternative to extend the shelf life of AOF by acting as barriers to water vapor, oxygen, and carbon dioxide and as carriers of substances to inhibit pathogenic and spoilage microorganisms. Natural antimicrobial agents may be incorporated into the corresponding suspensions, adding functionality to edible films and coatings, leading to the antimicrobial edible films and coatings (AEFC) obtaining.

There is increased interest in development and use of AEFC to preserve meat quality for longer shelf life periods while maintaining food safety, which is based on consumers demand for natural and safe products. Industry is concerned about these issues, while keeping competitive production costs [9]. Other key issues are sustainability through the use of biodegradable packaging materials and applications of by-products from the food industry that can generate added value [5].

Due to similar properties of edible films and coatings this review discusses characteristics of both types of coverings applied to meat products. This work focuses on a critical discussion of issues raised by recent research findings on the effectiveness of antimicrobial films and coatings and their potential application to enhance safety and quality of meat products.

2. Meat Products

2.1. Meat Importance and Consumption. Meat (including poultry and fish) is the first-choice source of animal protein for many people all over the world [10]. According to the Codex Alimentarius [11] meat is defined as "all parts of an animal that are intended for, or have been judged as safe and suitable for human consumption." Worldwide meat production is expected to be >250 million tons in 2014, with pork as the main product (108.9 million tons), whereas poultry production is expected around 87 million tons. Fish and seafood is also an important market, since world production in 2008 reached 142,287 tons [12]. Meat industry represents a significant share of national economies and therefore production and marketing systems should follow meat sanitation practices and additionally emerging preservation technologies such as the AEFC to extend shelf life and to avoid economic losses.

Nutritionally, meat importance is derived from its high quality protein containing all essential amino acids and its highly bioavailable minerals and vitamins [13]. In 2010, the average annual red meat consumption per capita in developing countries was 32.4 kg, whereas in industrialized countries was 79.2 kg, increasing to 124 kg/capita in the USA. Global annual poultry consumption rose from 11.1 to 13.6 kg/capita

between 2000 and 2012. In 2009, fish accounted for 16.6% of world population intake of AOF and 6.5% of all protein consumed [13].

2.2. Meat Spoilage. Meat quality is highly dependent on preslaughter handling of livestock and postslaughter handling of meat [10]. Among the main factors affecting meat quality is pH, which is determined by the glycogen content of the muscle and varies from 5.4 to 5.7 in postrigor muscle; another important factor is temperature, which must be quickly decreased from 37°C to refrigeration temperatures (4–8°C) [14].

There are three mechanisms involved in meat and meat products deterioration during processing and storage: microbial spoilage, lipid oxidation, and enzymatic autolysis. Microbial population may arrive from native microflora of the intestinal tract and skin of the animals or through environmental, human, handling, and storage conditions associated to the production chain [15]. Microbial growth in meat can result in slime formation, structural components degradation, decrease in water holding capacity, off odors, and texture and appearance changes [10]. Lipid oxidation depends on fatty acids composition, vitamin E concentration, and prooxidants such as free iron in muscles. Oxidation products, such as hydroperoxides, aldehydes, and ketones, can cause loss of color and nutritive value due to degradation of lipids, pigments, proteins, carbohydrates, and vitamins [10, 16]. Enzymatic autolysis of carbohydrates, fats, and proteins of the tissues results in softening and greenish discoloration of meat and may lead to microbial decomposition. Proteolytic enzymes are active even at low temperatures (5°C) leading to microbial growth, loss of water holding capacity, and biogenic amines production [17].

2.3. Meat Related Outbreaks. Meat products outbreaks are often due to inadequate cooking or cross-contamination from other foods. However, contamination may occur while meat is processed, cut, packaged, transported, sold, or handled. Pathogenic microorganisms do not survive thorough meat cooking, but several of their toxins and spores do [10].

Red meat is frequently involved in outbreaks, mainly due to the presence of *Salmonella* spp., *Listeria* spp., *Clostridium* spp., and *Staphylococcus* spp. [3]. Most outbreaks reported in the EU in 2010 were due to meat and meat products consumption in which *Salmonella* was the main pathogen involved [18]. *Listeria* infection is often considered as the most lethal; for instance, in 1998 hot dogs consumption caused 21 deaths and >100 illnesses [3]. Recently, an outbreak of *Salmonella typhimurium* was linked to the consumption of ground beef which caused hospitalization of seven people [19].

In the case of poultry *Salmonella* and *Campylobacter* account for most of the cases of food poisoning associated with chicken [3, 20]. In 2010 turkey contaminated with *Clostridium perfringens* caused 135 illnesses in Kansas (USA), whereas in 2011 ground turkey contaminated with *Salmonella* Heidelberg infected 136 people in 34 USA states [3].

Most outbreaks caused by fish and fish products are caused by natural toxins (scombrotoxin and ciguatoxin), rather than by bacteria or viruses. However, outbreaks caused mainly by *Vibrio parahaemolyticus* and *V. cholerae* in raw oysters have been reported; additionally *Clostridium botulinum*, *Staphylococcus aureus*, *Salmonella enterica*, and *Escherichia coli* were also involved in illnesses due to fish consumption [3, 21].

3. Edible Films and Coating Types

EFC act as barrier between food and the surrounding environment to enhance the quality of food products protecting them from physical, chemical, and biological deterioration. Design and application of EFC on meat products arises from the search of new preservation methods, the need to add value to by-products from renewable sources, the desire to give food products a more natural or ecological image, and reduction of environmental impact of using oil-derived plastic packaging materials [22]. Additionally, they may provide moisture loss reduction during storage of fresh or frozen meats, prevention of juice dripping, and decrease in myoglobin oxidation of red meats. There are two commercially available edible films, New Gem™, which contains spices and bilayer protein films that are used to enhance ham glaze and Coffi™, that is made from collagen nettings used to wrap boneless meat products [23]. Antimicrobials or antioxidant compounds incorporated into the polymer matrix may prevent growth of spoilage and pathogenic microorganisms, delay of meat fat rancidity, discoloration prevention, and even improvement of the nutritional quality of coated foods [24, 25].

3.1. Composition and Properties of Lipid-Based Films and Coatings. A wide range of hydrophobic compounds has been used to produce EFC, including animal and vegetable oils and fats (peanut, coconut, palm, cocoa, lard, butter, fatty acids, and mono-, di-, and triglycerides), waxes (candelilla, carnauba, beeswax, jojoba, and paraffin), natural resins (chicle, guarana, and olibanum), essential oils and extracts (camphor, mint, and citrus fruits essential oils), and emulsifiers and surface active agents (lecithin, fatty alcohols, and fatty acids) [26]. In meat products, emulsifiers and surface active agents are sometimes used as gas and moisture barriers. However, pure lipids can be combined with hydrocolloids such as protein, starch, cellulose, and their derivatives providing a multicomponent system able to be applied as meat coatings [27]. In fresh and processed meats, lipid incorporation into EFC can improve hydrophobicity, cohesiveness, and flexibility, making excellent moisture barriers, leading to prolongation of freshness, color, aroma, tenderness, and microbiological stability [24].

Palmitoylated alginate is the only lipid-containing material of AEFC recently reported to wrap beef muscle and ground beef [28] (Table 1). However, essential oil extracts have been widely used to promote antimicrobial activity of AEFC (column 3, Tables 1–3).

3.2. Composition and Properties of Protein-Based Films and Coatings. Film-forming proteins are derived from animals (casein, whey protein concentrate and isolate, collagen,

gelatin, and egg albumin) or plant sources (corn, soybean, wheat, cottonseed, peanut, and rice). Protein-based films adhere well to the meat hydrophilic surfaces and provide barrier for oxygen and carbon dioxide but do not resist water diffusion [27]. Plasticizers, such as polyethylene glycol or glycerol, are added to improve flexibility of the protein network, whereas water permeability can be overcome by adding hydrophobic materials such as beeswax or oils like oleic that can affect films properties such as crystallinity, hydrophobicity, surface charge, and molecular size, improving films characteristics and their application [6, 31, 34]. Despite their advantages, protein films may be susceptible to proteolytic enzymes present in meat products or allergenic protein fractions may cause adverse reactions to susceptible people [24].

3.3. Composition and Properties of Polysaccharides-Based Film and Coatings. Polysaccharide coatings are generally poor moisture barriers, but they have selective permeability to O_2 and CO_2 and resistance to fats and oils [25]. Polysaccharide films can be made of cellulose, starch (native and modified), pectins, seaweed extracts (alginates, carrageenan, and agar), gums (acacia, tragacanth, and guar), pullulan, and chitosan. These compounds impart hardness, crispness, compactness, viscosity, adhesiveness, and gel-forming ability to a variety of films [24, 44, 45]. Polysaccharide films and coatings can be used to extend the shelf life of muscle foods by preventing dehydration, oxidative rancidity, and surface browning. When applied to wrapped meat products and exposed to smoke and steam, the polysaccharide film actually dissolves and becomes integrated into the meat surface resulting in higher yields, improved structure and texture, and reduced moisture loss [27].

Materials recently used to obtain AEFC in meat and meat products, poultry, and fish and fish products are shown in column 2 of Tables 1, 2, and 3, respectively. Chitosan based AEFC were the most commonly reported in recent years and have been used to wrap pork meat hamburgers and sausages [39, 42] (Table 1), as films and coatings on roasted and sliced turkey [46, 47] (Table 2) and as films on cod fillets [48] (Table 3). Chitosan was used as both polymeric material and antimicrobial agent, for roast beef coating [32] (Table 1) and chicken breast fillets (wrapping and coating) [49, 50] (Table 2) and as coating of Atlantic cod and herring [51] and as films on sea bass fillets [52] (Table 3). WPI and cellulose (or its acetate salt), despite being less reported, are also materials used in AEFC for meat and meat products [29, 30, 33, 35] (Table 1): turkey frankfurters [53] (Table 2) and smoked salmon [54] (Table 3). Several reports mention pectin for production of AEFC to wrap cooked ham and bologna [40] (Table 1) and chicken breast [55, 56] (Table 2), whereas other reports show gelatin based antimicrobial films placed on top or between slices of fish products [48, 57–59] (Table 3).

4. Common Antimicrobials Used in EFC

Incorporation of antimicrobial compounds into EFC as an alternative to their direct application onto the meat surface has the advantage of gradual release of the antimicrobial

TABLE 1: Use of antimicrobial films and coatings in meat and meat products.

Product	Coating material	Antimicrobial compound	Target microorganism	Inoculation technique	Conditions	Results	Reference
Sliced bologna and summer sausage	Whey protein isolate (WPI, pH 5.2) films	0.5 to 1.0% p-aminobenzoic acid (PABA) and/or sorbic acid (SA)	L. monocytogenes, E. coli O157:H7, and S. enterica typhimurium (10^6 CFU/g)	0.1 mL of inoculum spread onto both surfaces	4°C, 21 days (d) Slices placed in plates covered with edible film Stored in aerobiosis	WPI films with SA or PABA reduced L. monocytogenes, E. coli, and S. typhimurium populations by 3.4–4.1, 3.1–3.6, and 3.1–4.1 log CFU/g, respectively, on both products	[29]
Hot dogs (beef 60%, pork 40%)	Whey protein isolate films (casings)	p-aminobenzoic acid (PABA) 1%	L. monocytogenes (10^3 CFU/g)	Immersion in L. monocytogenes culture for 1 min and dried in a safety cabinet	4°C, 42 d Vacuum-packaged samples	Growth inhibition for 42 d in refrigeration but no population reduction. Controls increased around 2.5 CFU/g	[30]
Beef muscle slices	Milk protein films	Oregano essential oil (OR) 1.0% (w/v), Pimento essential oil (PI) 1.0% (w/v), or 1% OR-PI (1:1)	Escherichia coli O157:H7 or Pseudomonas spp. (10^3 CFU/cm^2)	Spreading over meat surface Samples placed in plates, covered on either side with the corresponding film	4°C, 7 d Meat sterilized by radiation and then inoculated Samples in plates hermetically sealed	Film with OR was the most effective against both bacteria Reduction of 0.95 log of Pseudomonas spp. and 1.12 log reduction of E. coli O157:H7	[31]
Sterile beef muscle slices or ground beef	Palmitoylated alginate films Activated alginate beads	Covalently immobilized nisin (N) to activated alginate beads (AAB) (0–1000 IU/mL), or ground beef mixed with 0–1000 IU/mL of N	Staphylococcus aureus (10^4 CFU/g)	Inoculated using a sterile spoon and placed in sterile plates	4°C, 14 d Covered with immobilized nisin film or mixed with nisin solution	Reduction of 0.91 and 1.86 log CFU/cm^2 on samples covered with film (500 or 1000 IU/mL, resp.) After 14 days: N solution (500 or 1000 IU/g) mixed with ground beef reduced to 2.2 and 2.81 log CFU/g, respectively; N (500 or 1000 IU/g) in AAB reduced to 1.77 and 1.93 log CFU/g, respectively	[28]
Roast beef	Chitosan (CH, high or low molecular weight) coatings dissolved in lactic or acetic acid	Chitosan, lactic, or acetic acid (0.5 and 1%; w/v)	Listeria monocytogenes (10^6 CFU/g)	1 mL culture onto 5 g cubed meat, air dried 10 min Then dipped in chitosan for 30 s, dried for 1 h and placed into sterile Whirl-Pack bags	4°C, 28 d 5 g cubed roast beef samples placed into sterile bags	Reduction of 1–3 log CFU/g for low molecular weight chitosan in acetic and lactic acids, respectively, after 28 d	[32]
Frankfurters	Cellulose (produced by G. xylinus) films	Nisin (N), 625 and 2500 IU/mL	L. monocytogenes Scott A serotype 4b (10^6 CFU/mL)	Dipping in 0.85% saline sln. containing L. monocytogenes for 2 s	4°C, 14 d Samples wrapped in a single layer of film and vacuum-sealed for 2.5 s	Films containing 625 IU/mL N not significantly reduced L. monocytogenes populations. Films with 2500 IU/mL N decreased 2 log CFU/g compared to the control	[33]
Pork loins	Gelidium corneum–gelatin (GCG) films	Grapefruit seed extract (GFSE, 0.08% w/v) or green tea extract (GTE, 2.80% w/v)	E. coli O157:H7 (NCTC12079) and L. monocytogenes (KCTC 3710) (10^5 CFU/g each one)	Spread with a sterile glass rod and allowed to drain for 10 min	4°C, 10 d Samples were packed in direct contact to films and stored in sterile polystyrene trays	Samples packed with the GCG film containing GFSE or GTE decreased population of E. coli O157:H7 and L. monocytogenes in 1 and 2 log CFU/g, respectively, compared to the control	[34]

TABLE 1: Continued.

Product	Coating material	Antimicrobial compound	Target microorganism	Inoculation technique	Conditions	Results	Reference
Ham	Cellulose acetate films	Pediocin (ALTA 2351) (25% and 50%, w/v)	L. innocua (10⁶ CFU/mL) and Salmonella sp. (10⁶ CFU/mL)	Immersion in a 0.1% w/v peptone solution of L. innocua or Salmonella sp. for 10 min	12°C, 15 d Samples sterilized under UV (5 min each side) Films intercalated with ham slices packed in plastic bags and vacuum sealed	The 50% pediocin-film reduced L. innocua 2 log relative to the control. The 25% and 50% pediocin-films had similar performance on Salmonella sp. inhibition, both presenting 0.5 log reduction relative to the control	[35]
Fresh ground beef patties	Soy protein films	Oregano (OR), thyme (TH), or OR-TH essential oils (5%)	E. coli O157:H7, Staphylococcus aureus, Pseudomonas aeruginosa, and Lactobacillus plantarum	No inoculation	4°C, 12 d Film applied to the upper and bottom surfaces of patties and vacuum packaged in plastic bags	Pseudomonas spp. in samples coated with TH and OR films decreased in 1.13 and 1.27 log CFU/g, respectively. Coliforms were reduced by 1.6, 1.9 and 2.0 log CFU/g with addition of OR, OR-TH, and TH, respectively	[36]
Salami	Sodium caseinate (SC) films and coatings	Chitosan (CH) (2%)	Mesophilic and psychrotrophic aerobic bacteria and yeast and mold	No inoculation	10°C, 5 d, 65% RH. Film added by immersion and as wrapper Immersed slices air dried at 30°C and 50% RH for 50 min All food faces were contacted with wrapping film	CH and SC/CH films applied as both, coatings and wrappers, exerted a strong bactericidal action on 3 microbial populations analyzed, with reductions of 2 to 4.5 log CFU/g	[37]
Ground beef patties	Zein films	Lysozyme (LY) (43 mg/g) and disodium Ethylene diamine tetra acetic acid (Na₂EDTA, 19 mg/g)	Mesophilic microorganisms (TVC) and coliforms (TCC)	No inoculation	4°C, 7 d Films at both sides of each piece Wrapped with stretch plastic film and with aluminum foil	After 5 and 7 d, TVC of patties with LY and Na₂EDTA films were significantly lower (0.75–1.9 log CFU/g) than control films. After 5 d, TCC of patties with LY and Na₂EDTA films were significantly lower than the control but after 7 d, no significant difference in TCC of patties was found	[38]
Pork meat hamburgers	High molecular weight chitosan (1% w/v), acetic acid (1% w/v), lactic acid (1% w/v) films	Sunflower oil (1%)	Mesophilic bacteria, coliforms	No inoculation	5°C, 8 d Surface of both sides of the hamburgers coated with the films placed in PET trays	Reduction of 0.5–1 log for mesophilic microorganisms; 1 log CFU/g for coliforms	[39]

TABLE 1: Continued.

Product	Coating material	Antimicrobial compound	Target microorganism	Inoculation technique	Conditions	Results	Reference
Cooked ham and bologna	High methoxyl pectin + Apple, carrot or hibiscus puree films	Carvacrol (CV) or cinnamaldehyde (CM) (0.5%, 1.5%, and 3.0%, w/v)	*L. monocytogenes* (101M; serotype 4b) 10^6 CFU/mL	Dispersed on the surface as droplets. Inoculated samples dried under the biohood (30 min), flipped over, and inoculated on the other side Inoculated samples were dried again (30 min) and then surface wrapped with one of the test films	4°C, 7 d Samples were kept frozen and thawed before use. Sample wrapped in 2 pieces of circular films and parts of the films were not directly in contact with the meat surface	Films containing 3% CV showed 3 log reductions on ham at day 7. Bologna, films with 3% CV reduced 2 log CFU/g at day 7. Reductions with 1.5% CV were 0.5–1, 1–1.5, and 1–2 logs at day 0, 3, and 7, respectively. Films containing 3% CM, only 0.5–1.5 and 0.5–1.0 log CFU/g reductions were seen at day 7 on ham and bologna, respectively. Limited reduction (0.2–0.3 log CFU/g) was observed with 1.5% CM films	[40]
Bacon	Red algae (RA) films	1% w/v, grapefruit seed extract (GFSE)	*Escherichia coli* O157:H7 (10^6 CFU/g) and *L. monocytogenes* (10^7 CFU/g)	Spread separately on the surface of bacon with a sterile glass rod and allowed to rest for 30 min	4°C, 15 d Packed by wrapping	*E. coli* O157:H7 decreased 0.45 log CFU/g and *L. monocytogenes* decreased by 0.76 log CFU/g respect to the controls	[41]
Pork sausages	Chitosan films	Green tea extract 20% (w/v) in the chitosan film-forming solution	Mesophilic bacteria, yeasts and molds, lactic acid bacteria (LAB), not inoculated	No inoculation	4°C, 20 d Sausages wrapped with films, packaged into a pouch of low density polyethylene coated with polyamide plastic bag and heat-sealed	On day 12, faster growth in control samples for total viable count and molds and yeast was found; no difference for LAB	[42]
Cooked cured ham	Polylactic acid (PLA) films	Lauric arginate (LAE) (0% to 2.6%, w/w)	*Listeria monocytogenes* and *Salmonella enterica* serovar typhimurium (10^5 CFU/mL)	Inoculum of both *L. monocytogenes* and *S. typhimurium* onto surface of the sliced ham	4°C, 7 d Slices sterilized with UV on each side prior to inoculation Inoculated samples wrapped with LAE-coated PLA film and stored in closed plates	LAE-coated PLA film (2.6%) showed a significantly greater antibacterial activity, with *L. monocytogenes* and *S. typhimurium* levels reduced to <2 log CFU/film after 24 h exposure and remaining at this low level for the next 6 d	[43]

TABLE 2: Use of antimicrobial films and coatings in poultry.

Product	Coating material	Antimicrobial compound	Target microorganism	Inoculation technique	Conditions	Results	Ref.
Chicken breast (ready-to-eat cooked chicken cubes)	Zein coatings dissolved in propylene glycol (ZP) or ethanol (ZE)	Nisin (N) (1000 IU/g) and/or calcium (CP) propionate (1% w/v)	L. monocytogenes V7 (low and high inoculum 2.67 log CFU/g and 6.89 log CFU/g, respectively, at 4°C)	Cubes immersed in 24 h broth cultures for 30 s, allowed to drip free of excess inoculum, and dried. Frozen samples were irradiated (3.0 kGy) and kept frozen until used	4°C or 8°C, 24 d. Cubes boiled in water bath for 20 min, were inoculated, then dried, followed by dipping in edible ZP or ZE, with and without antimicrobials. Air dried samples (20 min) stored in sterile bags	L. monocytogenes was reduced by 4.5–5 log CFU/g relative to the control after high dose and 16 d at 4°C, with more significant effect when N was added to the films. Low inoculum dose and using ZPNCP film caused complete inhibition from 4 to 24 d, either at 4° or at 8°C	[18]
Turkey frankfurter	WPI coatings	Grape seed extract (GSE, 1.0–3.0% w/v), nisin (N, 6–18 kIU/g), malic acid (MA 1.0–3.0%; w/v), EDTA (1.6 mg/mL), and their combinations	L. monocytogenes, E. coli O157:H7, and Salmonella typhimurium (10^6 CFU/g)	Samples were defrosted and dipped into cultures of 10^6 CFU/mL of L. monocytogenes, E. coli O157:H7, or S. typhimurium for 1 min at room temperature. Inoculated samples were then air dried under laminar flow conditions	4°C, 28 d Samples were dipped in film-forming solutions (1 min) and dried (10 min, room temperature). Samples were then packed individually in sterile bags, and stored	L. monocytogenes decreased to 2.3 log/g (N, 6000 IU/g; GSE 0.5%; MA 1.0%). S. typhimurium decreased to 5 log CFU/g using any antimicrobial, whereas E. coli decreased to 4.6 log cycles using N, MA and EDTA. All reductions were relative to the control	[53]
Chicken breast	High methoxyl pectin 11400 with apple puree films	Carvacrol (C) or cinnamaldehyde (CM) at 0.5–3% (w/w).	Salmonella enterica serovar Enteritidis or E. coli O157:H7 (ATCC 35150) (10^7 CFU/g)	Inoculum was dispersed on the surface as droplets	23°C or 4°C, 3 d Samples were dipped in boiling water (40 s), plated and exposed in a bio-hood for drying (30 min). Sample was flipped over and inoculated in a similar way. Meat was wrapped using appropriate edible films	At 23°C, films with 3% antimicrobials showed the highest reductions (4.3–6.8 log CFU/g) of both S. Enteritidis and E. coli O157:H7. At 4°C, C exhibited greater activity than CM. Relative to control samples, films with 0.5–3% C reduced S. Enteritidis by 1.6–3 log CFU/g, whereas 1–3% CM films reduced its population by 1.2–2.8 log CFU/g. Films with 0.5–3% C reduced 1–3 log E. coli whereas 1–3% CM films inhibited 0.2–1.2 log CFU. Treatments were at 4°C	[55]

TABLE 2: Continued.

Product	Coating material	Antimicrobial compound	Target microorganism	Inoculation technique	Conditions	Results	Ref.
Chicken breast	k-carrageenan (kCF) films	Ovotransferrin, OTf (25 mg), EDTA (5 mM), and potassium sorbate (PS, 10 mg/g of k-carrageenan)	E. coli	No inoculation	5°C, 7 d. Samples were wrapped with k-carrageenan-based films and packed in plastic bags	Samples wrapped with kCF added with 5 mM EDTA alone or mixed with 25 mg OTf allowed 2.7 log CFU/g reduction of E. coli, compared to the control, at day 7. Addition of 25 mg of OTf or 10 mg PS slightly inhibited microbial growth	[60]
Turkey bologna	Gelatin films	Nisaplin based films (GNF) (0.025–0.5 %; w/v nisin) and Guardian CS1-50 based films (GGF) (0.5–4 %; w/v).	L. monocytogenes (10^6 cfu/mL)	Inoculated by surface spreading. Samples were thawed at 4°C for 18 h and then inoculated and covered with antimicrobial film. Each sample was vacuum-sealed	4°C, 56 d. Samples were irradiated at 4°C (2.4 mrad for 521 min) and stored at −70°C	Both 0.5% GNF and 1% GGF inhibited L. monocytogenes by 4 log CFU/cm^2 and 3 log CFU/cm^2, respectively, relative to the control, during storage at 4°C for 56 d. GGF inhibited L. monocytogenes by 2.17 log CFU/cm^2 at 7 d	[61]
Roasted turkey	Starch, chitosan, alginate, or pectin coatings	Sodium lactate (SL) and sodium diacetate (SD), OptiForm PD4 (OF4), NovaGARDCB1 (NG1), Protect-M (PM), and Guardian NR100 (GN)	A cocktail of five strains of L. monocytogenes (PSU1 serotype 1/2a, F5069 serotype 4b, ATCC19115 (serotype 4b), PSU9 serotype 1/2b, and Scott A serotype 4b) 10^3 UFC/mL	Spreading on both sides of the turkey surface, 10^3 CFU/cm^2. After inoculation, turkey samples were kept at 4°C for 20 min	4°C, 8 weeks. Coatings on each side were dried in a laminar-flow hood for 20 min each side. All samples were inserted into nylon/polyethylene pouches and vacuum sealed	OF4 (2.5%) alone or mixed with PM (0.12%) in films made from alginate, chitosan or pectin were the most effective, reaching L. monocytogenes reduction by 3.5 log/cm^2 relative to the control after storage	[46]
Chicken breasts	3% solution of high methoxyl pectin added with golden delicious apple puree films	Carvacrol (C) and cinnamaldehyde (CM) (0.5–3.0 %; w/v)	Campylobacter jejuni (D28a, H2a and A24a), 10^7 CFU/mL	Samples dipped in boiling water for 40 s and dried in a biohood for 1 h. Chicken was dip-inoculated for 5 min	23°C and 4°C, 72 h in anaerobiosis. Samples placed in sterile plates, dried in a 42°C, 10% CO$_2$ incubator for 1 h and then wrapped with apple films and stored	Films with ≥1.5% CM reduced populations of both strains to below detection at 23°C at 72 h. Films with 3% C reduced populations of A24a and H2a to below detection. Using 3% C, films reduced to 0.5 log CFU/g of both strains A24a and D28a and 0.9 logs for H2a at 4°C	[56]

TABLE 2: Continued.

Product	Coating material	Antimicrobial compound	Target microorganism	Inoculation technique	Conditions	Results	Ref.
Chicken breast fillets	Chitosan (CH) coating, deacetylation degree of 75–85%	Chitosan (1.5 % w/v) and/or oregano oil 0.25% v/w (OO)	Mesophilic microorganisms, Pseudomonas spp. and Brochothrix thermosphacta	No inoculation	12, 18 y 21 d, 4°C Samples were dipped into the chitosan solution (1.5 min) and drained. Sterile OO was added to the surface. Fillets were packed in plastic pouches, and stored in a modified atmosphere	Shelf life of chicken fillets can be extended using either OO and/or CH, by approximately 6–21 d	[49]
Chicken breasts fillets	Chitosan (CH) films	CH or CH-LAE (1, 5 or 10%, by weight)	Mesophiles, psychrophiles, yeast, moulds, Pseudomonas, coliforms, LAB, and hydrogen sulfide-producing bacteria	No inoculation	0, 2, 6 and 8 d, 4°C. Slices wrapped with CH or CH-5% LAE films, then packed in polyethylene films	CH films reduced 0.47–2.96 log population of fillets, depending on time and microbial group studied. Incorporation of LAE (5%) increased antimicrobial activity to 1.78–5.81 log reduction, and maintaining the initially low microbial fillets load for 8 d	[50]
Sliced turkey deli meat	Chitosan (CH, 2–5% w/w) films and coatings added with 2% solution of either acetic, lactic or levulinic acids	Lauric arginate (LAE, 50–200 mL/mL) and nisin (NIS, 25 mg/mL) alone or in combination	L. innocua (6-7 log CFU/cm²)	Even spread over the meat surface (3 × 3 cm²) using sterile spreaders	48 h, 37°C. Films were placed on top of inoculated turkey; coatings applied by spreading. The product was vacuum packed and stored at 10°C for 24 h prior to microbiological analysis	High CH levels reduced 4.6 log CFU/cm². NIS addition (486 IU/cm²) reduced Listeria by 2 and 2.4 log CFU/cm² for 2% and 5% CH, respectively. Combination of CH, LAE and NIS had similar reductions as only CH with LAE, suggesting no additive or synergistic effect by NIS. Despite no statistical difference (P < 0.05), coatings showed more microbial reduction than films	[47]

TABLE 3: Use of antimicrobial films and coatings in fish and seafood.

Product	Coating material	Antimicrobial compound	Target microorganism	Inoculation technique	Conditions	Results	Ref.
Atlantic cod (Gadus morhua) and herring (Clupea harengus)	Chitosan coatings	Chitosan (CH) with different molecular weights and viscosities (14, 57 or 360 mPa s)	Psychrotrophic microorganisms (PT) and total plate count (TPC)	No inoculation	12 d, 4°C	Herring fillets treated with 57 and 360 cP CH showed lower PT population than 14 cP CH-treated fillets after 6 d. CH treatments reduced to 10^3 and 10^2 TPC of herring and cod samples, respectively, after 12 d	[51]
Smoked salmon	Whey protein isolate (WPI) coatings	Lactoperoxidase system (LPO) (0–0.5%, w/v)	L. monocytogenes (V7 serotype 4b, LCDC 81-861 serotype 4b, Scott A serotype 4b, 101M and 108M) (10^2–10^4 CFU/g).	Spotted directly onto the salmon (I + C) or on top of applied coating (C + I) and spread with a hockey stick	4°C and 10°C, 35 d. I + C samples were inoculated and then dried for 0.5 h and coated. C + I samples dried for 1h and then inoculated	Samples coated by LPO-WPI showed <1.0 log CFU/g of L. monocytogenes at 4°C for 35 d (both treatments). L. monocytogenes was completely inhibited in C + I samples stored during 35 d at 10°C	[54]
Cold-smoked sardine (Sardina pilchardus)	Gelatin (G) films	Oregano extract (OE) (Origanum vulgare 1.5%, v/v) or rosemary (RM) (Rosmarinus officinalis, 20% w/v) or Chitosan (CH) (1.5% w/v), high pressure (300 MPa/20°C/15 min) (HP).	Total viable count (TVC), H2S-reducing organisms, luminescent bacteria, and Enterobacteriaceae	No inoculation	5°C, 20 d Fish slices were placed between two layers of edible films and were stored in clean bags	Fish coated with OE-G and RM-G films reduced TVC by 1.99 and 1.54 log CFU/g respectively, on d 16. H2S-reducing bacteria followed a similar pattern. OE and RM had no effect, but CH reduced to ≤10^3 CFU/g in all cases. Pressurized samples produced undetectable levels of all microorganisms for 20 d, except uncoated sample whose TVC was 10^5 CFU/g at d 20	[58]
Cod (Gadus morhua)	Gelatin (G) or in combination with chitosan (CH) films	Clove essential oil (CO)	Total bacterial count (TVC), H2S-producers organisms, luminescent organisms, Pseudomonas, Enterobacteriaceae (EB), and lactic acid bacteria (LAB)	No inoculation	2°C, 11 d Fillets were covered with the G-CH film containing CO, and vacuum-packed in plastic bags	TVC count was 6.1 log CFU/g. Luminescent bacteria reached 6 log CFU/g after 3 d, but later were undetected. H2S producers were completely inhibited from d 3 onwards. LAB and EB increased during storage despite storage temperature	[48]

TABLE 3: Continued.

Product	Coating material	Antimicrobial compound	Target microorganism	Inoculation technique	Conditions	Results	Ref.
Cold smoked salmon (CSS) slices and fillets	Alginate (AL), κ-carrageenan, pectin, gelatin, or starch coatings	Sodium lactate (SL, 0–2.4% w/v) and sodium diacetate (SD, 0–0.25% w/v), OptiForm (OF, 2.5% w/v)	Mixture of *L. monocytogenes* strains: PSU1, PSU9, F5069, ATCC 1915, and Scott A. 10^3 CFU/g	Surface-inoculated	4°C, 30 d Samples coated with AL incorporating SL/SD or OF, dried 20 min and stored at 4°C in vacuum sealed bags	Al coatings with 2.4% SL/0.25% SD and OF reduced *L. monocytogenes* by 3.2 and 4 log CFU/g (slices) and 2.4 and 3 log CFU/g (fillets), respectively, relative to control sample	[62]
Bream (*Megalobrama amblycephala*)	Alginate (AL) coatings	Vitamin C (VC, 5% w/v) and tea polyphenols (TP, 0.3% w/v)	TVC	No inoculation	21 d, 4°C Bream was dipped in AL-antimicrobial solutions (1 min), air-dried (1 min) and immersed in 2% (w/v) $CaCl_2$ (1 min) to obtain gels. Samples were packed and stored	After 4 d of storage. The TVC of VC and TP decreased by 1.6 and 1.5 log CFU/g, respectively, on day 21	[63]
Sea bass slices	Gelatin extracted from the skin of unicorn leatherjacket (*Aluterus monoceros*) films	Lemongrass essential oil (LEO) 25% (w/w)	Mesophilic (TVC) and psychrophilic (PS) microorganisms, enterobacteria (EB), and H_2S-producing bacteria LAB	No inoculation	12 d, 4°C For each slice, films were placed on both sides. Subsequently, the samples were placed in polystyrene trays wrapped with extensible polypropylene film	TVC of unwrapped sample increased to 7.2 log CFU/g at d 4 reaching 7.9 log CFU/g at d 12. TVC of LEO-film wrapped samples was 5.6 log CFU/g at d 12. PS count for control, G and LEO films was 6.0, 5.5 and 4.0 log CFU/g, respectively. LAB increased to 7.2, 6.7 and 5.9 log CFU/g at the end of storage. LEO-film showed the lowest EB counts (2.2 log CFU/g), as compared to control	[57]
Sea bass (*Dicentrarchus labrax*)	Chitosan (CH) films	CH with vacuum packaging	Total mesophilic aerobic bacteria (TVC) and psychrotrophic (PS) aerobic bacteria	No inoculation	4°C until end of shelf life. Fish fillets were covered using CH films, wrapped and vacuum packaged using polyethylene bags	The acceptable limit of 6 log CFU/g and 7 log CFU/g for PS and TVC bacteria, respectively, was reached after 25 d at 4°C. Control samples reached this limit after 5 d	[52]

TABLE 3: Continued.

Product	Coating material	Antimicrobial compound	Target microorganism	Inoculation technique	Conditions	Results	Ref.
Indian oil sardine (*Sardinella longiceps*)	Chitosan (CH) (1 and 2% w/v) coatings	Chitosan (1 and 2% w/v)	Mesophilic microorganisms (TVC)	No inoculation	11 d, 1-2°C. Fillets were dipped in 1 and 2% CH at 1-2°C for 10 min, drained for 5 min and placed in trays for 24 h, then sealed using HDPE	Eating quality was maintained for 8 and 10 d for 1 and 2% CH respectively, whereas untreated samples lasted 5 d. The limit of 10^7 CFU/g of TVC was exceeded after 7, 9 and 11 d for untreated, 1% and 2% CH treated samples, respectively	[64]
Salmon	Barley bran protein and gelatin (BBG) films	Grapefruit seed extract (GFSE) (0.5–1.2% w/v)	*E. coli* O157:H7 and *L. monocytogenes* (10^6 CFU/mL)	*E. coli* O157:H7 and *L. monocytogenes* were spread individually on sample surface and allowed to rest for 30 min	4°C, 15 d Samples wrapped using the BBG film. Samples packed in polyethylene terephthalate film were used as control	After 15 d, populations of *E. coli* and *L. monocytogenes* inoculated salmon with the BBG film containing GSE decreased by 0.53 and 0.50 log CFU/g, respectively, compared to the control	[59]
Cold-smoked salmon	Potato processing waste (PPW) films	Oregano essential oil (OO) 0.97% and 1.92% (185 and 289 mg oil/g film)	*L. monocytogenes* V7 (6.7–6.9 log CFU/g)	*L. monocytogenes* Overnight culture (100 μL) was spotted at 25–30 locations on salmon fillet, spread and dried in a biological hood (30 min)	4°C, 28 d Salmon samples were wrapped with edible films and were vacuum packed	Coated samples with PPW-OO, reduced *Listeria* population by 0.4–2.4 log CFU/g as compared to control samples, after storage period	[65]

compound from the AEFC leading to a reduction of added antimicrobial and to reduced sensory changes. Antimicrobial compounds within AEFC are less exposed to interaction with meat surface components than those added directly to the surface and thus maintaining their activity [66–68].

Antimicrobial agents recently incorporated in AEFC for meat and meat products, poultry, and fish and fish products are shown in column 3 of Tables 1, 2, and 3, respectively. Target microorganisms aimed by recently developed AEFC as well as inoculation technique for meat and meat products, poultry, and fish and fish products are shown in columns 4 and 5 of Tables 1, 2, and 3, respectively.

The characteristics and mode of action of most common antimicrobials used to promote meat safety are described below.

4.1. Organic Acids. The antimicrobial effect of organic acids depends on concentration of undissociated form, which can penetrate the bacterial cell membrane. Inside the cell, their dissociation leads to interference with membrane transport and disruption of proton motive force [30]. Organic acids incorporated into EFC include lactate and acetate [46], propionate [18], and p-aminobenzoic acid [30]. WPI coatings added with malic acid, nisin, and grape seed extract applied on turkey frankfurters decreased to 2.3 log CFU/g of *L. monocytogenes* and 5 log CFU/g *S. typhimurium* after 28 d of storage at 4°C [53] (Table 1). Zein based AEFC using calcium propionate combined with nisin, reduced up to 5 log CFU/g of *L. monocytogenes* after 14 d at 4°C, when used to coat chicken breast [18]. Sodium lactate combined with other commercial antimicrobials reduced to 3.5 log/cm^2 of this pathogen when roasted turkey was stored at 4°C for 8 weeks [46] (Table 2). Thus, organic acids, especially when acting combined with other antimicrobial agents, have an important role in maintaining microbiological quality of meat and meat products.

4.2. Essential Oils and Plant Extracts. Essential oils are complex mixtures of volatile compounds obtained from plants, which mainly include terpenes, terpenoids, and aliphatic chemicals, all characterized by low molecular weight [69]. Oils containing phenols such as thymol, carvacrol, and eugenol exhibit the highest activity against all kind of microorganisms. Essential oils usually show higher antibacterial activity than mixtures of their major antimicrobial components, suggesting that minor components are critical for enhanced activity [69]. The antimicrobial mechanism is attributed to the disturbance of the cytoplasmic membrane disrupting the proton motive force; active transport and coagulation of cell contents may occur [70]. Direct incorporation of essential oils in the formulation of AEFC applied to meat products is expected to reduce bacterial population but may alter their sensory characteristics [68]. Microencapsulation of essential oils or their ingredients may be an alternative to protect them from interaction with environmental factors, avoiding their oxidation or volatilization while exerting their antimicrobial effect. Moreover, encapsulation increases the oil solubility in water, prevents its release at an undesired stage, and makes it easier to handle [71, 72]. Essential oils or

their constituents that may be incorporated in AEFC on AOF include those extracted from lemongrass, oregano, pimento, thyme, or cinnamon [40, 57, 65]. Oregano essential oil has been the most commonly reported in recent years including a 1.5% extract (v/v), successfully used to reduce total viable count by 2 log CFU/g of cold smoked sardine covered with an AEFC after 20 d storage at 5°C [58], whereas at 1.9% it achieved *L. monocytogenes* population reduction by 2.4 log CFU/g after 28 d, at 4°C in wrapped cold smoked salmon [65] (Table 3). Oregano essential oil combined with thyme extract, was incorporated into a film placed on top and bottom of fresh ground beef patties reducing *Pseudomonas* spp. and coliforms populations [36], whereas mixed with pimento essential oil, the films covering beef muscle slices reduced to 1 log of *E. coli* O157:H7 after 7 d of storage at 4°C [31] (Table 1). Grapefruit seed extract (GSE) incorporated into AEFC was found to inhibit *E. coli* O157:H7 and *L. monocytogenes* from pork loins [34], bacon [41], and salmon [59] (Tables 1 and 3). However, some commercial GSE is adulterated with synthetic preservatives such as benzalkonium and benzethonium chlorides, which are solely responsible for the antimicrobial activity of GSE. These compounds show toxicity and allergenicity to humans, and it is unlikely that they are formed during any extraction and/or processing of grapefruit seeds and pulp [73, 74].

4.3. Bacteriocins. Bacteriocins from lactic acid bacteria are peptides produced by bacteria that inhibit or kill other related and unrelated microorganisms [75]. These agents are generally heat-stable, apparently hypoallergenic and readily degraded by proteolytic enzymes in the human intestinal tract [68]. Class I bacteriocins, such as nisin, bind to plasma membranes via nonspecific electrostatic interactions and have a dual mode of action. The antibacterial activity results from pore formation in the bacterial plasma membrane, leading to dissipation of the transmembrane potential and vital solute gradients. The high efficiency of pore formation is the result of a second mechanism involving the cell wall precursor Lipid II which increases the affinity of nisin for the membrane, stabilizes a transmembrane orientation of nisin, and forms and integral part of the nisin pore. The pore structure involves a complex made up of four lipid II and 8 nisin molecules, which interferes with peptidoglycan biosynthesis [76, 77]. Other bacteriocins such as pediocin have been widely studied in food systems, but nisin remains the only one approved by European Union (EU) and the USA where it enjoys GRAS status [68, 78]. The effect of nisin incorporation into AEFC is the most studied, either to protect beef and turkey frankfurters, or turkey bologna against *L. monocytogenes* [33, 53, 61] (Tables 1 and 2); but pediocin has also been tested [35].

4.4. Proteins. Lysozyme is a naturally produced enzyme active against gram-positive bacteria, by hydrolyzing N-glycosidic bonds connecting N-acetyl muramic acid with the fourth carbon atom of N-acetyl glucosamine of the peptidoglycan molecule in the cell wall. This antimicrobial has been formulated in whey protein isolate (WPI) films and tested for its diffusivity and antimicrobial effect on salmon slices [79] and also tested in ground beef patties using zein films [38] (Table 1).

4.5. Chitosan. Chitosan is a linear polysaccharide composed of randomly distributed β-(1-4)-linked D-glucosamine and N-acetyl-D-glucosamine. Chitosan is believed to chelate certain ions from the lipopolysaccharide (LPS) layer of the outer membrane of bacteria or to exhibit electrostatic interactions among its NH_3^+ groups and the negative charges of microbial cell membrane. In both cases cell permeability increases releasing key cellular components of bacteria. The antimicrobial action of chitosan is influenced by type of chitosan, degree of polymerization, and environmental conditions. Chitosan coatings act as barrier against oxygen transfer leading to growth inhibition of aerobic bacteria [42]. In addition to the functionality of chitosan as polymeric material and antimicrobial agent (Section 3.3), it has been used as coating and wrapper in salami [37] and as film and coating combined with lauric arginate and nisin to reduce *L. monocytogenes* population in sliced turkey deli meat [47] (Tables 1 and 2) and also in seafood and fish [48, 52].

4.6. Lauric Arginate. Lauric arginate (LAE) is a food-grade cationic surfactant that is highly active against a wide range of food pathogens and spoilage microorganisms including bacteria, yeasts, and molds. It is obtained through the reaction of L-arginine, hydrochloric acid, ethanol, thionyl chloride, sodium hydroxide, lauryl chloride, and deionized water [80]. LAE affects cells viability by disturbing membrane potential and causing structural changes, although no disruption of cells is detected. In gram-negative cells, LAE alter both the cytoplasm membrane and the external membrane, while in gram-positive cells, alterations were observed in the cell membrane and in the cytoplasm. However, in both cases, cells remained intact and cell lysis is not observed [81]. LAE is nontoxic and is metabolized to naturally occurring amino acids, mainly arginine and ornithine, after consumption. Effectiveness of LAE, alone or in combination with other antimicrobials, has been tested against *L. monocytogenes*, *S. enterica*, and *L. innocua* in cooked ham and sliced turkey deli meat producing 2 log reductions in all cases [43, 47] (Tables 1 and 2).

Antimicrobial agents recently incorporated in AEFC for meat and meat products, poultry, and fish and fish products are shown in column 3 of Tables 1, 2, and 3, respectively, whereas application conditions and effect of AEFC are shown in columns 6 and 7 of the same tables, respectively.

5. Migration of Antimicrobial Agents from Films

Few reports have considered the migration extent of antimicrobial agents from edible films to the food surface. A study showed the effect of film thickness, solution pH, and temperature on nisin migration from an active WPI edible film to an aqueous solution. Results indicated that nisin is able to migrate from the film where diffusivity increased at lower pH and thickness, while it increased at higher temperatures [82]. Sorbic acid migration from an active cellulose film into pastry

dough was evaluated for 40 days and it was not significantly affected by film thickness, achieving a migration of 0.07%, (w/v) [83]. Nisin release measured from low density polyethylene film was unpredictable but it was affected by temperature and pH [84]. Migration of lysozyme from WPI-glycerol films indicated that the diffusion coefficient decreased as the WPI-glycerol ratio increased or storage temperature decreased [79]. Chitosan-glycerol films incorporated with 1–10% (w/v) lauric arginate showed full release of the agent and followed a Fickian behavior in a few hours at 4° and 28°C. Films were active in liquid and solid media against bacteria, yeast and fungi achieving 1.8–5.8 log reductions [50]. These findings lead us to consider that antimicrobial agents incorporated into AEFC may prevent microbial contamination of food surfaces.

6. Application and Effect of AEFC on Meat Products

Antimicrobial packaging can be a promising tool for protecting meat from pathogens contamination by preventing microbial growth by direct contact of the package with its surface. The gradual release of an antimicrobial substance from a packaging film to the food surface for extended period of time may be more advantageous than incorporating the antimicrobial into foods [85].

Studies using chitosan films incorporated in meat products demonstrated that lipid oxidation is reduced, suggesting that it may be due to the antioxidant activity of chitosan [52], as well as its low oxygen permeability characteristic [42]. Similar results have been obtained when other compounds were incorporated such as essential oils [57], grapefruit extracts [41, 59], and lysozyme [38]. In all cases, the oxidation rates decreased maintaining an acceptable quality in meat, poultry, or fish products. However, even when the coating may confer protection against lipid oxidation, other characteristics may have changed, leading to modified sensory attributes that made the food unacceptable for consumers. Application of films on meat surface in some cases could increase the stability of the red meat color [57], but if coatings act as gas barriers undesirable color changes may occur [38]. Sensory studies on fish indicated that not only bacterial number is critical for fish acceptance, but other factors such as bacterial types, autolytic activity, biochemical properties of fish, and storage conditions are significant [76]. In other studies, using chitosan film incorporated with oregano essential oil did not negatively influence the taste of chicken samples, extending the shelf-life of chicken fillets by 14 days, maintaining acceptable sensory characteristics [49]. Therefore, each particular application should be evaluated to establish the conditions leading to maintain meat safety without altering sensory characteristics.

Potential benefits of using AEFC for the meat industry are prevention of moisture loss, avoiding texture, flavor, and color changes, producing a significant economic impact by increasing saleable weight of products. Other advantages include reduction of dripping enhancing products presentation and reduced use of absorbent pads at the bottom of trays. Low oxygen permeability leads to decreased lipids oxidation and brown color-causing myoglobin oxidation, reduced load of spoilage and pathogenic microorganisms, and partial inactivation of deteriorative proteolytic enzymes at the surface of coated meat. Volatile flavor loss and foreign odors pick-up by meat, poultry, or seafood could be restricted by using edible films and coatings and incorporation of additives such as antimicrobial agents can be used for direct treatment of meat surface. There are, however, some factors that may represent disadvantages of using AEFC; there is wide diversity of meat products whose characteristics may vary making it difficult to standardize a single application procedure. Composition and properties of AEFC will provide different functionality and may affect scaling up of application methods for coatings.

Selection of the appropriate AEFC for a specific meat product will depend on its nature, characteristics, specific needs, costs, and benefits that this technology can offer to the manufacturers and the consumer. Thus, more research is needed to improve production and application processes of AEFC intended for the meat industry to be economically feasible and appropriate for each product.

7. Conclusions

The application and effects of AEFC of different nature have been investigated in several AOF. Effectiveness shown by each one depends on meat source, polymer used, film barrier properties, target microorganism, antimicrobial substance, and conditions of storage among others. EFC are a good alternative to improve the quality and safety of food and also to add value to food industry by-products. However, some challenges remain such as the need to improve and standardize coating procedures according to industry requirements aiming to reduce costs and increase shelf life to meet consumer demands without altering sensory characteristics of meat and meat products.

Conflict of Interests

The authors declare that there is not conflict of interests regarding the publication of this paper.

Acknowledgments

The authors are grateful to PROMEP for a PhD grant to ISO and to CONACYT for financial support to project no. 166751.

References

[1] P. D. Warriss, *Meat Science: An Introductory Text*, CAB International Publishers, New York, NY, USA, 2010.

[2] R. L. Scharff, "Economic burden from health losses due to food-borne illness in the united states," *Journal of Food Protection*, vol. 75, no. 1, pp. 123–131, 2012.

[3] CDC, 2012, http://www.cdc.gov/features/dsFoodborneOutbreaks/.

[4] J. C. Buzby and T. Roberts, "Economic costs and trade impacts of microbial foodborne illness," *World Health Statistics Quarterly*, vol. 50, no. 1-2, pp. 57–66, 1997.

[5] J. H. Han and A. Gennadios, "Edible films and coatings: a review," in *Innovations in Food Packaging*, J. H. Han, Ed., pp. 239–262, Elsevier Science, New York, NY, USA, 2005.

[6] A. Cagri, Z. Ustunol, and E. T. Ryser, "Antimicrobial edible films and coatings," *Journal of Food Protection*, vol. 67, no. 4, pp. 833–848, 2004.

[7] A. E. Pavlath and W. Orts, "Edible films and coatings: why, what, and how?" in *Edible Films and Coatings for Food Applications*, M. E. Em buscado and K. C. Huber, Eds., pp. 57–112, Springer, New York, NY, USA, 2009.

[8] E. A. Baldwin and R. Hagenmaier, "Introduction," in *Edible Coatings and Films to Improve Food Quality*, E. A. Baldwin, R. Hagenmaier, and J. Bai, Eds., pp. 1–12, CRC Press, Boca Raton, Fla, USA, Second edition, 2012.

[9] Z. Ustunol, "Edible films and coatings for meat and poultry," in *Edible Films and Coatings for Food Applications*, M. E. Embuscado and K. C. Huber, Eds., pp. 245–268, Springer, New York, NY, USA, 2009.

[10] D. Dave and A. E. Ghaly, "Meat spoilage mechanisms and preservation techniques: a critical review," *The American Journal of Agricultural and Biological Sciences*, vol. 6, no. 4, pp. 486–510, 2011.

[11] Codex Alimentarius, Code of hygienic practice for meat. Codex Alimentarius Commision/Recommended Code of Practice. 58-2005. New Zealand, FAO/WHO, 2005.

[12] USDA, United States Department of Agriculture, 2014, http://www.usda.gov/wps/portal/usda/usdahome.

[13] FAO, 2012, http://www.fao.org/index_en.htm.

[14] ICMSF, *Microorganisms in Foods 6. Microbial Ecology of Food Commodities*, 2nd edition, 2005.

[15] J. Cerveny, J. D. Meyer, and P. A. Hall, "Microbiological spoilage of meat and poultry products compendium of the microbiological spoilage, of foods and beverages," in *Food Microbiology and Food Safety*, W. H. Sperber and M. P. Doyle, Eds., pp. 69–868, Springer, NY, NY, USA, 2009.

[16] P. E. Simitzisand and S. G. Deligeorgis, "Lipid oxidation of meat and use of essential oils as antioxidants in meat products," 2010, http://www.scitopics.com/Lipid_Oxidation_of_Meat_and_Use_of_Essential_Oils_as_Antioxidants_in_Meat_Products.html.

[17] K. Kuwahara and K. Osako, "Effect of sodium gluconate on gel formation of Japanese common squid mantle muscle," *Nippon Suisan Gakkaishi (Japanese Edition)*, vol. 69, no. 4, pp. 637–642, 2003.

[18] M. E. Janes, S. Kooshesh, and M. G. Johnson, "Control of *Listeria monocytogenes* on the surface of refrigerated, ready-to-eat chicken coated with edible zein film coatings containing nisin and/or calcium propionate," *Journal of Food Science*, vol. 67, no. 7, pp. 2754–2757, 2002.

[19] CDC, 2013, http://www.cdc.gov/features/dsFoodborneOutbreaks/.

[20] EFSA (European Food Safety Authority), European Centre for Disease Prevention and Control, The European Union Summary Report on Trends and Sources of Zoonoses, Zoonotic Agents and Food-borne Outbreaks in 2010, http://www.efsa.europa.eu/efsajournal.

[21] C. S. DeWaal, X. A. Tian, and F. Bhuiya, Outbreak Alert! 2008 Center for Science in the Public Interest (CSPI) Washington, 2013, http://www.cspinet.org/.

[22] K. Dangaran, P. M. Tomasula, and P. Qi, "Structure and function of protein-based edible films and coatings," in *Edible Films and Coatings for Food Applications*, M. E. Embuscado and K. C. Huber, Eds., pp. 25–56, Springer, New York, NY, USA, 2009.

[23] T. H. McHugh and R. J. Avena-Bustillos, "Applications of edible films and coatings to processed foods," in *Edible Coatings and Films to Improve Food Quality*, E. A. Baldwin, R. Hagenmaier, and J. Bai, Eds., pp. 291–318, CRC Press, Boca Raton, Fla, USA, 2012.

[24] A. Gennadios, M. A. Hanna, and L. B. Kurth, "Application of edible coatings on meats, poultry and seafoods: a review," *LWT—Food Science and Technology*, vol. 30, no. 4, pp. 337–350, 1997.

[25] R. Soliva-Fortuny, M. A. Rojas-Graü, and O. Martín-Belloso, "Polysaccharide coatings," in *Edible Coatings and Films To Improve Food Quality*, E. Baldwin, R. Hagenmaier, and J. Bai, Eds., pp. 103–136, CRC Press, Boca Raton, Fla, USA, 2012.

[26] F. Debeaufort and A. Voilley, "Lipid based edible films and coatings," in *Edible Films and Coatings for Food Applications*, M. E. Embuscado and K. C. Huber, Eds., pp. 135–168, Springer, New York, NY, USA, 2009.

[27] C. N. Cutter, "Opportunities for bio-based packaging technologies to improve the quality and safety of fresh and further processed muscle foods," *Meat Science*, vol. 74, no. 1, pp. 131–142, 2006.

[28] M. Millette, C. le Tien, W. Smoragiewicz, and M. Lacroix, "Inhibition of *Staphylococcus aureus* on beef by nisin-containing modified alginate films and beads," *Food Control*, vol. 18, no. 7, pp. 878–884, 2007.

[29] A. Cagri, Z. Ustunol, and E. T. Ryser, "Inhibition of three pathogens on bologna and summer sausage using antimicrobial edible films," *Journal of Food Science*, vol. 67, no. 6, pp. 2317–2324, 2002.

[30] A. Cagri, Z. Ustunol, W. Osburn, and E. T. Ryser, "Inhibition of *Listeria monocytogenes* on hot dogs using antimicrobial whey protein-based edible casings," *Journal of Food Science*, vol. 68, no. 1, pp. 291–299, 2003.

[31] M. Oussalah, S. Caillet, S. Salmiéri, L. Saucier, and M. Lacroix, "Antimicrobial and antioxidant effects of milk protein-based film containing essential oils for the preservation of whole beef muscle," *Journal of Agricultural and Food Chemistry*, vol. 52, no. 18, pp. 5598–5605, 2004.

[32] R. L. Beverlya, M. E. Janes, W. Prinyawiwatkula, and H. K. No, "Edible chitosan films on ready-to-eat roast beef for the control of *Listeria monocytogenes*," *Food Microbiology*, vol. 25, no. 3, pp. 534–537, 2008.

[33] V. T. Nguyen, M. J. Gidley, and G. A. Dykes, "Potential of a nisin-containing bacterial cellulose film to inhibit *Listeria monocytogenes* on processed meats," *Food Microbiology*, vol. 25, no. 3, pp. 471–478, 2008.

[34] Y. H. Hong, G. O. Lim, and K. B. Song, "Physical properties of *Gelidium corneum*-gelatin blend films containing grapefruit seed extract or green tea extract and its application in the packaging of pork loins," *Journal of Food Science*, vol. 74, no. 1, pp. 6–10, 2009.

[35] P. Santiago-Silva, N. F. F. Soares, J. E. Nóbrega et al., "Antimicrobial efficiency of film incorporated with pediocin (ALTA 2351) on preservation of sliced ham," *Food Control*, vol. 20, no. 1, pp. 85–89, 2009.

[36] Z. K. Emiroğlu, G. P. Yemiş, B. K. Coşkun;, and K. Candoğan, "Antimicrobial activity of soy edible films incorporated with thyme and oregano essential oils on fresh ground beef patties," *Meat Science*, vol. 86, no. 2, pp. 283–288, 2010.

[37] M. D. R. Moreira, M. Pereda, N. E. Marcovich, and S. I. Roura, "Antimicrobial effectiveness of bioactive packaging materials

from edible chitosan and casein polymers: assessment on carrot, cheese, and salami," *Journal of Food Science*, vol. 76, no. 1, pp. M54–M63, 2011.

[38] I. U. Ünalan, F. Korel, and A. Yemenicioğlu, "Active packaging of ground beef patties by edible zein films incorporated with partially purified lysozyme and Na2EDTA," *International Journal of Food Science and Technology*, vol. 46, no. 6, pp. 1289–1295, 2011.

[39] M. Vargas, A. Albors, and A. Chiralt, "Application of chitosan-sunflower oil edible films to pork meat hamburgers," in *Proceedings of the 11th International Congress on Engineering and Food (ICEF '11)*, vol. 1, pp. 39–43, Procedia Food Science, 2011.

[40] S. Ravishankar, D. Jaroni, L. Zhu, C. Olsen, T. McHugh, and M. Friedman, "Inactivation of *Listeria monocytogenes* on ham and bologna using pectin-based apple, carrot, and hibiscus edible films containing carvacrol and cinnamaldehyde," *Journal of Food Science*, vol. 77, no. 7, pp. 377–382, 2012.

[41] Y. J. Shin, H. Y. Song, Y. B. Seo, and K. B. Song, "Preparation of red algae film containing grapefruit seed extract and application for the packaging of cheese and bacon," *Food Science and Biotechnology*, vol. 21, no. 1, pp. 225–231, 2012.

[42] U. Siripatrawan and S. Noipha, "Active film from chitosan incorporating green tea extract for shelf life extension of pork sausages," *Food Hydrocolloids*, vol. 27, no. 1, pp. 102–108, 2012.

[43] P. Theinsathid, W. Visessanguan, J. Kruenate, Y. Kingcha, and S. Keeratipibul, "Antimicrobial activity of lauric arginate-coated polylactic acid films against *Listeria monocytogenes* and *Salmonella typhimurium* on cooked sliced ham," *Journal of Food Science*, vol. 77, no. 2, pp. 142–149, 2012.

[44] M. B. Nieto, "Structure and function of polysaccharide gum-based edible films and coatings," in *Edible Films and Coatings for Food Applications*, M. E. Embuscado and K. C. Huber, Eds., pp. 57–112, Springer, New York, NY, USA, 2009.

[45] E. Eroglu, M. Torun, C. Dincer, and A. Topuz, "Influence of pullulan-based edible coating on some quality properties of strawberry during cold storage," *Packaging Technology and Science*, 2014.

[46] Z. Jiang, H. Neetoo, and H. Chen, "Efficacy of freezing, frozen storage and edible antimicrobial coatings used in combination for control of *Listeria monocytogenes* on roasted turkey stored at chiller temperatures," *Food Microbiology*, vol. 28, no. 7, pp. 1394–1401, 2011.

[47] M. Guo, T. Z. Jin, L. Wang, O. J. Scullen, and C. H. Sommers, "Antimicrobial films and coatings for inactivation of *Listeria innocua* on ready-to-eat deli turkey meat," *Food Control*, vol. 40, pp. 64–70, 2014.

[48] J. Gómez-Estaca, A. López de Lacey, M. E. López-Caballero, M. C. Gómez-Guillén, and P. Montero, "Biodegradable gelatin-chitosan films incorporated with essential oils as antimicrobial agents for fish preservation," *Food Microbiology*, vol. 27, no. 7, pp. 889–896, 2010.

[49] S. Petrou, M. Tsiraki, V. Giatrakou, and I. N. Savvaidis, "Chitosan dipping or oregano oil treatments, singly or combined on modified atmosphere packaged chicken breast meat," *International Journal of Food Microbiology*, vol. 156, no. 3, pp. 264–271, 2012.

[50] L. Higueras, G. López-Carballo, P. Hernández-Muñoz, R. Gavara, and M. Rollini, "Development of a novel antimicrobial film based on chitosan with LAE (ethyl-Nα-dodecanoyl-L-arginate) and its application to fresh chicken," *International Journal of Food Microbiology*, vol. 165, no. 3, pp. 339–345, 2013.

[51] Y. J. Jeon, J. Y. V. A. Kamil, and F. Shahidi, "Chitosan as an edible invisible film for quality preservation of herring and Atlantic

cod," *Journal of Agricultural and Food Chemistry*, vol. 50, no. 18, pp. 5167–5178, 2002.

[52] A. Günlü and E. Koyun, "Effects of vacuum packaging and wrapping with chitosan-based edible film on the extension of the shelf life of sea bass (*Dicentrarchus labrax*) fillets in cold storage (4°C)," *Food and Bioprocess Technology*, vol. 6, no. 7, pp. 1713–1719, 2013.

[53] V. P. Gadang, N. S. Hettiarachchy, M. G. Johnson, and C. Owens, "Evaluation of antibacterial activity of whey protein isolate coating incorporated with nisin, grape seed extract, malic acid, and EDTA on a turkey frankfurter system," *Journal of Food Science*, vol. 73, no. 8, pp. 389–394, 2008.

[54] S. Min, L. J. Harris, and J. M. Krochta, "*Listeria monocytogenes* inhibition by whey protein films and coatings incorporating the lactoperoxidase system," *Journal of Food Science*, vol. 70, no. 7, pp. 317–324, 2005.

[55] S. Ravishankar, L. Zhu, C. W. Olsen, T. H. McHugh, and M. Friedman, "Edible apple film wraps containing plant antimicrobials inactivate foodborne pathogens on meat and poultry products," *Journal of Food Science*, vol. 74, no. 8, pp. 440–445, 2009.

[56] R. M. Mild, L. A. Joens, M. Friedman et al., "Antimicrobial edible apple films inactivate antibiotic resistant and susceptible *Campylobacter jejuni* strains on chicken breast," *Journal of Food Science*, vol. 76, no. 3, pp. 163–168, 2011.

[57] M. Ahmad, S. Benjakul, P. Sumpavapol, and N. P. Nirmal, "Quality changes of sea bass slices wrapped with gelatin film incorporated with lemongrass essential oil," *International Journal of Food Microbiology*, vol. 155, no. 3, pp. 171–178, 2012.

[58] J. Gómez-Estaca, P. Montero, B. Giménez, and M. C. Gómez-Guillén, "Effect of functional edible films and high pressure processing on microbial and oxidative spoilage in cold-smoked sardine (*Sardina pilchardus*)," *Food Chemistry*, vol. 105, no. 2, pp. 511–520, 2007.

[59] H. Y. Song, Y. J. Shin, and K. B. Song, "Preparation of a barley bran protein-gelatin composite film containing grapefruit seed extract and its application in salmon packaging," *Journal of Food Engineering*, vol. 113, no. 4, pp. 1736–1743, 2012.

[60] K. H. Seol, D. G. Lim, A. Jang, C. Jo, and M. Lee, "Antimicrobial effect of κ-carrageenan-based edible film containing ovotransferrin in fresh chicken breast stored at 5°C," *Meat Science*, vol. 83, no. 3, pp. 479–483, 2009.

[61] B. J. Min, I. Y. Han, and P. L. Dawson, "Antimicrobial gelatin films reduce *Listeria monocytogenes* on turkey bologna," *Poultry Science*, vol. 89, no. 6, pp. 1307–1314, 2010.

[62] H. Neetoo, M. Ye, and H. Chen, "Bioactive alginate coatings to control *Listeria monocytogenes* on cold-smoked salmon slices and fillets," *International Journal of Food Microbiology*, vol. 136, no. 3, pp. 326–331, 2010.

[63] Y. Song, L. Liu, H. Shen, J. You, and Y. Luo, "Effect of sodium alginate-based edible coating containing different anti-oxidants on quality and shelf life of refrigerated bream (*Megalobrama amblycephala*)," *Food Control*, vol. 22, no. 3-4, pp. 608–615, 2011.

[64] C. O. Mohan, C. N. Ravishankar, K. V. Lalitha, and T. K. Srinivasa Gopal, "Effect of chitosan edible coating on the quality of double filleted Indian oil sardine (*Sardinella longiceps*) during chilled storage," *Food Hydrocolloids*, vol. 26, no. 1, pp. 167–174, 2012.

[65] N. Tammineni, G. Ünlü, and S. C. Min, "Development of antimicrobial potato peel waste-based edible films with oregano essential oil to inhibit *Listeria monocytogenes* on cold-smoked

salmon," *International Journal of Food Science and Technology*, vol. 48, no. 1, pp. 1–4, 2013.

[66] P. Appendini and J. H. Hotchkiss, "Review of antimicrobial food packaging," *Innovative Food Science and Emerging Technologies*, vol. 3, no. 2, pp. 113–126, 2002.

[67] S. Quintavalla and L. Vicini, "Antimicrobial food packaging in meat industry," *Meat Science*, vol. 62, no. 3, pp. 373–380, 2002.

[68] V. COMA, "Bioactive packaging technologies for extended shelf life of meat-based products," *Meat Science*, vol. 78, no. 1-2, pp. 90–103, 2008.

[69] I. H. N. Bassolé and H. R. Juliani, "Essential oils in combination and their antimicrobial properties," *Molecules*, vol. 17, no. 4, pp. 3989–4006, 2012.

[70] S. Burt, "Essential oils: their antibacterial properties and potential applications in foods—a review," *International Journal of Food Microbiology*, vol. 94, no. 3, pp. 223–253, 2004.

[71] A. Arana-Sánchez, M. Estarrón-Espinosa, E. N. Obledo-Vázquez, E. Padilla-Camberos, R. Silva-Vázquez, and E. Lugo-Cervantes, "Antimicrobial and antioxidant activities of Mexican oregano essential oils (*Lippia graveolens* H. B. K.) with different composition when microencapsulated in β-cyclodextrin," *Letters in Applied Microbiology*, vol. 50, no. 6, pp. 585–590, 2010.

[72] C. C. Liolios, O. Gortzi, S. Lalas, J. Tsaknis, and I. Chinou, "Liposomal incorporation of carvacrol and thymol isolated from the essential oil of *Origanum dictamnus* L. and *in vitro* antimicrobial activity," *Food Chemistry*, vol. 112, no. 1, pp. 77–83, 2009.

[73] G. R. Takeoka, L. T. Dao, R. Y. Wong, and L. A. Harden, "Identification of benzalkonium chloride in commercial grapefruit seed extracts," *Journal of Agricultural and Food Chemistry*, vol. 53, no. 19, pp. 7630–7636, 2005.

[74] G. Takeoka, L. Dao, R. Y. Wong, R. Lundin, and N. Mahoney, "Identification of benzethonium chloride in commercial grapefruit seed extracts," *Journal of Agricultural and Food Chemistry*, vol. 49, no. 7, pp. 3316–3320, 2001.

[75] E. M. Balciunas, F. A. Castillo Martinez, S. D. Todorov, B. D. G. D. M. Franco, A. Converti, and R. P. D. S. Oliveira, "Novel biotechnological applications of bacteriocins: A review," *Food Control*, vol. 32, no. 1, pp. 134–142, 2013.

[76] C. Chatterjee, M. Paul, L. Xie, and W. A. van der Donk, "Biosynthesis and mode of action of lantibiotics," *Chemical Reviews*, vol. 105, no. 2, pp. 633–683, 2005.

[77] E. Breukink, "A lesson in efficient killing from two-component lantibiotics," *Molecular Microbiology*, vol. 61, no. 2, pp. 271–273, 2006.

[78] FDA US Food and Drug Administration. Nisin preparation: affirmation of GRAS status as direct human ingredient, 21 Code of Federal Regulations Part 184, Federal Register, 53, 1988.

[79] S. Min, T. R. Rumsey, and J. M. Krochta, "Diffusion of the antimicrobial lysozyme from a whey protein coating on smoked salmon," *Journal of Food Engineering*, vol. 84, no. 1, pp. 39–47, 2008.

[80] N. Terjung, M. Loeffler, M. Gibis, H. Salminen, J. Hinrichs, and J. Weiss, "Impact of lauric arginate application form on its antimicrobial activity in meat emulsions," *Food Biophysics*, vol. 9, pp. 88–98, 2014.

[81] E. Rodríguez, J. Seguer, X. Rocabayera, and A. Manresa, "Cellular effects of monohydrochloride of L-arginine, Nα- lauroyl ethylester (LAE) on exposure to *Salmonella typhimurium* and *Staphylococcus aureus*," *Journal of Applied Microbiology*, vol. 96, no. 5, pp. 903–912, 2004.

[82] G. Rossi-Márquez, J. H. Han, B. García-Almendárez, E. Castaño-Tostado, and C. Regalado-González, "Effect of temperature, pH and film thickness on nisin release from antimicrobial whey protein isolate edible films," *Journal of the Science of Food and Agriculture*, vol. 89, no. 14, pp. 2492–2497, 2009.

[83] M. F. A. Silveira, N. F. F. Soares, R. M. Geraldine, N. J. Andrade, and M. P. J. Gonçalves, "Antimicrobial efficiency and sorbic acid migration from active films into pastry dough," *Packaging Technology and Science*, vol. 20, no. 4, pp. 287–292, 2007.

[84] G. Mauriello, E. de Luca, A. La Storia, F. Villani, and D. Ercolini, "Antimicrobial activity of a nisin-activated plastic film for food packaging," *Letters in Applied Microbiology*, vol. 41, no. 6, pp. 464–469, 2005.

[85] M. Ye, H. Neetoo, and H. Chen, "Control of *Listeria monocytogenes* on ham steaks by antimicrobials incorporated into chitosan-coated plastic films," *Food Microbiology*, vol. 25, no. 2, pp. 260–268, 2008.

6

Impact of Endophytic Microorganisms on Plants, Environment and Humans

Dhanya N. Nair and S. Padmavathy

Research Department of Botany, Nirmala College for Women, Coimbatore, Tamil Nadu 641018, India

Correspondence should be addressed to Dhanya N. Nair; dhanya.dnn@outlook.com

Academic Editors: A. El-Shibiny and S. Tan

Endophytes are microorganisms (bacteria or fungi or actinomycetes) that dwell within robust plant tissues by having a symbiotic association. They are ubiquitously associated with almost all plants studied till date. Some commonly found endophytes are those belonging to the genera *Enterobacter sp.*, *Colletotrichum sp.*, *Phomopsis sp.*, *Phyllosticta sp.*, *Cladosporium sp.*, and so forth. Endophytic population is greatly affected by climatic conditions and location where the host plant grows. They produce a wide range of compounds useful for plants for their growth, protection to environmental conditions, and sustainability, in favour of a good dwelling place within the hosts. They protect plants from herbivory by producing certain compounds which will prevent animals from further grazing on the same plant and sometimes act as biocontrol agents. A large amount of bioactive compounds produced by them not only are useful for plants but also are of economical importance to humans. They serve as antibiotics, drugs or medicines, or the compounds of high relevance in research or as compounds useful to food industry. They are also found to have some important role in nutrient cycling, biodegradation, and bioremediation. In this review, we have tried to comprehend different roles of endophytes in plants and their significance and impacts on man and environment.

1. Introduction

Endophytes are bacterial or fungal microorganisms that colonize healthy plant tissue intercellularly and/or intracellularly without causing any apparent symptoms of disease [1]. They are ubiquitous, colonize in all plants, and have been isolated from almost all plants examined till date. Their association can be obligate or facultative and causes no harm to the host plants. They exhibit complex interactions with their hosts which involves mutualism and antagonism [2–11]. Plants strictly limit the growth of endophytes, and these endophytes use many mechanisms to gradually adapt to their living environments [12]. In order to maintain stable symbiosis, endophytes produce several compounds that promote growth of plants and help them adapt better to the environment [13, 14].

Improvement of endophyte resources could bring us a variety of benefits, such as novel and effective bioactive compounds that cannot be synthesized by chemical reactions. For this, there should be a better understanding about endophytes, their significance and roles. Understanding the biology of plants and their microbial ecology becomes important. As evidenced by more number of publications on endophytes in recent years, many studies have been performed for evaluating their colonization pattern of vegetative tissues as well as their effects on plant growth. These publications indirectly suggest their importance to the hosts and to the environment. This review aims to provide an overview about endophytes, their role and importance in plants and subsequently to the environment and human beings with reference to recent developments in endophytic research.

2. Isolation and Identification

Endophytic organisms have been isolated from different parts of plant. They were isolated from scale primordia, meristem and resin ducts [15, 16], leaf segments with midrib and roots

[17] and from stem, bark, leaf blade, petiole [18], and buds [19]. Sequence-based approach was used for investigating the transmission of diverse fungal endophytes in seed and needles of *Pinus monticola*, western white pine [20]. They isolated 2003 fungal endophytes from 750 surface-sterilized needles. In contrast, only 16 endophytic isolates were obtained from 800 surface-sterilized seeds.

There are endophytic bacteria, fungi, and/or actinomycetes whose isolation from the plant tissues has been a challenge since the studies on endophytes started. Several researchers have reviewed extensively different methods of the isolation of bacterial endophytes [21, 22]. Endophytes are isolated by initial surface sterilization followed by culturing from ground tissue extract [23] or by direct culturing of plant tissues [18] on media suitable for bacteria or fungi or actinomycetes. The impact of different culture media on isolation of endophytic fungal flora from root and fruits of *Azadirachta indica* A. Juss was studied by Verma et al. [24]. According to them, mycological agar (MCA) medium yielded the highest number of isolates, with the greatest species richness.

An obligatory endophyte, *Enterobacter cloacae*, was found to be associated with the pollen of several Mediterranean pines [25]. Most fungal endophytes isolated from plants and algae are members of the Ascomycota or their anamorphs, with only a few reports of basidiomycetous endophytes, these often being orchid mycorrhizas [26]. Basidiomycetous morphotypes were isolated from healthy leaves, rachis, and petioles of the oil palm *Elaeis guineensis* in a Thai plantation which were further characterized by molecular analysis using ribosomal DNA sequences. For the first time ever, the microorganism species *Acremonium terricola*, *Monodictys castaneae*, *Penicillium glandicola*, *Phoma tropica* and *Tetraploa aristata* were reported as endophytic fungi [27]. Some of the common and more frequently isolated endophytic fungi from different plants are given in Table 1.

Endophytic fungi have been classified into two broad groups discriminated based on phylogeny and life history traits as clavicipitaceous (C) which infect some grasses and the nonclavicipitaceous endophytes (NC-endophytes), which can be recovered from asymptomatic tissues of nonvascular plants, ferns and allies, conifers, and angiosperms [28]. NC-endophytes represent three distinct functional classes based on host colonization and transmission, *in planta* biodiversity and fitness benefits conferred to hosts while the C group has just one class.

Conventionally, identification of endophytes was done based on morphological characteristics for bacteria, fungi, and actinomycetes and with the help of biochemical tests for bacteria and actinomycetes. With the development of molecular biology, ribosomal DNA Internal Transcribed Spacer (ITS) sequence analysis is widely used for the identification of microorganisms. Ribosomal DNA (rDNA) ITS was proved to be a valuable source of evidence to resolve phylogenetic relationships at lower levels, such as among genera or species [29]. It was also reported that ITS sequences analysis was especially effective in nonsporulating fungi identification which reduced the impact of biased judgement [30] and the Large Subunit (LSU) and ITS data are powerful tools

TABLE 1: Fungi those are commonly isolated as endophytes from different plants.

Endophytes	Plant species	Citation
Phomopsis sp	*Neolitsea sericea*	[18]
	Pasania edulis	[17]
	Ginkgo biloba L.	[33]
	Tectona grandis and *Samanea saman* Merr.	[32]
	Taxus chinensis	[58]
Cladosporium sp.	*Opuntia ficus indica*	[27]
	Cinnamomum camphora	[89]
C. herbarum	*Lycopersicum esculentum* Mill.	[99]
	Triticum aestivum	[100]
Colletotrichum sp.	*Triticum aestivum*	[100]
	Citrus plants	[37]
	Cinnamomum camphora	[89]
	Pasania edulis	[17]
	Ginkgo biloba L.	[33]
	Tectona grandis and *Samanea saman* Merr.	[32]
	Huperzia serrata	[62]
C. gloeosporiodes	*Cinnamomum camphora*	[89]
	Lycopersicum esculentum Mill.	[99]
Phyllosticta sp	*Citrus sp.*	[37]
	Pasania edulis	[17]
	Coffea arabica	[101]
	Quercus variabilis	[65]
	Centella asiatica	[102]
	Panax quinquefolium	[103]
	Ginkgo biloba L.	[33]
Penicillium sp.	*Lycopersicum esculentum* Mill.	[99]
	Huperzia serrata	[62]
Acremonium sp.	*Taxus chinensis*	[58]
	Huperzia serrata	[37]

to resolve the taxonomy of basidiomycetous endophytes [26]. *Pleurostoma, Chaetomium, Coniochaeta (Lecythophora), Daldinia, Xylaria, Hypoxylon, Nodulisporium, Cazia,* and *Phellinus* isolated as endophytes from *Huperzia serrata* were confirmed for the first time by rDNA ITS analysis [31].

3. Effect of Climate on Endophytic Population

Endophytic population varies from plants to plants and from species to species. Within the same species it not only varies from region to region but also differs with change in climatic conditions of the same region. Temporal changes in relative frequency of total endophytic fungi were studied by Chareprasert et al. [32]. They found that matured leaves of teak (*Tectona grandis* L.) and rain tree (*Samanea saman* Merr.) had greater number of genera and species, with higher colonization frequency, than those in the young leaves and their occurrence in leaves increased during rainy season. The endophytic population and frequency tended to differ among sampling dates for all the organs studied, namely, young leaves, petiole, and twigs of *Gingko biloba* L. [33]. They proved that the occurrence of *Phyllosticta* sp. in both leaves and petioles was first detected in August and peaked in October with none in the month of May. *Phomopsis* sp. was detected in twigs throughout the growing season. These results suggest that the distribution of the two dominant endophytic fungi was organ-specific and differed within seasons.

4. Endophytes and Molecular Studies

With recent advances and developments in biotechnology, more studies at the molecular level are done with endophytes, which include metagenomic studies, use of molecular markers, molecular cloning, and genetic expression studies. Denaturing gradient gel electrophoresis (DGGE) profiles of 16S rRNA gene fragments amplified from total plant DNA were used to detect some nonculturable endophytic bacteria by comparing the profile with the bands obtained from the culturable endophytes from Citrus plant [34]. Bacterial automated ribosomal intergenic spacer analysis (B-ARISA) and Pyrosequencing was used to examine bacterial endophyte community of potato (*Solanum tuberosum*) cultivar [35]. B-ARISA profiles revealed a significant difference in the endophytic community between cultivars and canonical correspondence analysis showed a significant correlation between the community structure and plant biomass. Pyrosequencing was used to determine the bacterial operational taxonomic units (OTUs) richness. Metagenomic approach is another method used to find the microorganisms from different environments which cannot be cultured easily. This approach was used to find the 1-aminocyclopropane-1-carboxylate deaminase gene (*acdS*) operon from an uncultured endophytic microorganism colonizing *Solanum tuberosum* L. [36]. The authors in [36] concluded that metagenomic analysis can complement PCR-based analysis and yield information on whole gene operons.

Little variation within the endophytic population diversity in *Festuca eskia* was found, regardless of provenance altitude and site and/or endophyte infection frequency using Sequence Tagged Sites (STS) and Simple Sequence Repeats (SSR) markers [37]. SSR marker was also used to study the genetic variation among two isolated endophytes *Neotyphodium sibiricum* and *N. gansuense* from the host plant *Achnatherum sibiricum* [38]. Significant linkage disequilibrium of fungal SSR loci suggested that both fungal species primarily propagate by clonal growth through plant seeds, whereas variation in genetic diversity and the presence of hybrids in both endophytic species revealed that, although clonal propagation was prevalent, occasional recombination might also occur. Based on molecular cloning and genetic expression of analysis of geranylgeranyl diphosphate (GGPP) synthase, it was proposed that *ltmG, ltmM*, and *ltmK* are members of a set of genes required for lolitrem (a potent tremorgen to mammals) biosynthesis in endophytes *Epichloe festucae* and *Neotyphodium lolii* of the perennial ryegrass [39].

Molecular studies in endophytes have gone to the extent that complete genome of *Enterobacter* sp. 638, an endophytic plant growth promoting gamma-proteobacterium, that was isolated from the stem of poplar, a potentially important biofuel feed stock plant, was sequenced [40]. Sequencing revealed that it has 4,518,712 bp chromosome and a 157,749 bp plasmid (pENT638-1). Different sets of genes specific to the plant niche adaptation of this bacterium were identified by genome annotation and comparative genomics. This includes genes that code for putative proteins involved in survival in the rhizosphere (to cope with oxidative stress or uptake of nutrients released by plant roots), root adhesion (pili, adhesin, hemagglutinin, and cellulose biosynthesis), colonization/establishment inside the plant (chemiotaxis, flagella, and cellobiose phosphorylase), plant protection against fungal and bacterial infections (siderophore production and synthesis of the antimicrobial compounds 4-hydroxybenzoate and 2-phenylethanol), and overall improved poplar growth and development (through production of the phytohormones indole acetic acid, acetoin, and 2,3-butanediol).

5. Roles and Applications of Endophytes

5.1. Phytostimulation. Plants require 16 essential elements like C, H, N, O, and P and 11 more. These essential elements are available to plants for their growth and development in chemical form, which they obtain from atmosphere, soil, water, and organic matter. Endophytes also play an important role in the uptake of these nutrients. They elicit different modes of action in tall fescue adaptation to P deficiency [41] and induce increased uptake of N [42]. Endophytic bacteria produce a wide range of phytohormones, such as auxins, cytokinins, and gibberellic acids. *Burkholderia vietnamiensis*, a diazotrophic endophytic bacterium isolated from wild cottonwood (*Populus trichocarpa*), produced indole acetic acid (IAA), which promotes the growth of the plant [43]. This was confirmed by comparison between uninoculated control and plants inoculated with *B. vietnamiensis* on nitrogen free media, in which inoculated plants gained more dry weight and more nitrogen content. A new strain of fungus *Cladosporium sphaerospermum* isolated from the roots of *Glycine max* (L) Merr. showed the presence of higher amounts of bioactive GA3, GA4, and GA7, which induced maximum plant growth in both rice and soybean varieties [44].

5.2. Pigment Production. An orange pigment identified as quercetin glycoside was isolated from an endophytic fungus belonging to *Penicillium sp.* [45]. This was the first report on quercetin glycoside produced by endophytic fungus. Endophytic fungus strain named SX01, later identified as *Penicillium purpurogenum*, from the twigs of *Ginkgo biloba* L, was able to produce abundant soluble red pigments which could be used as natural food colorant [46]. A pigment isolated from the endophytic fungus *Monodictys castaneae* was found to inhibit few human pathogenic bacteria *Staphylococcus aureus, Klebsiella pneumonia, Salmonella typhi,* and *Vibrio cholerae* and was proved to be more active than streptomycin [47].

5.3. Enzyme Production. Many commercially important enzymes are produced by several soil micro-organisms. The hunt for other potential sources had led to the discovery of a few vital enzymes being produced by endophytes. Endophytic fungi like *Acremonium terricola, Aspergillus japonicas, Cladosporium cladosporioides, Cladosporium sphaerospermum, Fusarium lateritium, Monodictys castaneae, Nigrospora sphaerica, Penicillium aurantiogriseum, Penicillium glandicola, Pestalotiopsis guepinii, Phoma tropica, Phomopsis archeri, Tetraploa aristata,* and *Xylaria sp.* and many other unidentified species in *Opuntia ficus-indica* Mill. have indicated their promising potential for deployment in biotechnological processes involving production of pectinases, cellulases, xylanases, and proteases [27]. An endophyte, *Acremonium zeae,* isolated from maize produced the enzyme hemicellulase extracellularly [48]. This hydrolytic enzyme from *A. zeae* may be suitable for application in the bioconversion of lignocellulosic biomass into fermentable sugars.

5.4. Antimicrobial Activity. Most of the endophytes isolated from plants are known for their antimicrobial activity. They help in controlling microbial pathogens in plants and/or animals. Endophytes isolated from medicinal plants showed bioactivity for broad spectrum of pathogenic microorganisms [49–51]. A total of 37 endophytes were isolated all together from *Tectona grandis* L. and *Samanea saman* Merr. of which 18 could produce inhibitory substances effective against *Bacillus subtilis, Staphylococcus aureus,* and *Escherichia coli* and 3 isolates inhibited growth of *Candida albicans in vitro* [32]. Kumar et al. [52] assayed the bioactivity of the endophytic microorganisms like *Dothideomycetes sp., Alternaria tenuissima, Thielavia subthermophila, Alternaria sp., Nigrospora oryzae, Colletotrichum truncatum,* and *Chaetomium sp.,* isolated from the medicinal plant, *Tylophora indica,* against *Sclerotinia sclerotiorum* and *Fusarium oxysporum* which were found to inhibit their growth.

5.5. Source of Bioactives and Novel Compounds. Endophytes are capable of synthesizing bioactive compounds that are used by plants for defence against pathogens and some of these compounds have proven to be useful for novel drug discovery. Recent studies have reported hundreds of natural products including alkaloids, terpenoids, flavonoids, and steroids, from endophytes. Most of the bioactive compounds isolated from endophytes are known to have functions of antibiotics, immunosuppressants, anticancer agents, biological control agents, and so forth [53].

Maytansinoids, like rifamycin (Figure 1(a)) and geldanamycin (Figure 1(b)), which structurally belong to the ansamycin family of polyketide macrolactams are products of three closely related plant families (Celastraceae, Rhamnaceae, and Euphorbiaceae), mosses, and certain bacteria such as *Actinosynnema pretiosum.* It was hypothesized that microbes in the rhizosphere might be involved in the biosynthesis of plant maytansinoids [54]. Several endophytic actinomycetes were isolated from *Trewia nudiflora,* of which *Streptomyces* sp. 5B and *Streptomyces* sp. M27m3 were proved to have the potential of producing ansamycins [42]. One novel

chlorine-containing ansamycin, namely, naphthomycin K (Figure 1(c)), which was isolated from the endophytic strain *Streptomyces* sp. CS of the maytansinoids producer medicinal plant *Maytenus hookeri,* showed evident cytotoxic activity against P388 and A-549 cell lines, but no inhibitory activities against *Staphylococcus aureus* and *Mycobacterium tuberculosis* [55].

Siderophores are biologically active compound with function of chelating iron ions in living organisms. They have found extensive applications in the field of agriculture and medicine. They are also a component of virulence of microorganisms infecting man, animals, and plants [56]. Five different strains of *Phialocephala fortinii,* a dark septate fungal, were studied and all of them excreted three siderophores namely, ferricrocin, ferrirubin and ferrichrome C, whose production was dependent on pH and iron(III) concentration of the culture medium [57]. *P. fortinii* can thus be used for large scale production of these siderophores.

Taxol (Figure 1(d)) is a drug used to cure breast cancer, ovarian cancer, and lung cancer. An endophytic microorganism *Metarhizium anisopliae,* isolated from *Taxus chinensis,* was found to produce taxol in abundance *in vitro* [58]. *Colletotrichum gloeosporioides* isolated from the leaves of a medicinal plant, *Justicia gendarussa,* also produces taxol [59].

Huperzine A (HupA) (Figure 1(e)), a lycopodium alkaloid was isolated originally from *Huperzia serrata.* It has attracted intense attention since its marked role as cholinesterase inhibitor was discovered [60, 61]. Over 120 endophytic fungi were recovered from *H. serrata* and when screened for Hup-A, nine of them produced it [62]. From the all screened fungi, *Shiraia sp.* was found to be the most significant producer of HupA.

Two new benzopyranones, diaportheone A and B, were obtained via bioassay-guided isolation of the secondary metabolites from the endophytic fungus *Diaporthe sp.* P133 isolated from *Pandanus maryllifolius* leaves. Compounds diaportheone A and B inhibited the growth of the virulent strain of *Mycobacterium tuberculosis* H37Rv with minimum inhibitory concentrations of 100.9 and 3.5 μM, respectively [63].

Brefeldin A (Figure 1(f)) is a macrocyclic lactone synthesized from palmitate by a variety of fungi, which inhibits the protein secretion in the cells [64]. The membrane protein of the cell is retained in endoplastic reticulum from where it is not transported to Golgi apparatus and also results in retrotransport of the proteins in golgi complex, which have been secreted before the treatment of cells with it, to endoplasmic reticulum. It was initially known as an antiviral drug but lately was used to study protein synthesis and secretion in cells. Brefeldin A was isolated from active endophytic strain I(R)9-2, *Cladosporium sp.* from *Quercus variabilis* [65].

Nine new biologically active secondary metabolites were isolated from endophytic fungus *Alternaria alternata* residing in *Maytenus hookeri* and were characterized by NMR as (i) alternariol, (ii) alternariol monomethyl ether, (iii) 5-epialtenuene, (iv) altenuene, (v) uridine, (vi) adenosine, (vii) ACTG toxin-E, (viii) ergosta-4,6,8,22-tetraen-3-one, and (ix)

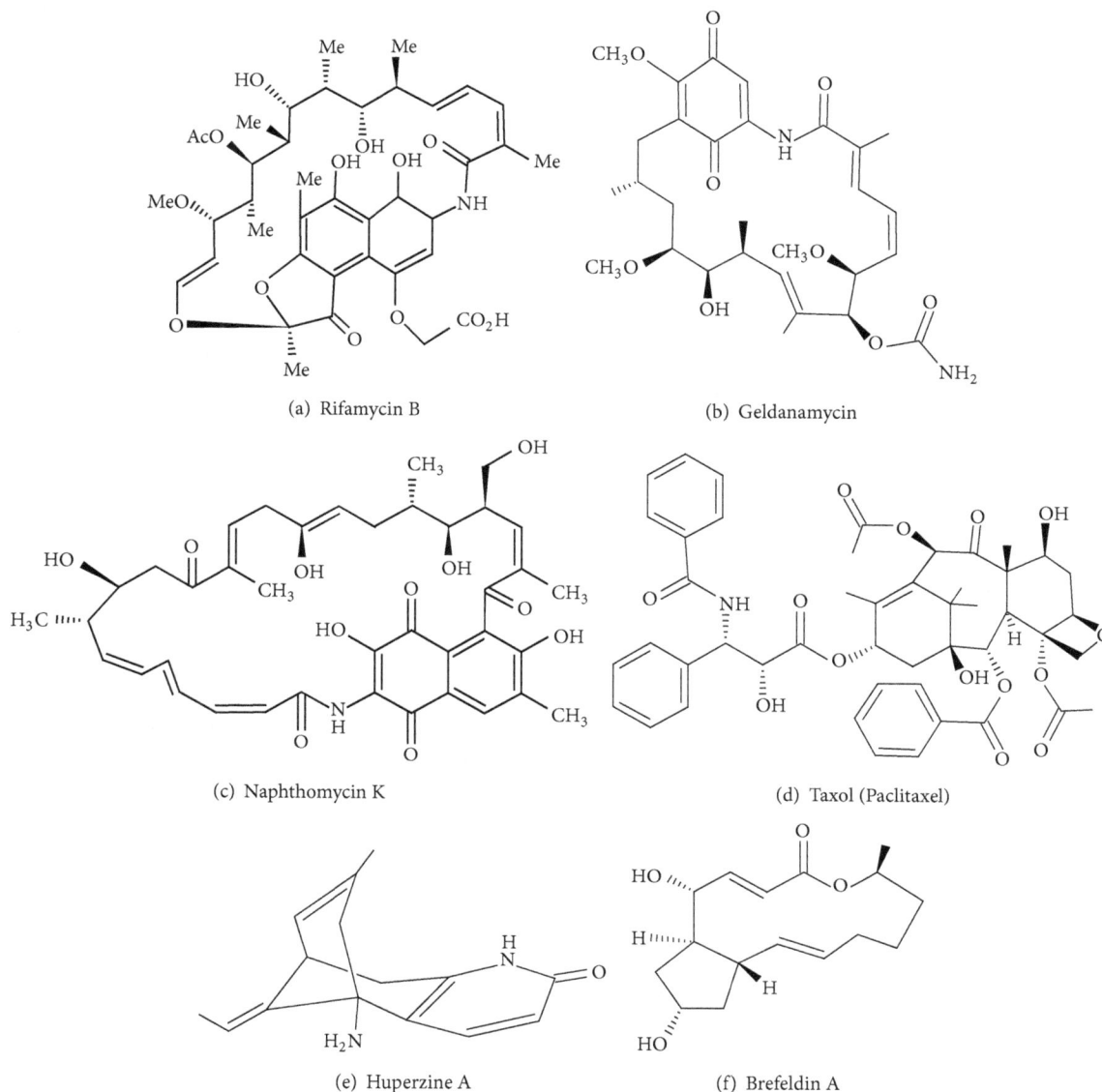

(a) Rifamycin B

(b) Geldanamycin

(c) Naphthomycin K

(d) Taxol (Paclitaxel)

(e) Huperzine A

(f) Brefeldin A

FIGURE 1: Chemical structures of some bioactive compounds produced by endophytic microorganisms.

ergosta-7,24(28)-dien-3-ol from the ethyl acetate-methanol-acetic acid extract of the solid-state fermentations of this fungus [66].

5.6. Reciprocal Interactions between Above- and Belowground Communities. The microbial community responses in soils conditioned by plants of the annual grass *Lolium multiflorum* with contrasting levels of infection with the endophyte *Neotyphodium occultans* were explored [67]. Soil conditioning by highly infected plants affected soil catabolic profiles and tended to increase soil fungal activity. A shift in bacterial community structures was detected while no changes were observed for fungi. Soil responses became evident even without changes in host plant biomass or soil organic carbon or total nitrogen content, suggesting that the endophyte modified host rhizo depositions during the conditioning phase.

A few researchers have reported changes in the rhizosphere chemistry and enzymatics activity mediated by endophyte presence in perennial host grasses [68, 69].

5.7. Biocontrol Agents. Endophytic microorganisms are regarded as an effective biocontrol agent, alternative to chemical control. Endophytic fungi have been described to play an important role in controlling insect herbivory not only in grasses [5] but also in conifers [70]. An endophytic fungi *Beauveria bassiana* known as an entomopathogen was found to control the borer insects in coffee seedlings [70] and sorghum [71]. The fungal pathogen *Botrytis cinerea* causes severe rotting on tomato fruits during storage and shelf life. The endophytic bacteria *Bacillus subtilis*, isolated from *Speranskia tuberculata* (Bge.) Baill, was found to be strongly antagonistic to the pathogen *B. cinerea* in *in vitro* studies[72].

A new strain of *Burkholderia pyrrocinia* JK-SH007 and *B. cepacia*, were identified as potential biocontrol agent against poplar canker [73].

Not only naturally occurring endophytes are used as biocontrol agents but also they are genetically engineered to express antipest proteins like lectins. Initial attempts were made to introduce heterologous gene into an endophytic microorganism for insect control [74, 75]. With the advent of time, many scientists have worked on this aspect, which has been one of the important studies in endophytes recently. Fungal endophyte of *Chaetomium globosum* YY-11 with antifungi activities, isolated from rape seedlings, and bacterial endophytes of *Enterobacter* sp. and *Bacillus subtilis* isolated from rice seedlings were used to express *Pinellia ternate* agglutinin (*PtA*) gene [76]. These recombinant endophytes expressing *PtA* gene were found to effectively control the population of sap sucking pests in several crop seedlings. Similarly, in a different study, recombinant endophytic bacteria *Enterobacter cloacae* expressing *PtA* gene proved to be a bioinsecticide against white backed planthopper, *Sogatella furcifera* [77]. Use of recombinant endophytes as biocontrol agents expressing different antipest proteins becomes a promising technique for control of plant pests as these endophytes can easily colonize within different crop plants successfully.

5.8. Nutrient Cycling. Nutrient cycling is a very important process that happens continuously to balance the existing nutrients existing and make it available for every component of the ecosystem. Biodegradation of the dead biomasses becomes one major step in it to bring back the utilized nutrients back to the ecosystem which in turn again becomes available to the organisms. This becomes a cyclic chain process. A lot of saprophytic organisms play a major role in it. Few studies have shown that endophytes have important role in biodegradation of the litter of its host plants [78–85]. During biodegradation of the litter, the endophytic microbes colonize initially within the plants [86] and facilitate the saprophytic microbes to act on through antagonistic interaction, and thus increasing the litter decomposition [87, 88]. In another study, it was demonstrated that all endophytes had the ability to decompose organic components, including lignin, cellulose, and hemicellulose however the preferences of various groups of endophytes with respect to organic compounds differed [89].

5.9. Bioremediation/Biodegradation. Endophytes have a powerful ability to breakdown complex compounds. Bioremediation is a method of removal of pollutants and wastes from the environment by the use of micro-organisms. It relies on the biological processes in microbes to breakdown these wastes. This is made possible due to the great microbial diversity. A group of researchers studied the role of endophytes in bioremediation in *Nicotiana tabaccum* plants [90]. Inoculation of *Nicotiana tabaccum* with endophytes resulted in improved biomass production under conditions of Cadmium (Cd) stress, and the total plant Cd concentration was higher compared to noninoculated plants. These results

demonstrated the beneficial effects of seed endophytes on metal toxicity and accumulation.

To explore the endophytic diversity for the breakdown of plastic, several dozen endophytic fungi were screened for their ability to degrade the synthetic polymer polyester polyurethane (PUR) [91]. Though several organisms demonstrated the ability to efficiently degrade PUR in both solid and liquid suspensions, robust activity was observed among several isolates in the genus *Pestalotiopsis*. Two *Pestalotiopsis microspora* isolates were uniquely able to grow on PUR as the sole carbon source under both aerobic and anaerobic conditions. Molecular characterization of this activity suggested that an enzyme serine hydrolase is responsible for degradation of PUR [91].

5.10. Production of Volatile Organic Compounds and Their Benefits. *Hypoxylon sp.* which is an endophytic fungus isolated from *Persea indica* produced an impressive spectrum of volatile organic compounds (VOCs), most notably 1,8-cineole, 1-methyl-1,4-cyclohexadiene, and tentatively identified alpha-methylene-alpha-fenchocamphorone, among many others, most of which are unidentified. It displayed maximal VOC antimicrobial activity against *Botrytis cinerea*, *Phytophthora cinnamomi*, *Cercospora beticola*, and *Sclerotinia sclerotiorum* suggesting that the VOCs may play some role in the biology of the fungus and its survival in its host plant [92]. They unequivocally demonstrated that 1,8-cineole (a monoterpene) is produced in addition by this *Hypoxylon sp.*, which represents a novel and important source of this compound. This monoterpene is an octane derivative and has potential use as a fuel additive as do the other VOCs of this organism. This study thus shows that fungal sourcing of this compound and other VOCs as produced by *Hypoxylon sp.* greatly expands their potential applications in medicine, industry, and energy production.

An unusual *Phomopsis sp.* was isolated as endophyte of *Odontoglossum sp.* (Orchidaceae), produced a unique mixture of volatile organic compounds (VOCs) including sabinene (a monoterpene with a peppery odor), 1-butanol, 3-methyl; benzeneethanol; 1-propanol, 2-methyl, and 2-propanone [93]. The gases of *Phomopsis sp.* possess antifungal properties and an artificial mixture of the VOCs mimicked the antibiotic effects of this organism with the greatest bioactivity against a wide range of plant pathogenic test fungi including *Pythium, Phytophthora, Sclerotinia, Rhizoctonia, Fusarium, Botrytis, Verticillium,* and *Colletotrichum*. As with many VOC-producing endophytes, this *Phomopsis sp.* did survive and grow in the presence of the inhibitory gases of *Muscodor albus*, an endophytic fungus. The authors in [93] had hypothesized that there was a possible involvement of VOC production by the fungus and its role in the biology/ecology of the fungus/plant/environmental relationship.

5.11. Endophytes with Multiple Roles. Many endophytes are known to have wide range of activity within hosts. Endophytic microbes were found to have herbicidal activity along with antimicrobial activity [94]. *Bacillus sp.* SLS18, known as a plant growth-promoting endophyte, was investigated for its role in the biomass production and manganese and cadmium uptake by *Sorghum bicolor* L., *Phytolacca acinosa* Roxb., and *Solanum nigrum* L. [95]. It displayed multiple heavy metals and antibiotics resistances. The strain also exhibited the capacity of producing indole-3-acetic acid, siderophores, and 1-aminocyclopropane-1-carboxylic acid deaminase.

5.12. Endophytes in Tissue Culture. Endophytes are largely useful to the host plants in many ways as discussed in earlier part of this review. But when it comes to plant tissue culture it is usually considered as a contaminant. The ultimate aim of tissue culture is to develop axenic plants. Though we surface sterilize the explants to be used for tissue culture, after few days, bacteria or fungi or actinomycetes or all of them start growing from tissue or the cultured explant. These contaminants are nothing but endophytic microbes resulting in complete loss of time, media, and explants, which sometimes may be of some rare and endangered species which needs to be conserved by tissue culture techniques.

Endophyte species composition and plant genotype together with tissue culture conditions are the key factors for gaining plant tissue cultures with high regeneration capacity. Interaction between the endophytes and specific secondary compounds may be an important factor for browning and cell death in the Scots pine calli [19]. These researchers examined the green, light brown, and dark brown calli by TEM with respect to presence of microbial cells in the tissues. The microbial cells were encountered more frequently in cells of the brown tissues than in the green, well growing tissues, which suggested that endophytes could either be involved with browning by inducing the senescence and release of tannins in the tissue or that the endophytes take over the otherwise senescing callus tissue.

Plant tissue culture, which commonly utilizes the meristems, which are considered to be sterile part of any plant, has however given numerous references to microbial existence in these tissues [96–98]. Elimination of these endophytes thus becomes a prime objective to develop axenic plants during tissue culture. Many protocols have been developed by many researchers to overcome this problem.

6. Conclusions

Recent years have seen great deal of interest among researchers in the studies on endophytic microorganisms, thanks to easier methods of isolation and identifications and current tools of molecular biology. Many bioactive compounds beneficial to pharmaceuticals, environment, agriculture, and industries are produced by endophytes. Due to their great importance to plants/human beings/environment, scientists have already started exploiting them very much for newer compounds and newer roles to the environment and human. It becomes sensible to review on past achievements in the field of endophytic research, opening up broader opportunities for the scientific community.

Endophytes can be either bacteria or fungi or actinomycetes. In this review, we have come to a conclusion that it is mostly the actinomycetes which are involved in production of pharmaceutically important compounds within the plants. Generally, the fungi are involved in the role of phytoremediation, biodegradation, and nutrient cycling and thus reduce the debris load on the environment in a better way. By and large, it is the bacterial community of endophytes which helps the plants in their better growth by producing different growth hormones. Application of different innovative biotechnological tools will help in strengthening the understanding of plant-endophyte interactions, producing new bioactive compounds, perk up the growth in plants, and improve biocontrol activity, reducing the debris and other wastes which are otherwise harmful to the ecosystem. Considering all these, definitely endophytes have proved to be a boon and have left good impact on plants, environment, and also human beings in several possible ways.

Conflict of Interests

The authors declare that there is no conflict of interests regarding the publication of this paper.

Acknowledgments

The authors are thankful to Mr. N. Prabhakaran, Mr. R. K. Selvakesavan, and Dr. S. Sivashankari for their critical reviewing, discussions, and discerning comments.

References

[1] D. Wilson, "Endophyte: the evolution of a term, and clarification of its use and definition," *Oikos*, vol. 73, no. 2, pp. 274–276, 1995.

[2] G. C. Carroll, "Fungal endophytes in stems and leaves: from latent pathogen to mutualistic symbiont," *Ecology*, vol. 69, no. 1, pp. 2–9, 1988.

[3] G. C. Carroll, "Fungal associates of woody plants as insect antagonists in leaves and stems," in *Microbial Mediation of Plant Herbivore Interactions*, P. Barbosa, V. A. Krischik, and C. G. Jones, Eds., pp. 253–272, John Wiley & Sons, New York, NY, USA, 1991.

[4] K. Clay, "Fungal endophytes of grasses: a defensive mutualism between plants and fungi," *Ecology*, vol. 69, no. 1, pp. 10–16, 1988.

[5] K. Clay, "Fungal endophytes of grasses," *Annual Review of Ecology, Evolution, and Systematics*, vol. 21, pp. 275–295, 1990.

[6] C. A. Gehring and T. G. Whitham, "Interactions between aboveground herbivores and the mycorrhizal mutualists of plants," *Trends in Ecology and Evolution*, vol. 9, no. 7, pp. 251–255, 1994.

[7] C. A. Gehring, N. S. Cobb, and T. G. Whitham, "Three-way interactions among ectomycorrhizal mutualists, scale insects, and resistant and susceptible pinyon pines," *The American Naturalist*, vol. 149, no. 5, pp. 824–841, 1997.

[8] N. C. Johnson, "Can fertilization of soil select less mutualistic mycorrhizae?" *Ecological Applications*, vol. 3, no. 4, pp. 749–757, 1993.

[9] N. C. Johnson, J. H. Graham, and F. A. Smith, "Functioning of mycorrhizal associations along the mutualism-parasitism continuum," *New Phytologist*, vol. 135, no. 4, pp. 575–586, 1997.

[10] M. A. Parker, "Plant fitness variation caused by different mutualist genotypes," *Ecology*, vol. 76, no. 5, pp. 1525–1535, 1995.

[11] M. A. Parker, "Mutualism in metapopulations of legumes and *Rhizobia*," *The American Naturalist*, vol. 153, supplement 5, pp. S48–S60, 1999.

[12] S. S. Dudeja, R. Giri, R. Saini, P. Suneja-Madan, and E. Kothe, "Interaction of endophytic microbes with legumes," *Journal of Basic Microbiology*, vol. 52, pp. 248–260, 2012.

[13] A. Das and A. Varma, "Symbiosis: the art of living," in *Symbiotic Fungi Principles and Practice*, A. Varma and A. C. Kharkwal, Eds., pp. 1–28, Springer, Berlin, Germany, 2009.

[14] S. Lee, M. Flores-Encarnación, M. Contreras-Zentella, L. Garcia-Flores, J. E. Escamilla, and C. Kennedy, "Indole-3-acetic acid biosynthesis is deficient in *Gluconacetobacter diazotrophicus* strains with mutations in *Cytochrome c* biogenesis genes," *Journal of Bacteriology*, vol. 186, no. 16, pp. 5384–5391, 2004.

[15] A. M. Pirttilä, H. Laukkanen, H. Pospiech, R. Myllylä, and A. Hohtola, "Detection of intracellular bacteria in the buds of scotch pine (*Pinus sylvestris* L.) by In situ hybridization," *Applied and Environmental Microbiology*, vol. 66, no. 7, pp. 3073–3077, 2000.

[16] A. M. Pirttilä, H. Pospiech, H. Laukkanen, R. Myllylä, and A. Hohtola, "Two endophytic fungi in different tissues of Scots pine buds (*Pinus sylvestris* L.)," *Microbial Ecology*, vol. 45, no. 1, pp. 53–62, 2003.

[17] K. Hata, R. Atari, and K. Sone, "Isolation of endophytic fungi from leaves of *Pasania edulis* and their within-leaf distributions," *Mycoscience*, vol. 43, no. 5, pp. 369–373, 2002.

[18] K. Hata and K. Sone, "Isolation of endophytes from leaves of *Neolitsea sericea* in broadleaf and conifer stands," *Mycoscience*, vol. 49, no. 4, pp. 229–232, 2008.

[19] A. M. Pirttilä, O. Podolich, J. J. Koskimäki, E. Hohtola, and A. Hohtola, "Role of origin and endophyte infection in browning of bud-derived tissue cultures of Scots pine (*Pinus sylvestris* L.)," *Plant Cell, Tissue and Organ Culture*, vol. 95, no. 1, pp. 47–55, 2008.

[20] R. J. Ganley and G. Newcombe, "Fungal endophytes in seeds and needles of *Pinus monticola*," *Mycological Research*, vol. 110, no. 3, pp. 318–327, 2006.

[21] J. Hallmann, A. Quadt-Hallmann, W. F. Mahaffee, and J. W. Kloepper, "Bacterial endophytes in agricultural crops," *Canadian Journal of Microbiology*, vol. 43, no. 10, pp. 895–914, 1997.

[22] B. Reinhold-Hurek and T. Hurek, "Life in grasses: diazotrophic endophytes," *Trends in Microbiology*, vol. 6, no. 4, pp. 139–144, 1998.

[23] R. Rai, P. K. Dash, B. M. Prasanna, and A. Singh, "Endophytic bacterial flora in the stem tissue of a tropical maize (*Zea mays* L.) genotype: isolation, identification and enumeration," *World Journal of Microbiology and Biotechnology*, vol. 23, no. 6, pp. 853–858, 2007.

[24] V. C. Verma, S. K. Gond, A. Kumar, R. N. Kharwar, L.-A. Boulanger, and G. A. Strobel, "Endophytic fungal flora from roots and fruits of an Indian neem plant *Azadirachta indica* A. juss., and impact of culture media on their isolation," *Indian Journal of Microbiology*, vol. 51, no. 4, pp. 469–476, 2011.

[25] A. Madmony, L. Chernin, S. Pleban, E. Peleg, and J. Riov, "*Enterobacter cloacae*, an obligatory endophyte of pollen grains of Mediterranean pines," *Folia Microbiologica*, vol. 50, no. 3, pp. 209–216, 2005.

[26] N. Rungjindamai, U. Pinruan, R. Choeyklin, T. Hattori, and E. B. G. Jones, "Molecular characterization of basidiomycetous endophytes isolated from leaves, rachis and petioles of the oil palm, *Elaeis guineensis*, in Thailand," *Fungal Diversity*, vol. 33, pp. 139–161, 2008.

[27] J. D. P. Bezerra, M. G. S. Santos, V. M. Svedese et al., "Richness of endophytic fungi isolated from *Opuntia ficus-indica* Mill. (Cactaceae) and preliminary screening for enzyme production," *World Journal of Microbiology and Biotechnology*, vol. 28, no. 5, pp. 1989–1995, 2012.

[28] R. J. Rodriguez, J. F. White Jr., A. E. Arnold, and R. S. Redman, "Fungal endophytes: diversity and functional roles," *New Phytologist*, vol. 182, no. 2, pp. 314–330, 2009.

[29] S. Youngbae, S. Kim, and C. W. Park, "A phylogenetic study of polygonum sect. tovara (polygonaceae) based on ITS sequences of nuclear ribosomal DNA," *Plant Biology*, vol. 40, pp. 47–52, 1997.

[30] Y. X. Chen, L. P. Zhang, and Z. T. Lu, "Analysis of the internal transcribed spacer (ITS) sequences in rDNA of 10 strains of *Fusarium* spp.," *Journal of Anhui Agricultural Sciences*, vol. 36, pp. 4886–4887, 2008.

[31] X. Y. Chen, Y. D. Qi, J. H. Wei et al., "Molecular identification of endophytic fungi from medicinal plant *Huperzia serrata* based on rDNA ITS analysis," *World Journal of Microbiology and Biotechnology*, vol. 27, no. 3, pp. 495–503, 2011.

[32] S. Chareprasert, J. Piapukiew, S. Thienhirun, A. J. S. Whalley, and P. Sihanonth, "Endophytic fungi of teak leaves *Tectona grandis* L. and rain tree leaves *Samanea saman* Merr.," *World Journal of Microbiology and Biotechnology*, vol. 22, no. 5, pp. 481–486, 2006.

[33] W. Thongsandee, Y. Matsuda, and S. Ito, "Temporal variations in endophytic fungal assemblages of *Ginkgo biloba* L.," *Journal of Forest Research*, vol. 17, no. 2, pp. 213–218, 2012.

[34] W. L. Araújo, J. Marcon, W. Maccheroni Jr., J. D. van Elsas, J. W. L. van Vuurde, and J. L. Azevedo, "Diversity of endophytic bacterial populations and their interaction with *Xylella fastidiosa* in citrus plants," *Applied and Environmental Microbiology*, vol. 68, no. 10, pp. 4906–4914, 2002.

[35] D. K. Manter, J. A. Delgado, D. G. Holm, and R. A. Stong, "Pyrosequencing reveals a highly diverse and cultivar-specific bacterial endophyte community in potato roots," *Microbial Ecology*, vol. 60, no. 1, pp. 157–166, 2010.

[36] B. Nikolic, H. Schwab, and A. Sessitsch, "Metagenomic analysis of the 1-aminocyclopropane-1-carboxylate deaminase gene (acdS) operon of an uncultured bacterial endophyte colonizing *Solanum tuberosum* L.," *Archives of Microbiology*, vol. 193, no. 9, pp. 665–676, 2011.

[37] C. Glienke-Blanco, C. I. Aguilar-Vildoso, M. L. C. Vieira, P. A. V. Barroso, and J. L. Azevedo, "Genetic variability in the endophytic fungus *Guignardia citricarpa* isolated from citrus plants," *Genetics and Molecular Biology*, vol. 25, no. 2, pp. 251–255, 2002.

[38] X. Zhang, A. Ren, H. Ci, and Y. Gao, "Genetic diversity and structure of *Neotyphodium* species and their host *Achnatherum sibiricum* in a natural grass-endophyte system," *Microbial Ecology*, vol. 59, no. 4, pp. 744–756, 2010.

[39] C. A. Young, M. K. Bryant, M. J. Christensen, B. A. Tapper, G. T. Bryan, and B. Scott, "Molecular cloning and genetic analysis of a symbiosis-expressed gene cluster for lolitrem biosynthesis from a mutualistic endophyte of perennial ryegrass," *Molecular Genetics and Genomics*, vol. 274, no. 1, pp. 13–29, 2005.

[40] S. Taghavi, D. van der Lelie, A. Hoffman et al., "Genome sequence of the plant growth promoting endophytic bacterium *Enterobacter* sp.," *PLoS Genetics*, vol. 6, no. 5, Article ID e1000943, 2010.

[41] D. P. Malinowski, G. A. Alloush, and D. P. Belesky, "Leaf endophyte *Neotyphodium coenophialum* modifies mineral uptake in tall fescue," *Plant and Soil*, vol. 227, no. 1-2, pp. 115–126, 2000.

[42] M. Arachevaleta, C. W. Bacon, C. S. Hoveland, and D. E. Radcliffe, "Effect of tall fescue endophyte on plant response to environmental stress," *Agronomy Journal*, vol. 81, pp. 83–90, 1989.

[43] G. Xin, G. Zhang, J. W. Kang, J. T. Staley, and S. L. Doty, "A diazotrophic, indole-3-acetic acid-producing endophyte from wild cottonwood," *Biology and Fertility of Soils*, vol. 45, no. 6, pp. 669–674, 2009.

[44] M. Hamayun, S. Afzal Khan, N. Ahmad et al., "*Cladosporium sphaerospermum* as a new plant growth-promoting endophyte from the roots of *Glycine max* (L.) Merr," *World Journal of Microbiology and Biotechnology*, vol. 25, no. 4, pp. 627–632, 2009.

[45] J. J. Liu, S. J. Chen, and H. X. Gong, "Study on endophytic fungi producing orange pigment isolated from *Ginkgo Biloba* L.," 2008, http://www.paper.edu.cn/en_releasepaper/content/23292.

[46] M. Qiu, R. Xie, Y. Shi et al., "Isolation and identification of endophytic fungus SX01, a red pigment producer from *Ginkgo Biloba* L.," *World Journal of Microbiology and Biotechnology*, vol. 26, no. 6, pp. 993–998, 2010.

[47] S. Visalakchi and J. Muthumary, "Antimicrobial activity of the new endophytic *Monodictys castaneae* SVJM139 pigment and its optimization," *African Journal of Microbiology Research*, vol. 3, no. 9, pp. 550–556, 2009.

[48] K. M. Bischoff, D. T. Wicklow, D. B. Jordan et al., "Extracellular hemicellulolytic enzymes from the maize endophyte *Acremonium zeae*," *Current Microbiology*, vol. 58, no. 5, pp. 499–503, 2009.

[49] L. D. Sette, M. R. Z. Passarini, C. Delarmelina, F. Salati, and M. C. T. Duarte, "Molecular characterization and antimicrobial activity of endophytic fungi from coffee plants," *World Journal of Microbiology and Biotechnology*, vol. 22, no. 11, pp. 1185–1195, 2006.

[50] K. A. Selim, A. A. El-Beih, T. M. AbdEl-Rahman, and A. I. El-Diwany, "Biodiversity and antimicrobial activity of endophytes associated with Egyptian medicinal plants," *Mycosphere*, vol. 2, no. 6, pp. 669–678, 2011.

[51] R. Devaraju and S. Sathish, "Endophytic Mycoflora of *Mirabilis jalapa* L. and studies on antimicrobial activity of its endophytic *Fusarium* sp.," *Asian Journal of Experimental Biological Sciences*, vol. 2, no. 1, pp. 75–79, 2011.

[52] S. Kumar, N. Kaushik, R. Edrada-Ebel, R. Ebel, and P. Proksch, "Isolation, characterization, and bioactivity of endophytic fungi of *Tylophora indica*," *World Journal of Microbiology and Biotechnology*, vol. 27, no. 3, pp. 571–577, 2011.

[53] B. Joseph and R. Mini Priya, "Bioactive compounds from endophytes and their potential in pharmaceutical effect: a review," *The American Journal of Biochemistry and Molecular Biology*, vol. 1, no. 3, pp. 291–309, 2011.

[54] C. B. Pullen, P. Schmitz, D. Hoffmann et al., "Occurrence and non-detectability of maytansinoids in individual plants of the genera Maytenus and Putterlickia," *Phytochemistry*, vol. 62, no. 3, pp. 377–387, 2003.

[55] C. H. Lu and Y. M. Shen, "A novel ansamycin, naphthomycin K from *Streptomyces* sp.," *Journal of Antibiotics*, vol. 60, no. 10, pp. 649–653, 2007.

[56] J. B. Neilands, "Siderophores," *Archives of Biochemistry and Biophysics*, vol. 302, no. 1, pp. 1–3, 1993.

[57] B. A. Bartholdy, M. Berreck, and K. Haselwandter, "Hydroxamate siderophore synthesis by *Phialocephala fortinii*, a typical dark septate fungal root endophyte," *BioMetals*, vol. 14, no. 1, pp. 33–42, 2001.

[58] K. Liu, X. Ding, B. Deng, and W. Chen, "Isolation and characterization of endophytic taxol-producing fungi from *Taxus chinensis*," *Journal of Industrial Microbiology and Biotechnology*, vol. 36, no. 9, pp. 1171–1177, 2009.

[59] V. Gangadevi and J. Muthumary, "Isolation of Colletotrichum gloeosporioides, a novel endophytic taxol-producing fungus from the leaves of a medicinal plant, *Justicia gendarussa*," *Mycologia Balcanica*, vol. 5, pp. 1–4, 2008.

[60] J. S. Liu, C. M. Yu, Y. Z. Zhou et al., "Study on the chemistry of huperzine-A and huperzine-B," *Acta Physico-Chimica Sinica*, vol. 44, pp. 1035–1040, 1986.

[61] J. S. Liu, Y. L. Zhu, C. M. Yu et al., "The structures of huperzine A and B, two new alkaloids exhibiting marked anticholinesterase activity," *Canadian Journal of Chemistry*, vol. 64, pp. 837–839, 1986.

[62] Y. Wang, Q. G. Zeng, Z. B. Zhang, R. M. Yan, L. Y. Wang, and D. Zhu, "Isolation and characterization of endophytic huperzine A-producing fungi from *Huperzia serrata*," *Journal of Industrial Microbiology and Biotechnology*, vol. 38, no. 9, pp. 1267–1278, 2011.

[63] M. E. Bungihan, M. A. Tan, M. Kitajima et al., "Bioactive metabolites of *Diaporthe* sp. P133, an endophytic fungus isolated from *Pandanus amaryllifolius*," *Journal of Natural Medicines*, vol. 65, no. 3-4, pp. 606–609, 2011.

[64] R. D. Klausner, J. G. Donaldson, and J. Lippincott-Schwartz, "Brefeldin A: insights into the control of membrane traffic and organelle structure," *Journal of Cell Biology*, vol. 116, no. 5, pp. 1071–1080, 1992.

[65] F. W. Wang, R. H. Jiao, A. B. Cheng, S. H. Tan, and Y. C. Song, "Antimicrobial potentials of endophytic fungi residing in *Quercus variabilis* and brefeldin A obtained from *Cladosporium* sp.," *World Journal of Microbiology and Biotechnology*, vol. 23, no. 1, pp. 79–83, 2007.

[66] Y.-T. Ma, L.-R. Qiao, W.-Q. Shi, A.-L. Zhang, and J.-M. Gao, "Metabolites produced by an endophyte *Alternaria alternata* isolated from *Maytenus hookeri*," *Chemistry of Natural Compounds*, vol. 46, no. 3, pp. 504–506, 2010.

[67] C. Casas, M. Omacini, M. S. Montecchia, and O. S. Correa, "Soil microbial community responses to the fungal endophyte *Neotyphodium* in Italian ryegrass," *Plant and Soil*, vol. 340, no. 1, pp. 347–355, 2011.

[68] D. P. Malinowski, G. A. Alloush, and D. P. Belesky, "Evidence for chemical changes on the root surface of fall fescue in response to infection with the fungal endophyte *Neotyphodium coenophialum*," *Plant and Soil*, vol. 205, no. 1, pp. 1–12, 1998.

[69] M. M. van Hecke, A. M. Treonis, and J. R. Kaufman, "How does the fungal endophyte *Neotyphodium coenophialum* affect tall fescue (*Festuca arundinacea*) rhizodeposition and soil microorganisms?" *Plant and Soil*, vol. 275, no. 1-2, pp. 101–109, 2005.

[70] F. Posada and F. E. Vega, "Inoculation and colonization of coffee seedlings (*Coffea arabica* L.) with the fungal entomopathogen *Beauveria bassiana* (Ascomycota: Hypocreales)," *Mycoscience*, vol. 47, no. 5, pp. 284–289, 2006.

[71] T. Tefera and S. Vidal, "Effect of inoculation method and plant growth medium on endophytic colonization of sorghum by the entomopathogenic fungus *Beauveria bassiana*," *BioControl*, vol. 54, no. 5, pp. 663–669, 2009.

[72] S. Wang, T. Hu, Y. Jiao, J. Wei, and K. Cao, "Isolation and characterization of *Bacillus subtilis* EB-28, an endophytic bacterium strain displaying biocontrol activity against *Botrytis cinerea* Pers," *Frontiers of Agriculture in China*, vol. 3, no. 3, pp. 247–252, 2009.

[73] J. H. Ren, J. R. Ye, H. Liu, X. L. Xu, and X. Q. Wu, "Isolation and characterization of a new *Burkholderia pyrrocinia* strain JK-SH007 as a potential biocontrol agent," *World Journal of Microbiology and Biotechnology*, vol. 27, no. 9, pp. 2203–2215, 2011.

[74] J. W. Fahey, "Endophytic bacteria for the delivery of agrochemicals to plants," in *Biologically Active Natural Products*, H. G. Cutler, Ed., pp. 120–128, American Chemical Society, Washington, DC, USA, 1988.

[75] J. W. Fahey, M. B. Dimock, S. F. Tomasino, J. M. Taylor, and P. S. Carlson, "Genetically engineered endophytes as biocontrol agents: a case study in industry," in *Microbial Ecology of Leaves*, J. H. Andrews, Ed., pp. 402–411, Springer, New York, NY, USA, 1991.

[76] X. Zhao, G. Qi, X. Zhang, N. Lan, and X. Ma, "Controlling sap-sucking insect pests with recombinant endophytes expressing plant lectin," *Nature Precedings*, vol. 21, article 21, 2010.

[77] X. Zhang, J. Li, G. Qi, K. Wen, J. Lu, and X. Zhao, "Insecticidal effect of recombinant endophytic bacterium containing *Pinellia ternata* agglutinin against white backed planthopper, *Sogatella furcifera*," *Crop Protection*, vol. 30, no. 11, pp. 1478–1484, 2011.

[78] M. M. Müller, R. Valjakka, A. Suokko, and J. Hantula, "Diversity of endophytic fungi of single Norway spruce needles and their role as pioneer decomposers," *Molecular Ecology*, vol. 10, no. 7, pp. 1801–1810, 2001.

[79] V. Kumaresan and T. S. Suryanarayanan, "Endophyte assemblages in young, mature and senescent leaves of Rhizophora apiculata: evidence for the role of endophytes in mangrove litter degradation," *Fungal Diversity*, vol. 9, pp. 81–91, 2002.

[80] T. Osono, "Effects of prior decomposition of beech leaf litter by phyllosphere fungi on substrate utilization by fungal decomposers," *Mycoscience*, vol. 44, no. 1, pp. 41–45, 2003.

[81] T. Osono, "Role of phyllosphere fungi of forest trees in the development of decomposer fungal communities and decomposition processes of leaf litter," *Canadian Journal of Microbiology*, vol. 52, no. 8, pp. 701–716, 2006.

[82] T. Korkama-Rajala, M. M. Müller, and T. Pennanen, "Decomposition and fungi of needle litter from slow- and fast-growing Norway spruce (*Picea abies*) clones," *Microbial Ecology*, vol. 56, no. 1, pp. 76–89, 2008.

[83] Y. Fukasawa, T. Osono, and H. Takeda, "Effects of attack of saprobic fungi on twig litter decomposition by endophytic fungi," *Ecological Research*, vol. 24, no. 5, pp. 1067–1073, 2009.

[84] T. Osono and D. Hirose, "Effects of prior decomposition of *Camellia japonica* leaf litter by an endophytic fungus on the subsequent decomposition by fungal colonizers," *Mycoscience*, vol. 50, no. 1, pp. 52–55, 2009.

[85] I. Promputtha, K. D. Hyde, E. H. C. McKenzie, J. F. Peberdy, and S. Lumyong, "Can leaf degrading enzymes provide evidence that endophytic fungi becoming saprobes?" *Fungal Diversity*, vol. 41, pp. 89–99, 2010.

[86] M. N. Thormann, R. S. Currah, and S. E. Bayley, "Succession of microfungal assemblages in decomposing peatland plants," *Plant and Soil*, vol. 250, no. 2, pp. 323–333, 2003.

[87] S. C. Fryar, T. K. Yuen, K. D. Hyde, and I. J. Hodgkiss, "The influence of competition between tropical fungi on wood colonization in streams," *Microbial Ecology*, vol. 41, no. 3, pp. 245–251, 2001.

[88] V. A. Terekhova and T. A. Semenova, "The structure of micromycete communities and their synecologic interactions with basidiomycetes during plant debris decomposition," *Microbiology*, vol. 74, no. 1, pp. 91–96, 2005.

[89] X. He, G. Han, Y. Lin et al., "Diversity and decomposition potential of endophytes in leaves of a *Cinnamomum camphora* plantation in China," *Ecological Research*, vol. 27, no. 2, pp. 273–284, 2012.

[90] C. Mastretta, S. Taghavi, D. van der Lelie et al., "Endophytic bacteria from seeds of *Nicotiana tabacum* can reduce cadmium phytotoxicity," *International Journal of Phytoremediation*, vol. 11, no. 3, pp. 251–267, 2009.

[91] J. R. Russell, J. Huang, P. Anand et al., "Biodegradation of polyester polyurethane by endophytic fungi," *Applied and Environmental Microbiology*, vol. 77, no. 17, pp. 6076–6084, 2011.

[92] A. R. Tomsheck, G. A. Strobel, E. Booth et al., "*Hypoxylon* sp., an endophyte of *Persea indica*, producing 1,8-Cineole and other bioactive volatiles with fuel potential," *Microbial Ecology*, vol. 60, no. 4, pp. 903–914, 2010.

[93] S. K. Singh, G. A. Strobel, B. Knighton, B. Geary, J. Sears, and D. Ezra, "An endophytic *Phomopsis* sp. possessing bioactivity and fuel potential with its volatile organic compounds," *Microbial Ecology*, vol. 61, no. 4, pp. 729–739, 2011.

[94] J. Li, G.-Z. Zhao, H.-Y. Huang et al., "Isolation and characterization of culturable endophytic actinobacteria associated with *Artemisia annua* L.," *Antonie van Leeuwenhoek*, vol. 101, no. 3, pp. 515–527, 2012.

[95] S. Luo, T. Xu, L. Chen et al., "Endophyte-assisted promotion of biomass production and metal-uptake of energy crop sweet sorghum by plant-growth-promoting endophyte *Bacillus* sp. SLS18," *Applied Microbiology and Biotechnology*, vol. 93, no. 4, pp. 1745–1753, 2012.

[96] M. A. Holland and J. C. Polacco, "PPFMs and other covert contaminants: is there more to plant physiology than just plant?" *Annual Review of Plant Physiology and Plant Molecular Biology*, vol. 45, pp. 197–209, 1994.

[97] C. Leifert, C. E. Morris, and W. M. Waites, "Ecology of microbial saprophytes and pathogens in tissue culture and field-grown plants: reasons for contamination problems in vitro," *Critical Reviews in Plant Sciences*, vol. 13, no. 2, pp. 139–183, 1994.

[98] P. Tanprasert and B. M. Reed, "Detection and identification of bacterial contaminants from strawberry runner explants," *In Vitro Cellular and Developmental Biology*, vol. 33, no. 3, pp. 221–226, 1997.

[99] S. Larran, C. Mónaco, and H. E. Alippi, "Endophytic fungi in leaves of *Lycopersicon esculentum* Mill," *World Journal of Microbiology and Biotechnology*, vol. 17, no. 2, pp. 181–184, 2001.

[100] S. Larran, A. Perelló, M. R. Simón, and V. Moreno, "Isolation and analysis of endophytic microorganisms in wheat (*Triticum aestivum* L.) leaves," *World Journal of Microbiology and Biotechnology*, vol. 18, no. 7, pp. 683–686, 2002.

[101] J. Santamaría and P. Bayman, "Fungal epiphytes and endophytes of coffee leaves (*Coffea arabica*)," *Microbial Ecology*, vol. 50, no. 1, pp. 1–8, 2005.

[102] E. F. Rakotoniriana, F. Munaut, C. Decock et al., "Endophytic fungi from leaves of *Centella asiatica*: occurrence and potential interactions within leaves," *Antonie van Leeuwenhoek*, vol. 93, no. 1-2, pp. 27–36, 2008.

[103] X. Xing, S. Guo, and J. Fu, "Biodiversity and distribution of endophytic fungi associated with *Panax quinquefolium* L. cultivated in a forest reserve," *Symbiosis*, vol. 51, no. 2, pp. 161–166, 2010.

7

Evaluation of Different Culture Media for Improvement in Bioinsecticides Production by Indigenous *Bacillus thuringiensis* and Their Application against Larvae of *Aedes aegypti*

Patil Chandrashekhar Devidas,[1] Borase Hemant Pandit,[1] and Patil Satish Vitthalrao[1,2]

[1] *School of Life Sciences, North Maharashtra University, P.O. Box 80, Jalgaon, Maharashtra 425001, India*
[2] *North Maharashtra Microbial Culture Collection Centre (NMCC), North Maharashtra University, P.O. Box 80, Jalgaon, Maharashtra 425001, India*

Correspondence should be addressed to Patil Satish Vitthalrao; satish.patil7@gmail.com

Academic Editors: S. R. Gerrard, S. Narayanan, R. Tofalo, and A. I. Vela

Production of indigenous isolate *Bacillus thuringiensis sv*2 (*Bt sv*2) was checked on conventional and nonconventional carbon and nitrogen sources in shake flasks. The effects on the production of biomass, toxin production, and spore formation capability of mosquito toxic strain were determined. Toxicity differs within the same strain depending on the growth medium. *Bt sv*2 produced with pigeon pea and soya bean flour were found highly effective with LC_{50} < 4 ppm against larvae of *Aedes aegypti*. These results were comparable with bacteria produced from Luria broth as a reference medium. Cost-effective analyses have revealed that production of biopesticide from test media is highly economical. The cost of production of *Bt sv*2 with soya bean flour was significantly reduced by 23-fold. The use of nonconventional sources has yielded a new knowledge in this area as the process development aspects of biomass production have been neglected as an area of research. These studies are very important from the point of media optimization for economic production of *Bacillus thuringiensis* based insecticides in mosquito control programmes.

1. Introduction

Microorganisms and microbial product with potential insecticidal activity can play an important role in controlling diseases by interrupting transmission mechanism by killing insect vectors at community level [1]. Worldwide efforts to screen effective entomopathogenic microorganisms for control of agriculturally and medically important insect pests have yielded many *Bacillus thuringiensis* (*Bt*) isolates with various insecticidal properties [2]. The Gram-positive bacterium *Bacillus thuringiensis* is well known for its ability to form spores and crystal proteins with insecticidal activity against a wide variety of lepidopteran, coleopteran, and dipteran insects [3]. The use of *B. thuringiensis* as commercial bioinsecticides was due to the remarkable ability of this bacterium to produce large quantities of insecticidal crystal proteins during large-scale fermentation. In many countries including developed ones, where mosquito borne diseases are still a serious problem, there is a need for large quantities of such microbial insecticides. Recently, more attention has been drawn to the low-cost production of *B. thuringiensis* which can be achieved through the optimization of culture conditions using appropriate media [4, 5]. Few published reports are available on use of low-cost ingredients for development of media for *B. thuringiensis* production. Agroindustrial residues and byproducts like cheese whey, soya bean milk, molasses, chicken feather waste, and paddy husk waste have been used as ingredients [6–8]. Edible seeds like mung beans (*Vigna radiata* (L.) R. Wilczek) were used as major sources of protein together with different combinations of soluble starch and/or sugarcane molasses as major carbohydrate sources for the production of delta-endotoxin [9]. Similarly, soya bean flour, groundnut cake powder, and wheat bran extract (*Triticum aestivum* L.) were separately used in large-scale production of *B. thuringiensis* bioinsecticide [10]. Media formulation and optimization are key considerations

Evaluation of Different Culture Media for Improvement in Bioinsecticides Production by Indigenous Bacillus thuringiensis and Their Application against Larvae of Aedes aegypti

59

in development of bioprocesses that can produce affordable biological agents, yet limited progress has been made in this area to satisfy market opportunities for affordable commercial biological insecticide products. It has been well documented that nutrient sources like carbon, nitrogen sources, and macronutrients strongly influence the growth, spore production, and the toxicity associated with parasporal proteinaceous crystalline inclusions during sporulation and synthesis of commercially useful metabolites in *Bacillus* species [11, 12]. Commonly used nutrient sources include a wide range of peptones, extracts, and hydrolysates, many of which are expensive for industrial-scale manufacture of large-volume products and have negative market acceptance as animal byproducts [13, 14]. The study of growth medium components affecting significantly the production of biomass, toxin production, and spore formation is a step required to advance in the design of a low-cost culture medium for the efficient production of all above responses.

Therefore, in the present work, an attempt has been made to determine the effects of several conventional and nonconventional carbon and nitrogen sources on the production of biomass, toxin production, and spore formation capability of mosquito toxic strain *Bt sv*2. This study also determines the cost effectiveness of potential substrates in production of *Bt* insecticide.

2. Material and Methods

2.1. Bacteria and Growth. *Bacillus thuringiensis sv*2 was locally isolated in India and tested for potential mosquito toxic activity [1]. *Bt sv*2 was maintained on nutrient agar slopes (HiMedia, Mumbai) at 4°C throughout the study. The organism was grown in 50 mL of nutrient broth with shaking for 24 hrs at 30°C. This culture was further used as an inoculum (1%v/v) for a basal medium composed of 1 L water and NaCl (0.25%), Na_2HPO_4 (0.1%), ($MgSO_4$ (0.02%), and $MnCl_2$ (0.005%) and (pH 7.2)). The medium was autoclaved for 30 min at 121°C.

2.2. Media Preparation. Conventional and nonconventional carbon and nitrogen sources were used as test materials.

Carbon sources tested were glucose, sucrose, fructose, corn starch, mannitol, beet root pulp (*Beta vulgaris* L.), banana (*Musa paradisiaca* L. var. Grand Naine) fruit pulp, and mahua (*Madhuca indica* L.) flower extract. For carbon sources study, basal medium was supplemented with yeast extract (0.5%) and various carbon sources were incorporated so as to correspond to 10 g/L carbon concentration.

Nitrogen sources tested were soya bean flour, pigeon pea flour, yeast extract, malt extract, beef extract, egg albumin powder, casein powder, ammonium sulphate, and sodium nitrate. For nitrogen sources studies, basal medium was supplemented with glucose (10 g/L) and various nitrogen sources were incorporated at a final concentration of 0.5 g/L.

The *Bt sv*2 cells were grown for 48 hrs at 30°C with shaking at 150 rpm in a basal medium supplemented with test carbon and/or nitrogen source. The carbon and nitrogen sources added to basal media constituents were investigated to get maximum dry cell mass and to determine yield coefficient, protein content, and toxicity to mosquito larvae.

2.3. Preparation of Nonconventional Nutrients

2.3.1. Flower Extract of Mahua. Flowers of *M. latifolia* L. were collected and dried in shade for 8 days at room temperature (28 ± 2°C); 100 g dry flowers were soaked in 200 mL hot distilled water (95°C) and incubated in shaker with 220 rpm at 29°C for 2 h. The extract filtered through muslin cloth was used as a source of sugar in medium.

2.3.2. Banana Pulp. Banana pulp was prepared by blending the ripe banana in a mixer and reduced to puree. Distilled water was added in puree at ratio of 3 : 1 to make a final puree which could be poured.

2.3.3. Beet Root. Fresh beet roots were purchased from local market. Pulp was prepared by blending the beet roots in a mixer and reduced to puree. Distilled water was added in puree at ratio of 3 : 1 to make a final puree which could be poured. This pulp was further used as source of sugar in nutrient medium.

2.3.4. Pigeon Pea and Soya Bean Flour. Pigeon pea and Soya bean flour were prepared by separately grinding beans finely enough to pass through a 100-mesh.

2.4. Cell Mass. After fermentation was completed, two samples (50 mL) were taken from each fermenter and then centrifuged at 8000 rpm for 15 min. The supernatants were discarded and the cell pellets were lyophilized. Dry weight was calculated and expressed in grams per liter (g/L). The same sample was used for the toxicity test.

2.4.1. Yield Coefficient. The yield coefficient was expressed as the ratio of carbon in the newly formed biomass to the carbon in the respective sugar source. Total carbon estimation was measured by the 3-5-dinitrosalicylic acid (DNS) modified method [15].

2.4.2. Protein Determination. 1 mL of culture medium was centrifuged for 10 min at 10000 g and the resulting pellets were washed twice with NaCl (1 mL) and twice with distilled water. These pellets were then suspended in 1 mL of NaOH (50 mM/L, pH 12.5) in order to solubilize protein crystals. After 2 h of incubation at 37°C, total proteins in the supernatant were measured by using the method by Bradford [16].

2.4.3. Bacterial Growth. *Bt sv*2 was inoculated in all the test media and allowed to grow under constant agitation in the shaker (200 rpm at 30°C for 72 h). Culture samples from the respective media were drawn (2.5 mL) every 6 h, till the end of the bacterial growth (72 h). The density of culture media was measured (at 600 nm) using the UV-Vis spectrophotometer (Shimadzu, Japan). The bacterial developmental stages (vegetative to sporulation) were monitored (Labomed Microscope, India).

2.5. Toxicity Assay. For the laboratory trial, early fourth instars larvae of *Aedes aegypti* were collected from city area of Jalgaon (21°2′54″N, 76°32′3″E; elevation, 209 m). The identified larvae were kept in plastic enamel trays containing dechlorinated tap water. They were maintained, and all the experiments were carried out at $28 \pm 2°$C and 75–85% relative humidity under 14 : 10 light and dark cycles. Larvae were fed with a diet of finely ground brewer's yeast and dog biscuits (3 : 1) [1]. Dry cell mass produced in different media was assayed against early fourth instars larvae of *Aedes aegypti*. Bioassay was performed by dissolving lyophilized *Bt sv2* powder in distilled water to get final different concentrations of 15, 10, 7.5, 5, 2.5, 1, and 0.5 ppm for 20 larvae in 50 mL of distilled water. Larval mortality was checked after 24 hrs of incubation. Each treatment was performed in three replicates each. In all cases, the mortality of control larvae, reared on a bacterial cell free diet (or water medium) and under the same environmental conditions as the experimental larvae, was recorded and calculated by Abbott [17] formula.

3. Statistical Analysis

The larvicidal activity of cell mass produced in each medium at different concentrations of 15, 10, 7.5, 5, 2.5, 1 and 0.5 ppm was subjected to probit regression analysis. The lethal concentrations in ppm (LC_{50}, LC_{90}) and the 95% confidence intervals of LC_{50} (upper confidence limit and lower confidence limit) were calculated. All conclusions are based on experiments that are repeated in time to ensure repeatability of results. Costs of the culture media were determined based on the ingredient prices in the western region of India. Media were compared based on their cost and potency against *Aedes aegypti* larvae.

Experimental data were analysed by one-way analysis of variance (ANOVA) using statistical software Minitab for Windows version 13. Treatment means were separated by Tukey's multiple comparison test at $\alpha = 0.05$.

4. Result and Discussion

The present study aimed to maximize production of *Bt* based bioinsecticide by ensuring high level of biomass, protein, and spore production from an indigenous isolate *Bt sv2* with relatively cheap nutrient sources. Initially, we screened 8 carbon substrates (glucose, sucrose, fructose, starch, mannitol, banana, beet root, and mahua) based on biomass production capability at laboratory scale. Amount of biomass produced by *Bt sv2* on different carbon source was found to vary (Table 1). Out of eight test carbon sources and reference medium (LB), only medium with glucose had the highest biomass yield (6.30 ± 0.03 g/L) in the basal production medium (Table 1). The second highest biomass yield (5.87 ± 0.08 g/L) was obtained in a medium with banana. Consumed sugars were determined to test rate of carbon utilization and to determine yield coefficient. Interestingly, more than 50% of initial sugars (glucose and banana) were remaining without being consumed even after prolonged fermentation studies (72 h). These results are particularly interesting because

remaining residue could be revalued as economic nutrient in next fermentation batches or to fed animals. We observed that addition of sucrose, fructose, starch, mannitol, and mahua as carbon source had no significant positive effect on biomass yield. These compounds were probably not preferred by the *Bt sv2* strain. On the other hand, glucose, banana, and beet root produced biomass at par with LB medium. According to our results, glucose should be added to the growth medium in order to obtain positive effects on the growth of *Bt sv2*, because glucose significantly stimulated biomass production. This type of response indicates that the inclusion of carbohydrates in a growth medium should be performed very carefully. Effect of these carbon sources was also analyzed by calculating their yield coefficient (Y); it showed that glucose is the best carbon source for biomass production (Table 1). We used glucose as an optimized carbon source in further studies considering its pure form compared to banana pulp and cost in purification of banana pulp. Our observation on banana pulp utilization is a new one in the production of *Bt* based biopesticides. Earlier, Shyam et al. [18] reported that waste ripe banana contains higher reducing sugars which helped in higher ethanol production using *Saccharomyces cerevisiae* fermentation. However, banana could also be used as an alternative to glucose after in detail factorial analysis of biomass production and cost analysis. Cell mass and product formation by microorganisms can be described quantitatively by yield coefficients expressed as the mass of cells or product formed per unit mass of substrate consumed. With the yield coefficients, the material balance equations for cells, substrate, and product can be straightforwardly formulated [19].

Since media components play a very important role in determining the yield and insecticidal activity of the spore crystal complex [20], effect of nitrogen sources on *Bt sv2* was evaluated for biomass yield, toxin content, and spore production. The usable form of *Bt* based product is in the form of spores. Thus, it is important to check spore production capacity of *Bt sv2* on test media. It was observed that natural nitrogen sources (pigeon pea and soya bean) support more biomass and toxin production than the synthetic nitrogen sources like ammonium sulphate and sodium nitrate (Table 2).

Spore counts are known to be more accurate than dry mass for yield determinations, because dry mass is affected by suspended solids in the media. This can be seen in medium supplemented with ammonium sulphate and sodium nitrate where a dry mass was relatively obtained at par with LB medium, while the spore count was lower than expected. Medium supplemented with pigeon pea and soya bean flour had the highest yield in terms of dry mass and spores per millilitre. No strong apparent relationship was found between spore production and biomass yield. The protein concentration, an indication of toxin production from all the tested media was quantified and there was a significant difference in the production of toxins (Table 2) between the media, which demonstrates that the toxins produced from each of the media was substrate specific.

To determine a close correlation between the growth and production of bacterial agents, we measured toxin to biomass

TABLE 1: Biomass production of *Bt sv*2 in different carbon sources.

Sugar source (g/L)	SC (g/L)	Biomass (g/L)	Y coefficient
BC + glucose (10)	4.70 (0.02)[e]	6.30 (0.03)[a]	1.34
BC + fructose (10)	7.12 (0.04)[b]	4.30 (0.09)[c]	0.60
BC + mannitol (10)	8.02 (0.03)[a]	1.27 (0.05)[e]	0.14
BC + sucrose (10)	5.28 (0.02)[d]	3.66 (0.07)[d]	0.70
BC + starch (10)	7.93 (0.02)[a]	1.07 (0.02)[e]	0.13
BC + banana (10)	4.82 (0.03)[e]	5.87 (0.08)[b]	1.22
BC + beet root (10)	6.32 (0.08)[c]	5.10 (0.11)[b]	0.80
BC + mahua (10)	6.76 (0.04)[c]	3.61 (0.04)[d]	0.53
LB medium	—	5.40 (0.03)[b]	—

SC: Sugar consumed at the end of fermentation; initial sugar concentration used in each test medium = 10 gm/L; BC: basal medium for carbon source study (NaCl 2.5 gm/L + Na_2HPO_4 1 gm/L + $MgSO_4$ 0.2 gm/L + $MnCl_2$ 0.05 gm/L + yeast extract 5 gm/L). Yield (Y) coefficient expressed as the ratio of carbon in the newly formed biomass to the carbon in the respective sugar source.
Data are presented as mean (SD).
Means within a given column followed by the same letter are not significantly different, Tukey's MRT (α = 0.05), $P < 0.0001$.

TABLE 2: Comparative analysis of *Bt sv*2 production in different nitrogen sources.

Nitrogen source (g/L)	Biomass yield (g/L)	Toxin content (protein mg/L)	Spores (CFU/mL)	Toxin/biomass (mg/g)
BN + ammonium sulphate (5)	5.44 (0.05)[c]	16.0 (0.20)[c]	7.30×10^7	2.94
BN + sodium nitrate (5)	4.90 (0.10)[d]	14.2 (0.30)[e]	9.10×10^6	2.89
BN + egg albumin (5)	4.96 (0.50)[d]	15.5 (0.20)[c]	1.50×10^7	3.12
BN + beef extract (5)	4.70 (0.03)[d]	11.3 (0.50)[d]	1.40×10^6	2.40
BN + casein (5)	6.67 (0.05)[b]	18.4 (0.60)[b]	1.69×10^8	2.75
BN + malt extract (5)	5.81 (0.06)[c]	15.4 (0.80)[c]	2.40×10^8	2.65
BN + soybean (5)	6.54 (0.04)[b]	19.7 (0.20)[b]	1.12×10^9	3.01
BN + pigeon pea (5)	7.45 (0.09)[a]	28.2 (0.70)[a]	2.24×10^9	3.78
BN + yeast extract (5)	6.26 (0.08)[b]	16.7 (0.40)[c]	9.10×10^8	2.66
LB medium (25)	5.40 (0.02)[c]	18.8 (0.20)[b]	1.32×10^9	3.48

BN: Basal medium for nitrogen source study (NaCl 2.5 gm/L + Na_2HPO_4 1 gm/L + $MgSO_4$ 0.2 gm/L + $MnCl_2$ 0.05 gm/L + glucose 10 gm/L);
data are presented as mean (SD). Means within a given column followed by the same letter are not significantly different, Tukey's MRT (α = 0.05), $P < 0.0001$.

ratio. Pigeon pea containing media had the highest toxin to biomass ratio (3.78), and spore production was also high. Similar trend was observed in all media with different nitrogen sources used. Ratio of toxin produced to the biomass yield could link well with the spore production. Thus, attention should therefore be directed not only towards fermentations with high yields and/or spore production; media should be selected to obtain the high toxins per volume of biomass produced.

In growth curve experiment, it was observed from the test culture media (pigeon pea and soya bean) that exponential phase of *Bt sv*2 was initiated from the sixth hour onwards (Figure 1). Rapid multiplication of bacterial cells followed by an increase in culture density was observed. This was extended up to 48 h after which the *Bt sv*2 entered into the stationary phase of growth (48 to 72 h). Sporulation started after 48 h of growth and complete sporulation was achieved in 54 h and by 72 h the spores were found to have been released from the cells. The *Bt sv*2 was able to digest the nutrients from all the culture media completely by 72 h. The growth pattern of *Bt sv*2 in pigeon pea and soya bean medium was higher than that from LB, corroborating the results of biomass and spore production in the present study (Table 2).

Similar enhanced growth pattern of *Bti* was reported with chicken feather, coconut cake, and manganese chloride based combination medium [21].

Toxicity tests (bioassays) with mosquito larvae and cost analysis were performed with *Bt sv*2 toxins produced from pigeon pea and soya bean media whereas LB was used as a reference medium. Table 3 represents comparative toxicities of *Bt sv*2 produced from LB, pigeon pea, and soya bean media against *Aedes aegypti* larvae. Here, the effect of toxin (as measured by LC_{50} and LC_{90} values) produced from pigeon pea media was increased by more than 2-fold compared to LB media. Toxins from soya bean media (LC_{50} 3.17) had effects close to LB media (LC_{50} 4.02).

The difference in bacteria mediated larvicidal efficacy may be due to the higher production of endotoxins (cry proteins). In the pigeon pea medium there is abundance of proteins and mixed salts which made it more suitable for growth and endotoxin production. The difference in growth and toxic activities of *Bt sv*2 in different media may be due to the differences in availability of growth nutrients. It is interesting to note that the toxicity obtained at the end of the fermentation depends on the culture medium and operating conditions [22]. Boulenouar et al. [23] also observed that the different

TABLE 3: Toxic effect of *Bt sv2* on different media against IVth instars larvae of *Aedes aegypti* and comparative cost analysis for production of toxin.

Culture media (g/L)	$LC_{50} \pm SE$ (LCL-UCL)	$LC_{90} \pm SE$ (LCL-UCL)	Regression equation	Cost per liter in USD	Net difference* (in ratio)
LB medium (25)	4.02 ± 0.26 (3.57–4.65)	7.59 ± 0.64 (6.56–9.25)	$Y = 0.438 + 2.37X$	1.61	—
BN + pigeon pea (5)	1.57 ± 0.09 (1.37–1.76)	3.20 ± 0.17 (2.89–3.62)	$Y = 3.66 + 3.69X$	0.07	23
BN + soyabean (5)	3.17 ± 0.16 (2.86–3.54)	5.94 ± 0.37 (5.31–6.85)	$Y = 0.167 + 3.12X$	0.07	23

LC_{50}: lethal concentration that kills 50% of the exposed larvae, LC_{90}: lethal concentration that kills 90% of the exposed larvae, SE: standard error, and LCL-UCL: 95% upper and lower fiducial limits. LC_{50} and LC_{90} expressed in ppm.
* Ratio of cost of LB medium required for preparation of 1 L medium with the test medium.

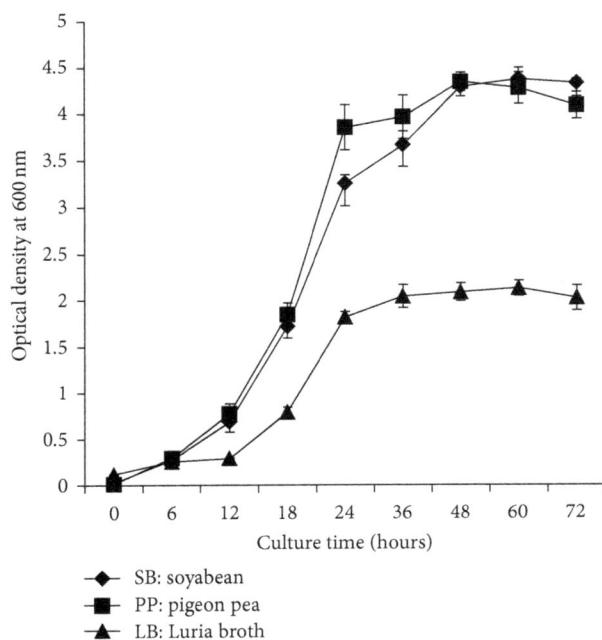

FIGURE 1: Growth pattern of *Bt sv2* produced in different media.

strains of the same bacteria may show different growth and toxic activities, which may be due to the differences in growth requirements of different strains. In the past, many attempts have been made to culture *Bacillus* to produce toxin at cheaper cost which has provided similar mosquito toxicity as observed in the present study [24–26].

The amount of raw materials used to prepare 1 liter of culture medium was 5 g, which is of negligible value on cost basis (Table 3). On the contrary, production on LB medium involved a cost of U.S. $ 1.61 per liter (source of price estimation was followed as per commercial preparation of fermentation media) which is 23 times more costly than pigeon pea and soya bean medium, respectively. However, transportation, electricity, and personnel cost are not included, because these expenditures are incurred commonly for both culture media. Moreover, pigeon pea and soya bean have grown in almost every part of India and most countries of the world. It has extensive usage, but storage has been its major problem as it is attacked by many pests. Damaged

seeds not suitable for cultivation or as food can be used for the production of fermentation media. An alternative use, for example, in fermentation medium, may provide additional revenue to farmers particularly in India and countries that have high levels of pigeon pea or soya bean production. In the choice of materials for the production of media for *B. thuringiensis*, three factors should be considered: availability, cost and how well the bacterium could utilize them. Some culture media selected in the present work could represent an economical benefit for biopesticide production, because they allowed maximum biomass, spore production, and ultimately higher toxin production levels at lower cost, compared to conventional nutrient source. The media preparations examined thus separately proved adequate for cultivation of *B. thuringiensis sv2*, at least on a laboratory scale. Medium with pigeon pea, which provided the best fermentation medium for the growth and sporulation of the *Bt sv2* strain in this study, can be considered for further development. If these low-cost *Bt sv2* preparations are as successful in field trials as they are in the laboratory, they could be an important tool for use in an integrated mosquito control programme in India and other developing countries where the bioinsecticides could be produced in regional laboratories in sufficient quantities to meet instant local demands during outbreak or to preserve for later use.

Conflict of Interests

The authors declare that there is no conflict of interests regarding the publication of this paper.

Acknowledgments

The authors are thankful to anonymous reviewers for critical evaluation of the paper. Dr. S. V. Patil is thankful to Rajiv Gandhi Science & Technology Commission, Mumbai, India for financial assistance (RGSTC/File-09/DPP-051/11).

References

[1] C. D. Patil, S. V. Patil, B. K. Salunke, and R. B. Salunkhe, "Insecticidal potency of bacterial species *Bacillus thuringiensis* SV2 and *Serratia nematodiphila* SV6 against larvae of mosquito species *Aedes aegypti*, *Anopheles stephensi*, and *Culex quinquefasciatus*," *Parasitology Research*, vol. 110, no. 5, pp. 1841–1847, 2012.

[2] J. S. Feitelson, J. Payne, and L. Kim, "*Bacillus thuringiensis*: insects and beyond," *Nature Biotechnology*, vol. 10, no. 3, pp. 271–275, 1992.

[3] E. Schnepf, N. Crickmore, J. Van Rie et al., "*Bacillus thuringiensis* and its pesticidal crystal proteins," *Microbiology and Molecular Biology Reviews*, vol. 62, no. 3, pp. 775–806, 1998.

[4] N. Zouari, A. Dhouib, R. Ellouz, and S. Jaoua, "Nutritional requirements of a strain of *Bacillus thuringiensis* subsp. *kurstaki* and use of gruel hydrolysate for the formulation of a new medium for δ-endotoxin production," *Applied Biochemistry and Biotechnology A*, vol. 69, no. 1, pp. 41–52, 1998.

[5] S. Ben Khedher, S. Jaoua, and N. Zouari, "Improvement of bioinsecticides production by sporeless *Bacillus thuringiensis* strains in response to various stresses in low cost medium," *Current Microbiology*, vol. 62, no. 5, pp. 1467–1477, 2011.

[6] L. F. A. Alves, S. B. Alves, R. M. Pereira, and D. M. F. Capalbo, "Production of *Bacillus thuringiensis* berliner var. *kurstaki* grown in alternative media," *Biocontrol Science and Technology*, vol. 7, no. 3, pp. 377–383, 1997.

[7] S. Poopathi, K. A. Kumar, L. Kabilan, and V. Sekar, "Development of low-cost media for the culture of mosquito larvicides, *Bacillus sphaericus* and *Bacillus thuringiensis* serovar. *israelensis*," *World Journal of Microbiology and Biotechnology*, vol. 18, no. 3, pp. 209–216, 2002.

[8] B. Paul, S. Paul, and M. A. Khan, "A potential economical substrate for large-scale production of *Bacillus thuringiensis* var. *kurstaki* for caterpillar control," *Biocontrol Science and Technology*, vol. 21, no. 11, pp. 1363–1368, 2011.

[9] S. G. Mummigatti and A. N. Raghunathan, "Influence of media composition on the production of δ-endotoxin by *Bacillus thuringiensis* var. *thuringiensis*," *Journal of Invertebrate Pathology*, vol. 55, no. 2, pp. 147–151, 1990.

[10] G. Prabakaran and K. Balaraman, "Development of a cost-effective medium for the large scale production of *Bacillus thuringiensis* var *israelensis*," *Biological Control*, vol. 36, no. 3, pp. 288–292, 2006.

[11] S. Yan, S. Mohammedi, R. D. Tyagi, R. Y. Surampalli, and J. R. Valéro, "Growth of four serovar of *Bacillus thuringiensis* (var. *kurstaki, israelensis, tenebrionis, and aizawai*) in wastewater sludge," *Practice Periodical of Hazardous, Toxic, and Radioactive Waste Management*, vol. 11, no. 2, pp. 123–129, 2007.

[12] M. K. Gouda, A. E. Swellam, and S. H. Omar, "Production of PHB by a Bacillus megaterium strain using sugarcane molasses and corn steep liquor as sole carbon and nitrogen sources," *Microbiological Research*, vol. 156, no. 3, pp. 201–207, 2001.

[13] Y. Nohata and R. Kurane, "Complete defined medium for large-scale production of polysaccharide bioabsorbent from *Alcaligenes latus* B-16," *Journal of Fermentation and Bioengineering*, vol. 83, no. 1, pp. 116–117, 1997.

[14] A. Vuolanto, N. Von Weymarn, J. Kerovuo, H. Ojamo, and M. Leisola, "Phytase production by high cell density culture of recombinant *Bacillus subtilis*," *Biotechnology Letters*, vol. 23, no. 10, pp. 761–766, 2001.

[15] G. L. Miller, "Use of dinitrosalicylic acid reagent for determination of reducing sugar," *Analytical Chemistry*, vol. 31, no. 3, pp. 426–428, 1959.

[16] M. M. Bradford, "A rapid and sensitive method for the quantitation of microgram quantities of protein utilizing the principle of protein dye binding," *Analytical Biochemistry*, vol. 72, no. 1-2, pp. 248–254, 1976.

[17] W. S. Abbott, "A method of computing the effectiveness of an insecticide," *Journal of Economic Entomology*, vol. 18, pp. 265–267, 1925.

[18] K. R. Shyam, M. I. Ganesh, R. Rajeswari, and H. Harikrishnan, "Utilization of waste ripe Banana, and peels for bio ethanol production using *Saccharomyces cerevisia*," *Journal of Bioscience and Research*, vol. 2, no. 2, pp. 67–71.

[19] J. Hong, "Yield coefficients for cell mass and product formation," *Biotechnology Bioengineering*, vol. 33, no. 4, pp. 506–507, 1989.

[20] H. T. Dulmage, "Production of the spore-δ-endotoxin complex by variants of *Bacillus thuringiensis* in two fermentation media," *Journal of Invertebrate Pathology*, vol. 16, no. 3, pp. 385–389, 1970.

[21] S. Poopathi and B. Archana, "Optimization of medium composition for the production of mosquitocidal toxins from *Bacillus thuringiensis* subsp. *israelensis*," *Indian Journal of Experimental Biology*, vol. 50, no. 1, pp. 65–71, 2012.

[22] M. Ramírez-Lepe and M. Ramírez-Suero, "Biological control of Mosquito larvae by *Bacillus thuringiensis* subsp. *israelensis*," in *Insecticides—Pest Engineering*, chapter 11, InTech Press, 2012.

[23] N. Boulenouar, F. Al-Quadan, and H. Akel, "Effect of various combinations of growth temperature, pH and NaCl on intracellular activities of G6PDH and 6PGDH from four *Bacillus* strains isolated from Jordanian hot springs," *Journal of Biological Sciences*, vol. 6, no. 3, pp. 586–590, 2006.

[24] M. Shojaaddini, S. Moharramipour, M. Khodabandeh, and A. Talebi, "Development of a cost effective medium for production of *Bacillus thuringiensis* bioinsecticide using food barley," *Journal of Plant Protection Research*, vol. 50, no. 1, pp. 9–14, 2010.

[25] K. Yadav, S. Dhiman, I. Baruah, and L. Singh, "Development of cost effective medium for production of *Bacillus sphaericus* strain isolated from Assam, India," *Microbiology Journal*, vol. 1, no. 2, pp. 65–70, 2011.

[26] S. Poopathi and S. Abidha, "Coffee husk waste for fermentation production of mosquitocidal bacteria," *Journal of Economic Entomology*, vol. 104, no. 6, pp. 1816–1823, 2011.

8

Medium Optimization for the Production of Fibrinolytic Enzyme by *Paenibacillus* sp. IND8 Using Response Surface Methodology

Ponnuswamy Vijayaraghavan and Samuel Gnana Prakash Vincent

International Centre for Nanobiotechnology, Centre for Marine Science and Technology, Manonmaniam Sundaranar University, Rajakkamangalam, Kanyakumari District, Tamil Nadu 629 502, India

Correspondence should be addressed to Ponnuswamy Vijayaraghavan; venzymes@gmail.com

Academic Editors: T. Betakova, Y. Mu, and L. Ramirez

Production of fibrinolytic enzyme by a newly isolated *Paenibacillus* sp. IND8 was optimized using wheat bran in solid state fermentation. A 2^5 full factorial design (first-order model) was applied to elucidate the key factors as moisture, pH, sucrose, yeast extract, and sodium dihydrogen phosphate. Statistical analysis of the results has shown that moisture, sucrose, and sodium dihydrogen phosphate have the most significant effects on fibrinolytic enzymes production ($P < 0.05$). Central composite design (CCD) was used to determine the optimal concentrations of these three components and the experimental results were fitted with a second-order polynomial model at 95% level ($P < 0.05$). Overall, 4.5-fold increase in fibrinolytic enzyme production was achieved in the optimized medium as compared with the unoptimized medium.

1. Introduction

Fibrin is the main protein component of the blood clot, and it is normally formed from fibrinogen by the action of thrombin (EC. 3. 4. 21. 5) after trauma or injury. Accumulation of fibrin in blood vessels usually increases thrombosis, leading to myocardial infarction and other cardiovascular diseases (CVDs). A variety of fibrinolytic enzymes such as tissue plasminogen activators (t-PA), urokinase (u-PA), and streptokinase were extensively studied and used as thrombolytic agents [1]. Although t-PA and u-PA are still widely used in thrombolytic therapy today, their expensive prices and undesirable side effects prompt researchers to search for cheaper and safer thrombolytic agents. In recent years, fibrinolytic enzymes from various sources, including microorganisms, worms, and animals, have been the subject of active researches because of their potential as novel agents in preventing or treating CVDs by dissolving fibrin blood clot [2, 3]. The genus Bacillus from traditional fermented food is an important one among the microorganisms that have been found to produce the fibrinolytic enzymes [4]. Fibrinolytic enzymes from these food-grade microorganisms can be promising alternatives for t-PA or streptokinase in thrombolytic therapy. Nattokinase, a potent fibrinolytic enzyme, has been reported to have a potent thrombolytic activity [5]. Based on its food origin and relatively strong fibrinolytic activity, nattokinase has advantages over other commercially used medicine, in preventative and prolonged effects, convenient oral administration, and stability in the gastrointestinal tract [6].

Many fibrinolytic enzymes have been isolated from various foods such as Korean *chungkookjang* [7], Chinese *douche* [8], soybean grits [9], and Indonesian *tempeh* [10]. Endophytic bacteria such as *Paenibacillus* produce biotechnologically important enzymes. The genus *Paenibacillus* was created by Ash et al. [11] to accommodate the "group 3" of the genus *Bacillus*. It comprises over 30 species of facultative anaerobes and endospore-forming, neutrophilic, periflagellated, heterotrophic, and low G + C Gram-positive bacilli. The name reflects this fact; in Latin *paene* means *almost*, and therefore the *Paenibacillus* is almost a *Bacillus*. Very few studies were carried out in the optimization and characterization of

Medium Optimization for the Production of Fibrinolytic Enzyme by Paenibacillus sp. IND8 Using Response
Surface Methodology

65

fibrinolytic enzymes from *Paenibacillus* sp. Recently, Lu et al. [12] purified and characterized fibrinolytic enzymes from *Paenibacillus polymyxa* EJS-3.

Moreover, very few studies were reported on statistical optimization of fibrinolytic enzymes production in solid-state fermentation (SSF). Tao et al. [13] optimized process parameters for the production of fibrinolytic enzymes by *Fusarium oxysporum*. Compared to submerged fermentation, SSF yields more enzyme and it could reduce the production cost of the enzyme. From an industrial point of view, around 30–40% of the production cost of enzymes is estimated to account the cost of the growth medium [14]. Therefore, optimization of the fermentation process parameters in SSF through a statistical approach is important for a significant improvement in yield as well as a decrease in the production cost of the enzyme. The selection of medium components is another critical factor for the production of fibrinolytic enzymes because each microbe requires unique nutrient components and environmental conditions for its growth and the production of fibrinolytic enzymes [9, 15]. Since wheat bran was recognized as the standard substrate for SSF, it was selected for the production of fibrinolytic enzymes.

The traditional one-at-a-time optimization strategy is simple, but it fails to locate the region of optimal response because the comprehensive effect of factors is not taken into consideration for the production of fibrinolytic enzymes [16]. The statistical experimental design provides a universal language with which experts from different areas such as academia, engineering, business, and industry can communicate for setting, performing, and analyzing experiments for research. Statistically designed experiments are more effective than other classical one-at-a-time optimization strategy because it can study many variables simultaneously with a low number of observations, saving time, and costs [17, 18]. The statistical method such as factorial design, central composite design and response surface methodology (RSM) were frequently used to optimize the process parameters for the production of antimicrobial metabolites [19], bio-surfactants [20], and fibrinolytic enzymes [16, 21]. The main objective of this study was to optimize the process parameters by statistical approach for enhancing fibrinolytic enzymes production by *Paenibacillus* sp. using two-level full factorial's design followed by RSM.

2. Materials and Methods

2.1. Isolation of Fibrinolytic Enzymes-Producing Strain. The fibrinolytic enzymes-producing strain-IND8 was isolated from the cooked Indian rice. Samples collected were plated onto skim-milk agar plates containing (g/L) peptone 5, yeast extract 5, NaCl 1.5, agar 15, and skim milk 10. These plates were incubated for 24–48 h at 37°C and a clear zone on skimmed milk hydrolysis gave an indication of protease-producing strains. These protease producing strains were subjected to fibrinolytic enzymes screening. Fibrinolytic enzymes production was carried out in the culture medium composed of (g/L) peptone 5, yeast extract 5, NaCl 1.5, and Casein 10. Medium was autoclaved at 121°C for 20 min and

a loopful culture of the selected organism was inoculated. Submerged fermentation was performed on a rotary shaker (150 rpm) for 48 h at 37°C, in 250 mL Erlenmeyer flasks. The cultures were centrifuged and the supernatants were used for determination of fibrinolytic activity using a fibrin plate. The fibrin plate was composed of 1% (w/v) agarose, 0.5% (w/v) fibrinogen, 1% (v/v), and thrombin (100 NIH units/mL) (pH 7.4) [22]. The fibrin plate was allowed to stand for 1 h at room temperature to form a fibrin clot layer. Ten microliters of crude enzyme was dropped into holes and incubated for 5 h at 37°C, fibrinolytic enzymes exhibited a clear zone of degradation of fibrin around the well indicating its activity. The single-strain IND8 showing the largest halo zone on the fibrin plate was selected and further identified.

2.2. 16S rDNA Sequencing. The genomic DNA was extracted from the cells of an 18 h culture using QIAGEN genomic DNA purification kit according to the manufacturer's instructions. The 16s rDNA gene was amplified by PCR (Peltier Thermal Cycler Machine, USA) using the upstream (P1: $^{5'}$AGAGTTTGATCMTGGCTAG$^{3'}$) and the downstream primers (P2: $^{5'}$ACGGGCGG TGTGTRC$^{3'}$) and DNA polymerase (Sigma, USA). The amplified product was sequenced and sequence comparison with the databases was performed using BLAST through the NCBI server [23]. The 830 bp 16S rDNA sequences of *Paenibacillus* IND8 strain were submitted to GenBank database under an accession number KF250416.

2.3. Assay of Fibrinolytic Enzymes Activity. The culture supernatant (0.1 mL) suitably diluted was mixed with 2.5 mL of 0.1 M Tris-HCl buffer (pH 7.8) containing 0.01 M calcium chloride. To this, 2.5 mL of fibrin (1.2%, w/v) was added and incubated for 15 min at 37°C. The reaction was stopped by adding 5.0 mL of 0.11 M trichloroaceticacid containing 0.22 M sodium acetate and 0.33 M acetic acid. The absorbance was measured at 275 nm against sample blank. A standard curve was performed using L-tyrosine. One unit of fibrinolytic activity was defined as the amount of enzyme which liberates 1 μg of tyrosine per minute under the experimental conditions used.

2.4. Fibrin Zymography and In Vitro Analysis of Blood Clot. Fibrin zymography was carried out in 12% SDS-polyacrylamide gel containing fibrinogen (0.12%, w/v) and 100 μL thrombin (10 NIH units/mL). After electrophoresis at 4°C, the gel was incubated in 0.05 M sodium phosphate buffer (pH 7.4) containing 2.5% triton X-100 for 30 min at room temperature. Further, the gel was washed with distilled water for 30 min and incubated in sodium phosphate buffer (pH 7.4, 0.05 M) for 5 h at 37°C. It was stained with coomassie brilliant blue R-250 for 1 h, after which it was destained and bands with fibrinolytic activities were visualized as the nonstained region of the gel. Clot lytic effects of fibrinolytic activities were studied with natural clot *in vitro*. The goat blood clot was cut into the same size, and crude fibrinolytic enzymes were added. The mixture was incubated at room temperature for 24 h and analyzed for its activities on fibrin blood clot [24].

2.5. *Primary Screening of Process Parameters Based on One-at-a-Time Strategy.* In the present study, SSF was carried out using wheat bran as a substrate. SSF was carried out separately in a 100 mL Erlenmeyer flask containing 2.0 g (w/w) of the substrate moistened with 2.0 mL buffer (pH 8.0, 0.1 M). The contents were sterilized and inoculated with 0.2 mL of 18 h grown (0.796 OD at 600 nm) culture broth under sterile conditions. The process parameters such as the fermentation period (24–96 h), pH (6.0–10.0), moisture content (60%–140%), inoculum size (3%–15%), carbon sources (1%, w/w) (maltose, sucrose, starch, glucose, xylose, and trehalose), nitrogen sources (1%, w/w) (casein, yeast extract, peptone, beef extract, gelatin, and urea), and inorganic salts (ammonium chloride, sodium dihydrogen phosphate (NaH_2PO_4), calcium chloride, sodium nitrate, disodium hydrogen phosphate (Na_2HPO_4), ammonium sulphate, and ferrous sulphate) were evaluated. Twenty milliliters of double distilled water was added with the fermented medium and enzyme extracted as described earlier [25]. All experiments were carried out in triplicate, and average values are reported.

2.6. *Evaluation of Significant Factors Affecting Fibrinolytic Enzymes Production by 2^5 Factorial Designs.* Two-level full factorial designs were carried out for screening the most significant factors affecting the fibrinolytic enzymes production by *Paenibacillus* IND8. Five factors, namely, sucrose (carbon source), yeast extract (nitrogen source), NaH_2PO_4 (inorganic salt), pH, and moisture content of the medium were selected for the analysis of significant factors. Based on two-level full factorial design each factor was examined at two-levels (– and +). The other factors such as fermentation period and inoculum were kept at middle level. Two-level full factorial designs were based on the following first-order polynomial model. Consider

$$Y = \alpha_0 + \sum_i \alpha_i x_i + \sum_{ij} \alpha_{ij} x_i x_j + \sum_{ijk} \alpha_{ijk} x_i x_j x_k, \quad (1)$$

where Y is the response (fibrinolytic activity). α_{ij} and α_{ijk} were the ijth and ijkth interaction coefficients; α_i was the ith linear coefficient and α_0 was an intercept.

Fibrinolytic activity assay was carried out in triplicates and the average of these experimental values was taken as response Y. ANOVA was used to estimate the statistical parameters and the values of "Prob > F" less than 0.05 indicated that the model terms are significant. Statistical software, Design-Expert 8.0.7.0 (StatEase Inc, Minneapolis, USA), was used to design the experiment. Experimental design and results of the 2^5 factorial designs were described in Table 1. The significant factors ($P < 0.05$) obtained from two-level full factorial designs were further optimized by RSM.

2.7. *Central Composite Design and Response Surface Methodology.* Central composite design was employed in the present investigation to estimate the main effects. The factors used were sucrose, NaH_2PO_4, and moisture for enhanced fibrinolytic enzymes production. Each factor in the design was studied at five levels ($-\alpha$, -1, 0, $+1$, and $+\alpha$) in a set of 20 experiments that included 8 factorial, 6 axial, and 6 center

points. All experiments were conducted in triplicates and the mean values of fibrinolytic activities (units/mL) were taken as the response (Y). The second-order polynomial equation was employed to fit the experimental data. For a three-factor system the second-order polynomial equation is as follows:

$$Y = \beta_0 + \sum_{I=1}^{3} \beta_i X_i + \sum_{i=1}^{3} \beta_{ii} X_i^2 + \sum_{ij=1}^{3} \beta_{ij} X_{ij}, \quad (2)$$

where Y was the response, β_0 was the offset term, and β_i, β_{ii}, and β_{ij} were the coefficients of linear terms, square terms, and coefficients of interactive terms, respectively. X_i's were A, B and C, X_{ij}'s were AB, AC, and BC.

Analysis of variance (ANOVA) was used to estimate the model. The values of "Prob > F" less than 0.05 indicated that the model terms were significant. The fitted polynomial equation was expressed as three-dimensional surface plots to visualize the relationship between the responses and the levels of each factors used in the design. The statistical software (Design-Expert 8.0.7.0, StatEase Inc, Minneapolis, USA) was used to plot the 3D graphs.

2.8. *Statistical Model Validation.* With the help of the special features of RSM and 3D graph and perturbation plot the optimum value of the combination of the three factors (sucrose, NaH_2PO_4, and moisture) were validated. Experiments were carried out in triplicates in Erlenmeyer flask under theoretically predicted conditions to validate the model.

3. Results and Discussion

3.1. *Bacterial Strain.* Among the bacterial isolates strain IND8 was selected for this study in light of exhibiting strong fibrinolytic activities. The isolated strain was Gram-positive rods, citrate-, oxidase-, and nitrate-positive. It was negative to urea-, indole- and gelatin-hydrolysis. It hydrolyzed casein and fermented carbohydrates. The strain-IND8 was identified as *Paenibacillus* sp. based on its 16s rDNA sequence and designated as *Paenibacillus* sp. IND8. The phylogenetic tree constructed from the sequenced data by the neighbor-joining method showed the detailed evolutionary relationship between the strain IND8 and other closely related *Paenibacillus* sp. (Figure 1).

3.2. *Fibrin Zymography and In Vitro Analysis of Fibrinolytic Activities.* Fibrin zymography revealed at least three major and two minor fibrinolytic proteases were determined from the crude extract (figure not shown). Fibrin clot degradation was observed within 24 h of incubation at room temperature (30°C) in the tube containing fibrinolytic enzymes. In the saline solution suspend tubes, the blood clot remained (figure not shown). These fibrinolytic enzymes may have wide application in pharmaceutical industry. Fibrinolytic enzymes from this kind of food grade organisms could effectively prevent and treat cardiovascular diseases [26].

3.3. *Initial Screening of Physical and Nutrient Factors for Statistical Optimization Process.* The physical and nutrient

Medium Optimization for the Production of Fibrinolytic Enzyme by Paenibacillus sp. IND8 Using Response
Surface Methodology

67

TABLE 1: Experimental design and results of the 2^5 fractorial design.

Run	Sucrose	Yeast extract	NaH$_2$PO$_4$	pH	Moisture	Response (Y)
1	−1	1	−1	1	1	1711
2	−1	−1	1	−1	1	2169
3	1	1	−1	1	−1	1931
4	1	1	1	−1	−1	2297
5	−1	1	1	−1	1	3413
6	−1	1	1	1	−1	1244
7	−1	−1	−1	−1	1	1867
8	1	−1	1	1	1	1995
9	1	1	1	1	−1	1976
10	−1	−1	−1	1	1	2873
11	1	−1	1	−1	−1	723
12	−1	1	−1	−1	−1	787
13	1	−1	−1	−1	1	2681
14	1	−1	−1	−1	−1	2315
15	−1	1	−1	1	−1	1400
16	1	−1	1	1	−1	1400
17	1	1	−1	1	1	3514
18	1	1	−1	−1	1	2086
19	−1	−1	1	1	1	2535
20	1	−1	−1	1	1	2251
21	1	−1	−1	1	−1	2910
22	1	1	−1	−1	−1	1171
23	−1	−1	1	1	−1	1345
24	−1	1	1	1	1	1940
25	1	−1	1	−1	1	3331
26	−1	−1	1	−1	−1	2535
27	−1	1	1	−1	−1	2544
28	−1	−1	−1	1	−1	1381
29	1	1	1	1	1	2123
30	−1	−1	−1	−1	−1	1894
31	−1	1	−1	−1	1	1995
32	1	1	1	−1	1	3768

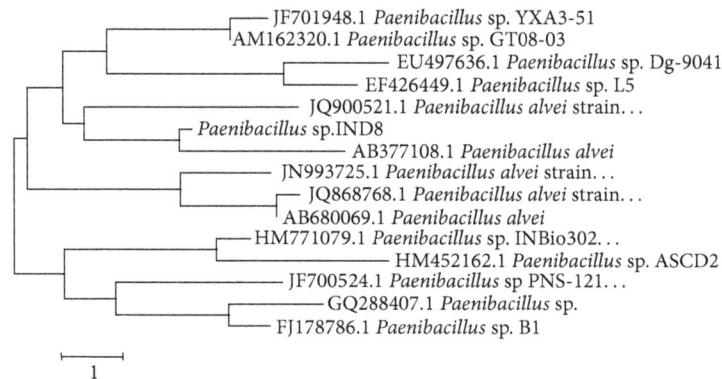

FIGURE 1: Phylogenetic relationships of stain IND8 and other closely related *Paenibacillus* based on 16S rDNA sequence. Bar = 1 substitution per site.

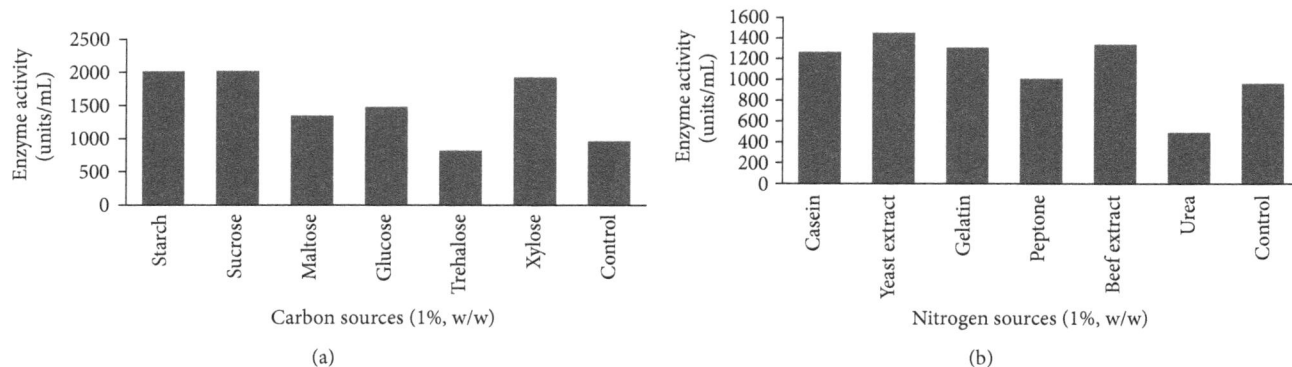

FIGURE 2: (a) Effect of different carbon sources on production of fibrinolytic enzymes. (b) Effect of different nitrogen sources on production of fibrinolytic enzymes.

factors were optimized by a traditional one-factor-at-a-time approach. In order to achieve the maximum yield of fibrinolytic enzymes, the following optimum process parameters are needed: fermentation period (72 h), pH (8.0), moisture (80–100%), and inoculum (6–9%). Among the carbon sources supplemented, sucrose supported more enzymes production (2017 units/mL) than other sources (Figure 2(a)). Among the nitrogen sources, yeast extract showed more enzymes production (1451 units/mL) (Figure 2(b)). Fibrinolytic enzymes production was high in the presence of sodium dihydrogen phosphate as the sole source of an inorganic salt (1630 units/mL). One-factor-at-a-time experiments revealed that sucrose, yeast extract, and sodium dihydrogen phosphate significantly increased fibrinolytic enzymes production. Hence, these nutrient factors were selected for statistical optimization. From the SSF point of view, moisture is one of the critical factors for enzyme production; thus it was selected for statistical optimization. The protease production by microbial strains strongly depends on the extracellular pH [27]. Hence, the pH of the medium was also considered for statistical medium optimization approach.

Conventional experimental approach used for media optimization employing "change-one-factor-at-a-time" is extremely time consuming and expensive and laborious for screening a large number of variables. Optimizing all the significant parameters by statistical experimental designs can eliminate these limitations of a single factor optimization process collectively. Several statistical designs are available such as full factorial, fractional factorial or Plackett-Burman designs, Taguchi's robust designs, and response surface methodology [28]. In this study, the important factors such as sucrose, yeast extract, NaH_2PO_4, pH, and moisture were optimized by two-level full factorial designs and response surface methodology.

3.4. Two-Level Factorial Designs on Elucidation of Medium Components. The fibrinolytic enzymes production varied from 723 to 3768 units/mL with different combinations of the components supplemented with wheat bran. Table 1 represented the two-level full factorial designs for five selected

variables (sucrose, yeast extract, NaH_2PO_4, pH, and moisture) and the corresponding response for fibrinolytic enzymes production. The nutrient factors such as, sucrose, yeast extract, and NaH_2PO_4 were positively correlated and this indicated that the further increase in the concentrations of these nutrients could increase the production of fibrinolytic enzymes. Among the all nutrient and physical factors, moisture content of the medium was significantly influenced by fibrinolytic enzymes production.

Analysis of variance (ANOVA) was used to evaluate the observed results. The model F value of 125.57 implied the model was significant. There is only a 0.01% chance that a "model F value" this large could occur due to noise. Values of "Prob > F" less than 0.05 indicate that model terms were significant. In this model A, C, D, E, AB, AC, AD, AE, BC, BD, BE, CD, CE, DE, ABC, ABD, ACE, ADE, BCD, BCE, BDE, CDE, ABCD, ABCE, ABDE, ACDE, and ABCDE were highly significant. The correlation coefficient of this model was 0.998. The "predicted R-squared" of 0.924 was in reasonable agreement with the "adjusted R-squared" of 0.99. "Adequate precision" measures the signal to noise ratio. A ratio greater than 4 is desirable. In this model the ratio of 45.11 indicates an adequate signal. For pH, the coefficient estimate was negative (-95.22) and it indicated that the reduction of pH could positively influence the enzymes production. Yeast extract did not influence the fibrinolytic enzymes production significantly ($P > 0.05$). Neglecting the insignificant variables, the model equation for fibrinolytic enzymes production can be written as

Enzyme activity = $+2128.28 + 151.22A + 80.34C - 95.22D + 387.47E + 88.28AB - 158.22AC + 78.22AD + 51.66AE + 214.03BC - 43.66BD + 62.53BE - 293.66CD + 63.16CE - 52.78DE + 46.59ABC + 88.41ABD + 100.34ACE - 178.09ADE - 159.84BCD - 115.28BCE - 55.09BDE - 69.34CDE - 48.28ABCD - 158.22ABCE + 204.97ABDE - 116.91ACDE - 52.84ABCDE.$

Enzyme production was found to be high (3768 units/mL) in the wheat bran containing 0.75% sucrose, 0.5% yeast extract, and 0.1% NaH_2PO_4 with 100% moisture content at pH 7.0 (run 32). Wheat bran was considered as a standard substrate for the production of proteolytic enzymes and

Medium Optimization for the Production of Fibrinolytic Enzyme by Paenibacillus sp. IND8 Using Response Surface Methodology

69

TABLE 2: ANOVA for selected factorial model.

Source	Sum of squares	df	Mean square	F Value	P value
Model	$1.788E + 007$	27	$6.621E + 005$	125.57	0.0001
A-sucrose	$7.317E + 005$	1	$7.317E + 005$	138.78	0.0003
C-NaH_2PO_4	$2.066E + 005$	1	$2.066E + 005$	39.18	0.0033
D-pH	$2.901E + 005$	1	$2.901E + 005$	55.02	0.0018
E-moisture	$4.804E + 006$	1	$4.804E + 006$	911.14	<0.0001
Residual	21091.13	4	5272.78		
Cor total	$1.790E + 007$	31			

TABLE 3: Variables and their levels for response surface methodology.

Variables	Symbol	Coded levels				
		−1.681	−1	0	+1	+1.681
Sucrose	A	0.079	0.25	0.5	0.75	0.92
NaH_2PO_4	B	0	0.01	0.055	0.1	0.13
Moisture	C	46.36	60	80	100	113.64

TABLE 4: Central composite designs matrix and results on the production of fibrinolytic enzymes.

Run	Factor A	Factor B	Factor C	Response (Y)
1	1	−1	−1	3398
2	1	1	1	4008
3	0	0	0	3450
4	−1.682	0	0	2903
5	1.682	0	0	2661
6	0	0	0	3360
7	−1	1	1	3335
8	0	−1.682	0	3661
9	0	0	0	3808
10	0	0	1.682	4418
11	1	1	−1	1513
12	0	0	0	3673
13	−1	−1	−1	1520
14	0	0	0	3755
15	1	−1	1	3829
16	0	0	0	4314
17	−1	−1	1	4241
18	−1	1	−1	450
19	0	1.682	0	1713
20	0	0	−1.682	1957

Chang et al. [29] used wheat bran medium for the production of fibrinolytic enzymes by *Bacillus subtilis* IMR-NK1. The two-level full factorial designs showed that the medium containing 100% moisture showed increased fibrinolytic activities. The moisture content of the fermentation medium is one of the main factors in SSF and often determines the success of a process [30]. Based on the experimental result, yeast extract is the best choice of nitrogen source for enzyme bioprocess. These observations were in accordance with the observations made with *Bacillus* sp. [31].

Fibrinolytic activities were found to be high in the presence sucrose with the wheat bran medium. These results were in accordance with reported protease production in the presence of different sugars [27]. Based on the calculated t values (Table 2), sucrose, NaH_2PO_4, and moisture were selected for further optimization by RSM.

3.5. Optimization by Central Composite Designs (CCD) and Statistical Analysis.
CCD was used for optimization of three variables: sucrose (A), NaH_2PO_4 (B), and moisture (C), each were studied at five coded levels, that is, ($−\alpha$, $−1$, 0, $+1$, $+\alpha$) as shown by Table 3 and the models were explained with twenty runs (Table 4).

In the recent years major research and development on the use of statistical methods involving various statistical software packages for the optimization studies with the aim of obtaining high yields of amylases, proteases, biosurfactants, neomycin, and so forth, [32]. Limitations and drawbacks of the single factor optimization can be eliminated by employing RSM which was used to explain the combined effects of all the factors in a fermentation process [33]. The commonly used response surface design was CCD [34] involving five levels, for each factor needed for quadratic terms to be estimable in the second-order model.

In this study the responses of the CCD were well fitted with a second-order polynomial equation.

Fibrinolytic activity (Y) = $+3728.21 + 204.66A − 509.50B + 927.80C + 33.75AB − 335.00AC + 278.50BC − 344.06A^2 − 377.65B^2 − 200$.

Here, A-sucrose; B-NaH_2PO_4; C-moisture.

The model F value of 10.74 implied that the model was significant. There is only a 0.05% chance that a "model F value" this large could occur due to noise. Values of "Prob > F" less than 0.05 indicated that the model terms were significant. In this model, the linear terms B and C ($P < 0.05$) and quadratic terms A^2 and B^2 ($P < 0.05$) were statistically significant (Table 5). The linear factor A, interaction terms (AB, AC, BC), and quadratic term (C^2) were insignificant ($P > 0.05$). The "lack of fit F value" of 2.85 implies the lack of fit is not significant relative to the pure error. There is a 13.75% chance that a "lack of fit F value" this large could occur due to noise. Nonsignificant lack of fit is good. "Adequate precision" measures the signal (response) to noise (deviation) ratio. A ratio greater than 4 is desirable. In this model the ratio 12.729 indicates an adequate signal and therefore the model is significant for the enzyme bioprocess.

The goodness-of-fit of the model was checked by comparing the coefficient of determination (R^2) with adjusted coefficient of determination (R^2), which are measures of the amount of the reduction in the variability of response obtained by using the repressor variables in the model [35]. In this model, the coefficient of determination (R^2) is 0.9063 and it explains 90.63% variability in the model and the adjusted R^2 was 0.82. The R^2 value closer to 1.0 shows a stronger model with better predictability [36]. Figures 3(a)–3(c) depicted the 3D plot of the response from the interaction among variables and determined the optimum concentrations of each factor

TABLE 5: ANOVA for response surface quadratic model.

Source	Sum of squares	df	Mean square	F Value	P value
Model	$2.109E + 007$	9	$2.344E + 006$	10.74	0.0005 significant
A-sucrose	$5.720E + 005$	1	$5.720E + 005$	2.62	0.1365
B-NaH$_2$PO$_4$	$3.545E + 006$	1	$5.545E + 006$	16.25	0.0024
C-moisture	$1.176E + 007$	1	$1.176E + 007$	53.88	<0.0001
AB	9112.50	1	9112.50	0.042	0.8422
AC	$8.978E + 005$	1	$8.978E + 005$	4.11	0.0700
BC	$6.205E + 005$	1	$6.205E + 005$	2.84	0.1226
A^2	$1.706E + 006$	1	$1.706E + 006$	7.82	0.0189
B^2	$2.055E + 006$	1	$2.055E + 006$	9.42	0.0119
C^2	$5.805E + 005$	1	$5.805E + 005$	2.66	0.1339
Residual	$2.182E + 006$	10	$2.182E + 005$		
Lack of fit	$1.616E + 006$	5	$3.232E + 005$	2.85	0.1373 not significant
Pure error	$5.662E + 005$	5	$1.132E + 005$		
Cor total	$2.328E + 007$	19			

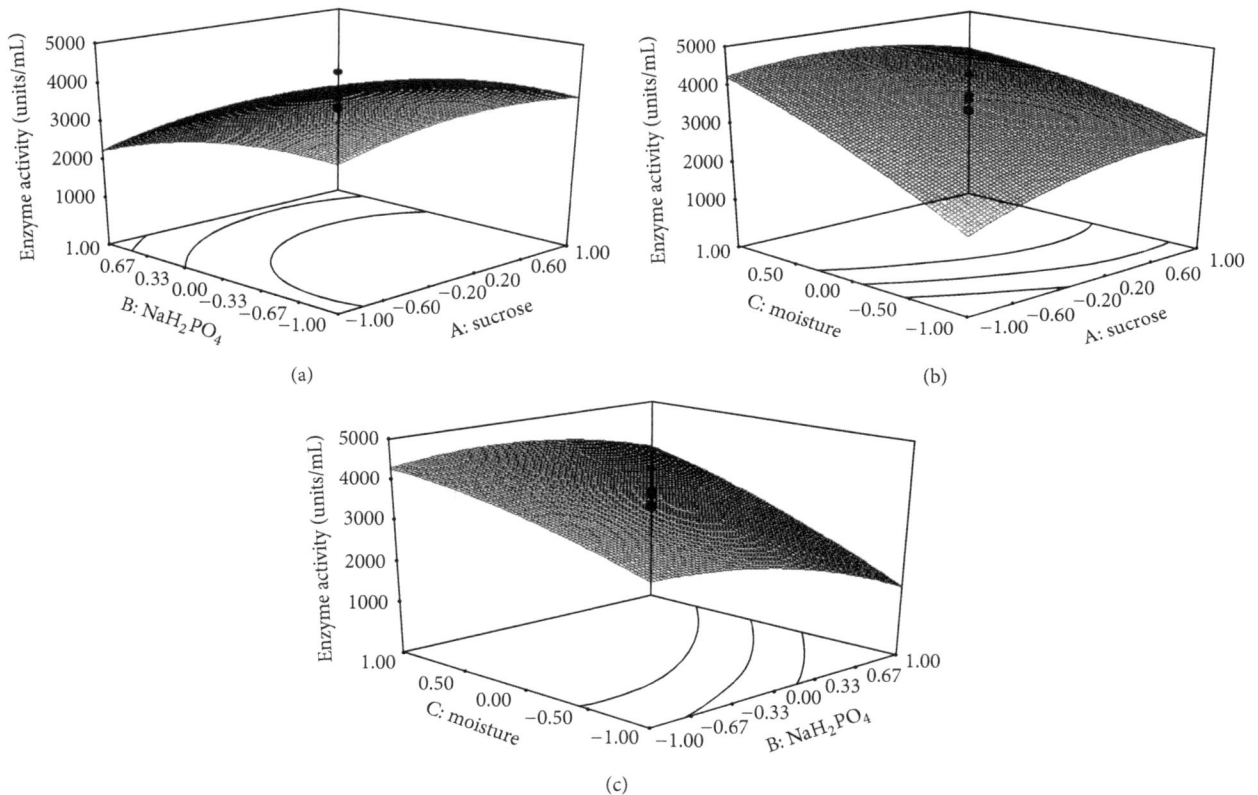

(a)

(b)

(c)

FIGURE 3: (a)–(c) Response surface plots showing the effect of interaction of sucrose and NaH$_2$PO$_4$ (a), sucrose and moisture (b), and moisture and NaH$_2$PO$_4$ (c).

for maximum fibrinolytic enzymes production by *Paenibacillus* sp. Fibrinolytic enzymes production varied significantly upon changing the initial concentrations of NaH$_2$PO$_4$ and moisture. The three-dimensional plots revealed that an increase in either moisture content of the medium, sucrose, or NaH$_2$PO$_4$ resulted in fibrinolytic enzymes production up to optimum level, whereas further increase in concentration decreased the enzyme yield. The perturbation plot (Figure 4)

showed that moisture had a significant effect on fibrinolytic enzymes production compared to other variables. The fibrinolytic enzymes production were 4418 units/mL in an optimized medium composed of (%) sucrose (0.5), NaH$_2$PO$_4$ (0.075), and moisture (113.64). Results revealed that the RSM optimized medium increased 4.5-fold of fibrinolytic enzymes production than the unoptimized medium. Fibrinolytic activities were comparatively higher than the earlier

Medium Optimization for the Production of Fibrinolytic Enzyme by Paenibacillus sp. IND8 Using Response Surface Methodology

71

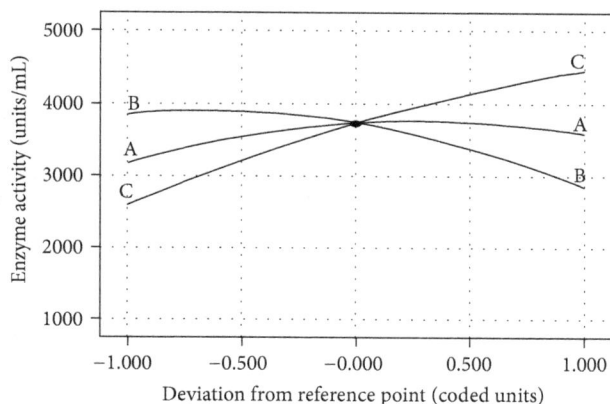

FIGURE 4: Perturbation graph summarizing the effect of sucrose (A), NaH_2PO_4 (B), and moisture (C) on fibrinolytic enzymes production.

report of RSM on *Bacillus subtilis* [21] and on *Bacillus* sp. strain AS-S20-1 [37].

3.6. Validation of the Experimental Designs. To validate the model equation, experiments were carried out in triplicates for fibrinolytic enzymes production in optimized conditions predicted by the experimental model and the fibrinolytic enzymes production was found to be 4683 units/mL. The theoretical predicted response for the model equation was 4720 units/mL. The experimental result was very close to the predicted response which validated this model experimentally.

4. Conclusion

A *Paenibacillus* sp. IND8 shows enhanced production of fibrinolytic enzymes in statistically optimized medium. The optimized medium showed high fibrinolytic activities of 4418 units/mL, which is 4.5-fold than that of the unoptimized medium. Purification and characterization of these fibrinolytic enzymes are in progress. This enzyme may be considered as a new source for thrombolytic agents.

Conflict of Interests

The authors declare that there is no conflict of interests regarding the publication of this paper.

Acknowledgment

The author Ponnuswamy Vijayaraghavan gratefully acknowledges the Council for Scientific and Industrial Research (CSIR), India, for providing a Senior Research Fellowship (ref. 09/652(0024)/2012 EMR-1).

References

[1] L. I. Moukhametova, R. B. Aisina, G. Y. Lomakina, and S. D. Varfolomeev, "Properties of the urokinase-type plasminogen activator modified with phenylglyoxal," *Russian Journal of Bioorganic Chemistry*, vol. 28, no. 4, pp. 278–283, 2002.

[2] F. Yan, J. Yan, W. Sun et al., "Thrombolytic effect of subtilisin QK on carrageenan induced thrombosis model in mice," *Journal of Thrombosis and Thrombolysis*, vol. 28, no. 4, pp. 444–448, 2009.

[3] J. Yuan, J. Yang, Z. Zhuang, Y. Yang, L. Lin, and S. Wang, "Thrombolytic effects of douchi fibrinolytic enzyme from *Bacillus subtilis* LD-8547 in vitro and in vivo," *BMC Biotechnology*, vol. 10, pp. 12–36, 2012.

[4] Y. Peng, X. Yang, and Y. Zhang, "Microbial fibrinolytic enzymes: an overview of source, production, properties, and thrombolytic activity in vivo," *Applied Microbiology and Biotechnology*, vol. 69, no. 2, pp. 126–132, 2005.

[5] H. Sumi, H. Hamada, H. Tsushima, H. Mihara, and H. Muraki, "A novel fibrinolytic enzyme (nattokinase) in the vegetable cheese natto; a typical and popular soybean food in the Japanese diet," *Experientia*, vol. 43, no. 10, pp. 1110–1111, 1987.

[6] H. Sumi, H. Hamada, K. Nakanishi, and H. Hiratani, "Enhancement of the fibrinolytic activity in plasma by oral administration of nattokinase," *Acta Haematologica*, vol. 84, no. 3, pp. 139–143, 1990.

[7] W. Kim, K. Choi, Y. Kim et al., "Purification and characterization of a fibrinolytic enzyme produced from *Bacillus* sp. strain CK 11-4 screened from Chungkook-Jang," *Applied and Environmental Microbiology*, vol. 62, no. 7, pp. 2482–2488, 1996.

[8] Y. Peng, Q. Huang, R.-H. Zhang, and Y.-Z. Zhang, "Purification and characterization of a fibrinolytic enzyme produced by *Bacillus amyloliquefaciens* DC-4 screened from douchi, a traditional Chinese soybean food," *Comparative Biochemistry and Physiology B*, vol. 134, no. 1, pp. 45–52, 2003.

[9] J.-H. Seo and S.-P. Lee, "Production of fibrinolytic enzyme from soybean grits fermented by *Bacillus firmus* NA-1," *Journal of Medicinal Food*, vol. 7, no. 4, pp. 442–449, 2004.

[10] S. Sugimoto, T. Fujii, T. Morimiya, O. Johdo, and T. Nakamura, "The fibrinolytic activity of a novel protease derived from a *Tempeh* producing fungus, *Fusarium* sp. BLB," *Bioscience, Biotechnology and Biochemistry*, vol. 71, no. 9, pp. 2184–2189, 2007.

[11] C. Ash, F. G. Priest, and M. D. Collins, "Molecular identification of rRNA group 3 bacilli (Ash, Farrow, Wallbanks and Collins) using a PCR probe test," *Antonie van Leeuwenhoek*, vol. 64, no. 3-4, pp. 253–260, 1994.

[12] F. Lu, Z. Lu, X. Bie et al., "Purification and characterization of a novel anticoagulant and fibrinolytic enzyme produced by endophytic bacterium *Paenibacillus polymyxa* EJS-3," *Thrombosis Research*, vol. 126, no. 5, pp. e349–e355, 2010.

[13] S. Tao, L. Peng, L. Beihui, L. Deming, and L. Zuohu, "Solid state fermentation of rice chaff for fibrinolytic enzyme production by *Fusarium oxysporum*," *Biotechnology Letters*, vol. 19, no. 5, pp. 465–467, 1997.

[14] H.-S. Joo, C. G. Kumar, G.-C. Park, S. R. Paik, and C.-S. Chang, "Oxidant and SDS-stable alkaline protease from *Bacillus clausii* I-52: production and some properties," *Journal of Applied Microbiology*, vol. 95, no. 2, pp. 267–272, 2003.

[15] J. Lee, S. Park, W.-A. Choi et al., "Production of a fibrinolytic enzyme in bioreactor culture by *Bacillus subtilis* BK-17," *Journal of Microbiology and Biotechnology*, vol. 9, no. 4, pp. 443–449, 1999.

[16] L. Xiao-Lan, D. Lian-Xiang, L. Fu-Ping, Z. Xi-Qun, and X. Jing, "Purification and characterization of a novel fibrinolytic enzyme from *Rhizopus chinensis* 12," *Applied Microbiology and Biotechnology*, vol. 67, no. 2, pp. 209–214, 2005.

[17] D. C. Montogomery, *Design and Analysis of Experiments*, John Wiley & Sons, New York, NY, USA, 5th edition, 2001.

[18] R. H. Myers and D. C. Montogomery, *Surface Methodology: Process and Product Optimization Using Designed Experiments*, John Wiley & Sons, New York, NY, USA, 2nd edition, 2002.

[19] X. Wang, L. Huang, Z. Kang, H. Buchenauer, and X. Gao, "Optimization of the fermentation process of actinomycete strain Hhs.015(T)," *Journal of Biomedicine and Biotechnology*, vol. 2010, Article ID 141876, 10 pages, 2010.

[20] M. A. Z. Coelho, G. C. Fontes, P. F. F. Amaral, and M. Nele, "Factorial design to optimize biosurfactant production by yarrowia lipolytica," *Journal of Biomedicine and Biotechnology*, vol. 2010, Article ID 821306, 8 pages, 2010.

[21] V. Deepak, K. Kalishwaralal, S. Ramkumarpandian, S. V. Babu, S. R. Senthilkumar, and G. Sangiliyandi, "Optimization of media composition for Nattokinase production by *Bacillus subtilis* using response surface methodology," *Bioresource Technology*, vol. 99, no. 17, pp. 8170–8174, 2008.

[22] T. Astrup and S. Müllertz, "The fibrin plate method for estimating fibrinolytic activity," *Archives of Biochemistry and Biophysics*, vol. 40, no. 2, pp. 346–351, 1952.

[23] S. F. Altschul, T. L. Madden, A. A. Schäffer et al., "Gapped BLAST and PSI-BLAST: a new generation of protein database search programs," *Nucleic Acids Research*, vol. 25, no. 17, pp. 3389–3402, 1997.

[24] J. H. Ko, J. P. Yan, L. Zhu, and Y. P. Qi, "Identification of two novel fibrinolytic enzymes from *Bacillus subtilis* QK02," *Comparative Biochemistry and Physiology C*, vol. 137, no. 1, pp. 65–74, 2004.

[25] P. Vijayaraghavan and S. G. P. Vincent, "Cow dung as a novel, inexpensive substrate for the production of a halo-tolerant alkaline protease by *Halomonas* sp. PV1 for eco-friendly applications," *Biochemical Engineering Journal*, vol. 69, pp. 57–60, 2012.

[26] Y. Mine, A. H. K. Wong, and B. Jiang, "Fibrinolytic enzymes in Asian traditional fermented foods," *Food Research International*, vol. 38, no. 3, pp. 243–250, 2005.

[27] P. Ellaiah, B. Srinivasulu, and K. Adinarayana, "A review on microbial alkaline proteases," *Journal of Scientific and Industrial Research*, vol. 61, no. 9, pp. 690–704, 2002.

[28] D. C. Montgomery, E. A. Peck, and G. G. Vining, *Introduction to Linear Regression Analysis*, John Wiley & Sons, New York, NY, USA, 3rd edition, 2000.

[29] C.-T. Chang, M.-H. Fan, F.-C. Kuo, and H.-Y. Sung, "Potent fibrinolytic enzyme from a mutant of *Bacillus subtilis* IMR-NK1," *Journal of Agricultural and Food Chemistry*, vol. 48, no. 8, pp. 3210–3216, 2000.

[30] B. K. Lonsane, N. P. Childyal, S. Budiatman, and S. V. Ramakrishna, "Engineering aspects of solid state fermentation," *Enzyme and Microbial Technology*, vol. 7, no. 6, pp. 258–265, 1985.

[31] R. S. Prakasham, C. S. Rao, and P. N. Sarma, "Green gram husk-an inexpensive substrate for alkaline protease production by *Bacillus* sp. in solid-state fermentation," *Bioresource Technology*, vol. 97, no. 13, pp. 1449–1454, 2006.

[32] M. Hajji, A. Rebai, N. Gharsallah, and M. Nasri, "Optimization of alkaline protease production by *Aspergillus clavatus* ES1 in *Mirabilis jalapa* tuber powder using statistical experimental design," *Applied Microbiology and Biotechnology*, vol. 79, no. 6, pp. 915–923, 2008.

[33] M. Elibol, "Optimization of medium composition for actinorhodin production by *Streptomyces coelicolor* A3(2) with response surface methodology," *Process Biochemistry*, vol. 39, no. 9, pp. 1057–1062, 2004.

[34] A. Dean and D. Voss, *Design and Analysis of Experiments*, Springer, New Delhi, India, 2006.

[35] D. Ba and I. H. Boyaci, "Modeling and optimization I: usability of response surface methodology," *Journal of Food Engineering*, vol. 78, no. 3, pp. 836–845, 2007.

[36] P. D. Ha, "Statistical problem solving," in *Experimental Design in Biotechnology*, P. D. Haaland, Ed., pp. 1–18, Marcel Dekker, New York, NY, USA, 1989.

[37] A. K. Mukherjee and S. K. Rai, "A statistical approach for the enhanced production of alkaline protease showing fibrinolytic activity from a newly isolated Gram-negative *Bacillus* sp. strain AS-S20-I," *New Biotechnology*, vol. 28, no. 2, pp. 182–189, 2011.

9

Biogeographical Variation and Population Genetic Structure of *Sporisorium scitamineum* in Mainland China: Insights from ISSR and SP-SRAP Markers

Liping Xu,[1] Yunhai Lu,[1] Qian You,[1] Xiaolan Liu,[1] Michael Paul Grisham,[2] Yongbao Pan,[2] and Youxiong Que[1]

[1] *Key Laboratory of Sugarcane Biology and Genetic Breeding, Ministry of Agriculture/Fujian Agriculture and Forestry University, Fuzhou 350002, China*
[2] *USDA-ARS, Sugarcane Research Unit, Houma, LA 70360, USA*

Correspondence should be addressed to Liping Xu; xlpmail@126.com and Youxiong Que; queyouxiong@hotmail.com

Academic Editors: V. Edwards-Jones, P. Jones, and L. Kong

A total of 100 *Sporisorium scitamineum* isolates were investigated by inter simple sequence repeat (ISSR) and single primer-sequence related amplified polymorphism (SP-SRAP) markers. These isolates were clearly assorted into three distinct clusters regardless of method used: either cluster analysis or by principal component analysis (PCA) of the ISSR, SP-SRAP, or ISSR + SP-SRAP data set. The total gene diversity (H_t) and gene diversity between subpopulations (H_s) were estimated to be 0.34 to 0.38 and 0.22 to 0.29, respectively, by analyzing separately the ISSR and SP-SRAP data sets, and to be 0.26–0.36 by analyzing ISSR + SP-SRAP data set. The gene diversity attributable to differentiation among populations (G_{st}) was estimated to be 0.35 and 0.22, and the gene flow (Nm) was 0.94 and 1.78, respectively, when analyzing separately ISSR and SP-SRAP data set, and was 0.27 and 1.33, respectively, when analyzing ISSR + SP-SRAP data set. Our study showed that there is considerable genetic variation in the analyzed 100 isolates, and the environmental heterogeneity has played an important role for this observed high degree of variation. The genetic differentiation of sugarcane smut fungus depends to a large extent on the heterogeneity of their habitats and is the result of long-term adaptations of pathogens to their ecological environments.

1. Introduction

China is the second largest consumer of sugar and the third largest producer internationally [1]. Sugarcane (*Saccharum* complex) is the most important sugar-yielding crop in China and accounts for about 92% of sugar output [1]. It has become a more important economic crop due to an increase in sugar demand. Furthermore, sugarcane offers the potential to be an ideal bioenergy crop because it can produce readily fermentable sugars, high fiber content, and very high yields of green biomass [1]. Sugarcane smut caused by *Sporisorium scitamineum* is one of the most intractable and devastating diseases of sugarcane in the world [2–7]. It causes not only considerable yield loss, but also leads to variety elimination due to susceptibility to the disease [8]. At present, the most economical and efficient way to control the smut disease is to plant resistant cultivars. Prior to the 1940s, sugarcane smut was limited to Asia, a few countries of South and East Africa (Natal, Mozambique, and Zimbabwe), and the Mascarenes (Madagascar, Mauritius, and Reunion) [9]. Since then, smut has spread to almost all sugarcane producing countries. In China, smut has been the most prevalent disease in the last ten years due to the susceptibility of the sugarcane cultivar ROC22 which occupies more than 50% of the total sugarcane planting area. The average of smut infection rate in ROC22 is over 8% in plant cane and 15% in ratoon cane, respectively (personal communication).

The smut-resistant mechanism of sugarcane has been widely investigated at the morphological, physiological, and mainly molecular level [1–3]. Several kinds of molecular techniques, including cDNA-amplified fragment length polymorphism (cDNA-AFLP) [10–13], differential-display reverse transcription-PCR (DDRT-PCR) [14], cDNA microarray [15], Solexa sequencing [16], and two-dimensional gel electrophoresis (2-DE) coupled with matrix-assisted laser desorption ionization-time of flight mass spectrometry (MALDI-TOF-TOF/MS) [17], have been occupied for the researches on the molecular interaction between sugarcane and *S. scitamineum*. The investigation of the genetic diversity of *S. scitamineum* is essential for a better understanding of the structure of pathogen populations and the interactions among host and pathogen and the environment and as a basis for developing smut-resistant sugacane varieties. Molecular detection of the smut pathogen in sugarcane was firstly realized by using polymerase chain reaction (PCR) and then TaqMan Real-time PCR to amplify the bE mating-type gene of *S. scitamineum* [5, 18, 19]. The intraspecies diversity within *S. scitamineum* isolates has been studied by the methods of random amplification of polymorphic DNA (RAPD) [6, 20], amplified fragment length polymorphism (AFLP) [4], and internal transcribed spacer (ITS) sequence analysis [21, 22]. Raboin et al. analyzed a collection of *S. scitamineum* populations from 15 sugarcane producing countries for polymorphisms at 17 microsatellite loci [9]. The authors found that the genetic diversity was extremely low among the American and African populations but was high among the Asian populations. They also found that the American and African *S. scitamineum* populations all belonged to a single lineage which was also found among some Asian populations. In China, Que et al. studied the molecular evolution of *S. scitamineum* by analyzing a set of 23 isolates collected from six primary sugarcane producing areas in Mainland China with RAPD and sequence-related amplified polymorphism (SRAP) markers [7]. The results of RAPD, SRAP, and combined RAPD-SRAP showed all that the molecular evolution of *S. scitamineum* was associated with its geographical origin. No evidence of coevolution between the pathogen and sugarcane was found. These analyses also did not provide information about the race differentiation of *S. scitamineum*. Furthermore, the numbers of host genotypes, smut isolates, and geographical origin of districts in China for testing were limited to 21, 23, and 6, respectively, in the study of Que et al. [7], and the smut isolates studied by Raboin et al. [9] and Singh et al. [6] did not include the isolates from Mainland China. Thus, to better understand the smut population structure and the influence of environmental and genotype heterogeneity on the population structure, analysis of more isolates representing more regions, more cultivar/genotypes, and a representative sugarcane variety in different regions is needed.

Population genetic analysis of *S. scitamineum* could help to identify potential sources of resistance, which are expected in areas where genetic diversity of both pathogen and host is maximal [9]. The genetic diversity of *S. scitamineum* populations was shown to be extremely low among 142 single teliospore isolates collected from 15 countries based on the analysis of AFLP and RAPD markers [9]. ISSR markers are more reliable than RAPD because of the longer primers and the higher annealing temperature of ISSR than RAPD [23]. Therefore, the ISSR markers have been widely used in marker-assisted selection, genetic diversity analysis, or

DNA fingerprinting [23–26]. Sequence-related amplification polymorphism (SRAP) technology was firstly developed from *Brassica* crops by Li and Quiros [27]. Currently, all SRAP markers are based on the use of a combination of upstream and downstream primer pairs [27]. In our previous research, the feasibility of single primer sequence related SRAP (SP-SRAP) markers have been proved for genetic diversity analysis (private bulletin).

In this study, we use ISSR and SP-SRAP techniques to study a more representative *S. scitamineum* isolate collection from Mainland China with the aim of understanding, at the molecular level, the differentiation, variation, and population structure of *S. scitamineum* in Mainland China and the effectiveness of the environmental and genotype heterogeneity on the influence of the pathogen population dynamics to provide the basis for *S. scitamineum* resistance breeding and the geographical distribution of varieties.

2. Materials and Methods

2.1. S. scitamineum Isolate Sampling and Isolation. One hundred pathogen strains were isolated in 2010 from 38 different sugarcane genotypes in 7 provinces (20 sugarcane planting districts) of China, including Guangxi (which accounts for 62% sugarcane planting area) (36 strains), Yunnan (26 strains) (accounts for about 20% sugarcane planting area), Guangdong (13 strains), Fujian (9 strains), Jiangxi (8 strains), Hainan (5 strains), and Sichuan (3 strains). More detailed information about these strains are given in Table 1 and Figure 1.

Single teliospores were isolated from whips of infected sugarcane with sterile tips or inoculation needle under the sterile conditions and separated by serial 10-fold dilution until only 1–5 single-teliospore(s) were observed in a single drop ($10 \mu L$) under the microscope (40×10). $10–30 \mu L$ of dilution was then evenly plated onto potato dextrose agar (PDA) medium, containing $75 \, mgl^{-1}$ streptomycin sulphate according to Yang et al. [28], then incubated for 1–3 days at $28°C$. Germinating single teliospores were transferred to new plates containing PDA medium.

2.2. DNA Extraction and S. scitamineum Isolate Identification. Genomic DNA from hyphal colonies developed from single teliospores were extracted by the modified SDS method described by Que et al. [20]then stored at $-20°C$. The quality and quantity of extracted DNA were controlled on a 0.6% agarose gel. Electrophoresis was performed in $1 \times$ TAE running buffer, pH 8.0, at 100 V. The agarose gel was checked with a NanoVue Spectrophotometer (GE Healthcare). DNA was quantified by fluorometry and adjusted to $20 \, ng/\mu L$ in distilled water. In order to understand the relationship between single teliospores from the same smut whip, we randomly selected five single teliospores from one smut whip collected from the sugarcane cultivar ROC22 and cloned and sequenced the rDNA internal transcribed spacer (rDNA-ITS) regions (ITS1, ITS2, and 5.8S rDNA) as described by Singh et al. [6]. To ensure that all the DNA samples were from *S. scitamineum*, we amplified these DNA samples by

Biogeographical Variation and Population Genetic Structure of Sporisorium scitamineum in Mainland China: Insights from ISSR and SP-SRAP Markers

75

TABLE 1: Information about 100 strains of *S. scitamineum*.

TABLE 1: Continued.

No.	Source	Genotype of host	No.	Source	Genotype of host
1	Futuo, Guangxi	ROC22	50	Lincang, Yunnan	LZ78-85
2	Futuo, Guangxi	ROC22	51	Lincang, Yunnan	ROC22
3	Futuo, Guangxi	ROC22	52	Lincang, Yunnan	LC03-182
4	Futuo, Guangxi	YL17	53	Lincang, Yunnan	YZ03-103
5	Nanning, Guangxi	ROC22	54	Longchuan, Yunnan	ROC22
6	Nanning, Guangxi	ROC22	55	Longchuan, Yunnan	GT94-119
7	Nanning, Guangxi	ROC22	56	Luxi, Yunnan	LZ78-85
8	Nanning, Guangxi	ROC22	57	Luxi, Yunnan	YL6
9	Chongzuo, Guangxi	ROC22	58	Rili, Yunnan	LK80-279
10	Chongzuo, Guangxi	ROC22	59	Rili, Yunnan	YZ95-128
11	Chongzuo, Guangxi	ROC22	60	Yingjiang, Yunnan	LC03-182
12	Chongzuo, Guangxi	ROC22	61	Lianghe, Yunnan	LC03-182
13	Chongzuo, Guangxi	ROC22	62	Lianghe, Yunnan	LZ78-85
14	Chongzuo, Guangxi	YC03-182	63	Nankang, Jiangxi	FN28
15	Chongzuo, Guangxi	ROC22	64	Nankang, Jiangxi	ROC16
16	Tianyang, Guangxi	YC03-182	65	Nankang, Jiangxi	YT91-600
17	Tianyang, Guangxi	ROC22	66	Nankang, Jiangxi	LC03-182
18	Tiandong, Guangxi	ROC22	67	Nankang, Jiangxi	ROC22
19	Tiandong, Guangxi	YT94-128	68	Nankang, Jiangxi	Co412
20	Tiandong, Guangxi	ROC22	69	Nankang, Jiangxi	YZ95-128
21	Tiandong, Guangxi	ROC22	70	Nankang, Jiangxi	GT02-351
22	Tiandong, Guangxi	LC03-182	71	Zhangzhou, Fujian	ROC22
23	Nongkesuo, Guangxi	ROC22	72	Zhangzhou, Fujian	YT93-158
24	Taocheng, Guangxi	ROC22	73	Zhangzhou, Fujian	YT93-158
25	Hepu, Guangxi	YL6	74	Zhangzhou, Fujian	ROC16
26	Hepu, Guangxi	YT00-236	75	Zhangzhou, Fujian	YT93-158
27	Yinhai, Guangxi	ROC22	76	Zhangzhou, Fujian	ROC16
28	Yinhai, Guangxi	GT02-901	77	Fuzhou, Fujian	GT98-296
29	Liucheng, Guangxi	ROC22	78	Fuzhou, Fujian	FN04-2861
30	Liucheng, Guangxi	LC03-182	79	Fuzhou, Fujian	NCo310
31	Liucheng, Guangxi	ROC22	80	Zhanjiang, Guangdong	YT89-113
32	Liucheng, Guangxi	LC03-1137	81	Zhanjiang, Guangdong	YT79-117
33	Laibing, Guangxi	LC03-182	83	Zhanjiang, Guangdong	YT00-236
34	Laibing, Guangxi	ROC22	84	Zhanjiang, Guangdong	ROC16
35	Laibing, Guangxi	ROC16	85	Zhanjiang, Guangdong	ROC16
36	Laibing, Guangxi	FN28	86	Zhanjiang, Guangdong	YT89-113
37	Baoshan, Yunnan	R6048	87	Zhanjiang, Guangdong	ROC22
38	Baoshan, Yunnan	ROC22	88	Zhanjiang, Guangdong	YT89-113
39	Baoshan, Yunnan	ROC22	89	Zhanjiang, Guangdong	ROC22
40	Baoshan, Yunnan	YT86-368	90	Zhanjiang, Guangdong	YZ03-194
41	Baoshan, Yunnan	GT12	91	Zhanjiang, Guangdong	ROC22
42	Nile, Yunnan	MT69-421	92	Zhanjiang, Guangdong	YT96-86
43	Nile, Yunnan	ROC16	93	Haikou, Hainan	ROC22
44	Kaiyuan, Yunnan	Q170	94	Haikou, Hainan	ROC22
45	Kaiyuan, Yunnan	GT11	95	Haikou, Hainan	ROC22
46	Kaiyuan, Yunnan	LC03-182	96	Danzhou, Hainan	ROC22
47	Kaiyuan, Yunnan	MT69-421	98	Miyi, Sichuan	CZ6
48	Xinping, Yunnan	MT69-422	99	Huili, Sichuan	CZ6
49	Xinping, Yunnan	ROC22	100	Dechang, Sichuan	CZ6

S. scitamineum specific primers *b*E4 and *b*E8 [18], cloned the amplicons by T-vector, sequenced the cloned fragments, and compared them by BLASTn to the NCBI databases (http://www.ncbi.nlm.nih.gov/).

2.3. PCR Reactions, ISSR, and SP-SRAP Analysis. To ensure the identity of our *S. scitamineum* samples, we first performed PCR amplifications using the universal primers ITS1 and ITS4 for detection of fungi [29] and the specific primers *b*E4 and *b*E8 for detection of *S. scitamineum* [18].

For ISSR analysis, the primers used (P5, P6, P10, P21, P25, P26, and P27) were commercially synthesis and *Ex Taq* polymerase purchased from the TaKaRa Biotechnology Co., Ltd. (Dalian, China). Amplification reactions were performed in a total volume of 20 μL containing 20 ng of DNA, 5 pmol of each primer, 4.5 nmol of each dNTP, 2.2 μL of 10 × *Ex Taq* Buffer, and 0.5 U *Ex Taq* polymerase. The final volume was adjusted to 20 μL with sterile distilled water [8, 24]. The amplification cycles consisted of an initial denaturation at 95°C for 4 min, followed by 35 cycles of 30 sec at 94°C, 30 sec at 48–56°C, and 30 sec at 72°C, and a final extension phase of 10 min at 72°C. PCR products were separated by electrophoresis in 3.0% agarose gel with 1 × TAE buffer at 60 V for 3 h. The gels were stained with ethidium bromide and photographed under ultraviolet light with NanoVue Spectrophotometer (GE Healthcare).

For SP-SRAP analysis, the amplification reactions were performed in a volume of 25 μL containing 2.4 μL of 10 × Ex Tap Buffer, 5 nmol of each dNTP, 20 pmol of primer, 1.0 U *Ex Taq* polymerase, and 40 ng of DNA template. The amplification cycles consisted of an initial denaturation at 95°C for 5 min, followed by 5 cycles of 1 min at 94°C, 1 min at 35°C, and 1 min at 72°C, 35 cycles of 1 min at 94°C, 1 min at 35°C, and 1 min at 72°C, and a final extension phase of 10 min at 72°C. PCR products were separated by electrophoresis in 3.0% agarose gel with 1 × TAE buffer at 60 V for 3 h. The gels were stained with ethidium bromide and then photographed under ultraviolet light as above.

2.4. Data Analysis. The softwares of NTSYS.PC (Numerical Taxonomy System, Applied Biostatistics, Inc.) [30] and POPGENE1.31 were used for analysis of ISSR, SP-SRAP, and ISSR + SP-SRAP data. A binary matrix was firstly generated by scoring the presence or absence of each individual band in all lanes. A similarity matrix was secondly generated by using Jaccard coefficient. The Jaccard coefficient takes into account the bands present in at least one of the two compared individuals. The absence of bands is not interpreted as a similar character between isolates. The similarity matrix was used for unweighted pair-group method with arithmetic average (UPGMA) cluster analysis and principal component analysis (PCA) [30]. Observed number of alleles (N_a), effective number of alleles (N_e), Nei's measure of gene diversity (h), and Shannon's Information index (I) were used to evaluate the genetic diversity within each population by POPGENE1.31 [31, 32]. Mean values of Nei's gene diversity of total populations (H_t), Shannon's diversity index between populations (H_s), proportion of gene diversity attributable to differentiation among populations (G_{st}), and estimates of gene flow parameter among populations $Nm = 0.5(1 - G_{st})/G_{st}$ were obtained across loci [31, 32].

2.5. Sequence Analysis of ITS Regions. Based on results of ISSR and SP-SRAP analyses, some pairwise comparisons among isolates with low similarity values were selected for further investigation of ITS sequences (ITS1, ITS2, and 5.8S rDNA). The sequences were analysed by using the software DNAMAN (http://www.lynnon.com/) and the consensus sequences were compared by BLASTn to the NCBI databases (http://www.ncbi.nlm.nih.gov).

3. Results

3.1. Confirmation of S. scitamineum Samples. The DNA extracts of 100 *S. scitamineum* isolates were firstly analyzed by NanoVue minim spectrophotometer. The ratios of OD_{260}/OD_{230} ranged from 1.90 to 2.20 and the ratios of OD_{260}/OD_{280} ranged from 1.60 to 2.10, indicating that all these DNA samples meet the requirement of the following experiments. To ensure the identity of our *S. scitamineum* samples, we performed PCR amplifications on these DNA samples using the universal primers ITS1 and ITS4 for specific detection of fungi [29] and the specific primer pair *b*E4 and *b*E8 for specific detection of *S. scitamineum* [18]. PCR amplification of the rDNA-ITS sequence using the specific primer pair ITS1 and ITS4 yielded a single 755 bp DNA fragment of expected size for each of the 100 DNA samples, extracted from the cultures of single-teliospores of the 100 *S. scitamineum* isolates. Sequencing of five randomly selected amplified products showed a sequence similarity of >99.93% among them. PCR analysis of the 100 samples by *S. scitamineum* specific primer pair *b*E4 and *b*E8 revealed a fragment with expected size of 420 bp, which is present on all the tested samples. These fragments were cloned in TA-vector and sequenced. Sequence analysis showed that this fragment was a sequence specific to *S. scitamineum*.

3.2. Genetic Diversity of S. scitamineum Isolates by ISSR Analysis. Among the 18 ISSR primers tested on a set of five DNA samples, seven, namely P5, P6, P10, P21, P25, P26, and P27, were selected to amplify the DNA extracts of the 100 *S. scitamineum* isolates. A total of 121 bands were obtained, of which, 105 were polymorphic among the 100 samples. The ISSR PCR amplification products and their levels of polymorphism are presented in Table 2. Each of these seven primers generated nine to 19 bands per isolate, which range in size from 250 bp to 5,000 bp. On an average, 17.3 bands were amplified by each primer, and the percentage of polymorphic bands was as high as 86.8% as estimated from the 100 isolates.

Based on the ISSR PCR amplification data, the genetic similarity indexes between the 100 *S. scitamineum* isolates were calculated using Jaccard coefficient [30], and their values ranged from 0.46 to 0.90, with an average of 0.66. The maximum similarity coefficient of 0.9008 was obtained between the isolates from the hosts ROC16 (number 76) and GT98-296 (number 77), both of which were isolated in

Biogeographical Variation and Population Genetic Structure of Sporisorium scitamineum in Mainland China: Insights from ISSR and SP-SRAP Markers

77

FIGURE 1: Geographic locations of the 100 *S. scitamineum* isolates isolated from 7 Provinces of China. Notes: The seven provinces where the *S. scitamineum* isolates were collected are marked in the red dots.

Fujian province. The minimum pairwise similarity coefficient of 0.46 was obtained between the isolates from the hosts GT11 (number 45) and CZ6 (number 98), of which the former was isolated in Yunnan province, and the latter was isolated in Sichuan province.

Based on the ISSR data set, a dendrogram was generated for the 100 *S. scitamineum* isolates by unweighted pair-group method with arithmetic average (UPGMA) cluster analysis and is presented in Figure 2. They were grouped into three clusters at the level of 0.006, I, II, and III, respectively. The cluster analysis showed that most of these strains were relatively concentrated. Of the 100 isolates, 58 isolates (58%) were grouped in cluster I, including all the isolates collected in Guangxi, Hainan, and Sichuan, together with four isolates collected in Jiangxi and 10 isolates collected in Yunnan. Nineteen isolates (19%) were grouped into the cluster II, including 16 from Yunnan and 3 from Jiangxi. The remaining 23 isolates fell into the cluster III, including the 13 isolates from Guangdong and eight isolates from Fujian, plus one isolate from Jiangxi. These results suggest that the smut pathogen isolates collected from same region (province) tended to cluster in the same group.

A principal component analysis (PCA) based on the ISSR data set was performed. The first three components explained 97% of the total variance and divided clearly the 100 isolates into three groups, A, B, and C, as shown in Figure 3 by the dimension 2-3 plot. Group A includes 45 isolates, of which 19 are from Yunnan and 36 are from Guangxi. Group B includes 32 isolates, of which seven are from Yunnan, seven are from Jiangxi, five are from Hainan, and three are from Sichuan. Group C includes the remaining 23 isolates, of which 13 are from Guangdong, nine are from Fujian, and one is from Jiangxi. Compared to the results of UPGMA analysis, the 45 isolates assigned to group A by PCA belong to the cluster I using UPGMA, the 19 isolates assigned to group B by PCA are the same as those in the cluster II using UPGMA and the 23 isolates assigned to group C by PCA are the same as those in the cluster III using UPGMA. From the above analyses, the smut pathogen strains from a same region (province) tend to group into the same group. For example, the 36 isolates from Guangxi are clustered into a single group A, the five isolates from Hainan and the three isolates from Sichuan cluster in

TABLE 2: List of the 7 selected ISSR primers and their PCR amplification results obtained on the 100 *S. scitamineum* isolates.

Primer	Sequence (5′-3′)	Bands		
		Total (No.)	Polymorphic (No.)	Polymorphic (%)
P5	$(AG)_8G$	18	14	77.8
P6	$(AG)_8C$	14	12	85.7
P10	$(AG)_8T$	20	18	85.0
P21	$(AG)_8YA$	20	19	95.0
P25	$(GA)_8YC$	18	17	94.4
P26	$(GA)_8YT$	19	17	89.4
P27	$(GA)_8YG$	12	9	75.0
Total		121	105	
Mean			17.3	86.8

group B, and the 13 isolates from Guangdong and the nine isolates from Fujian clustering group C. Of the remaining 34 isolates, 19 of the 26 from Yunnan cluster in group A and seven into group B and seven of the eight isolates from Jiangxi cluster in group B.

3.3. Genetic Diversity of S. scitamineum Isolates by SP-SRAP Analysis.
Among the 17 SP-SRAP primers tested on a set of five DNA samples, seven produced clear amplification patterns with bands ranging in size from 500 bp to 5,000 bps. The seven primers were used to amplify the DNA of the 100 *S. scitamineum* isolates and yielded a total of 153 bands, of which 152 (99.3%) were identified as polymorphic. The number of polymorphic bands generated by each primer ranged from 9 to 27 with an average of 21.7. Except the primer P2, which produced one monomorphic band (of 10 bands) among the 100 isolates, the other 6 primers produced 100% polymorphic bands (Table 3).

The genetic similarity indexes generated from SP-SRAP data for the 100 *S. scitamineum* isolates ranged from 0.44 to 0.79 with an average of 0.62. The maximum genetic similarity coefficient value was 0.7943 which was obtained from the following three compared isolate pairs: ROC22 (number. 7)

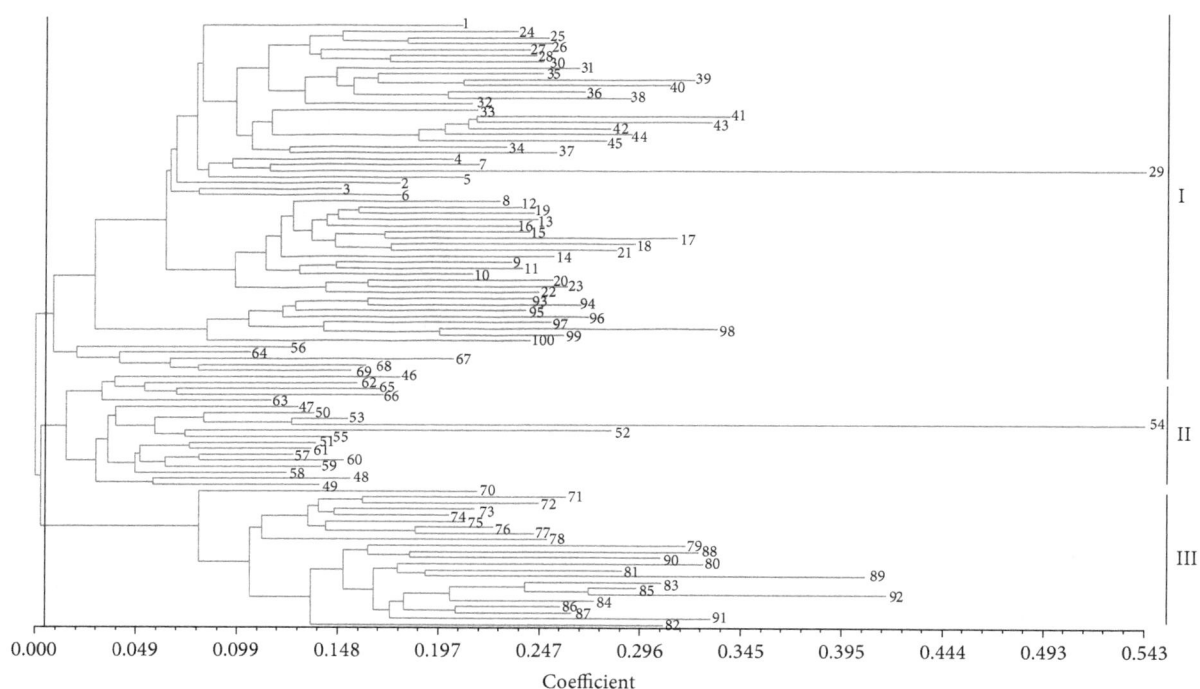

FIGURE 2: Dendrogram generated for the 100 *S. scitamineum* isolates (numbers 1–100) based on ISSR data.

TABLE 3: List of the 7 selected SP-SRAP primers and their PCR amplification results obtained on the 100 *S. scitamineum* isolates.

Primer	Sequence (5′-3′)	Bands		
		Total (No.)	Polymorphic (No.)	Polymorphic (%)
P1	TGAGTCCAAACCGGATA	23	23	100.0
P2	TGAGTCCAAACCGGAAG	10	9	90.0
P3	TGAGTCCAAACCGGACA	20	20	100.0
P4	TGAGTCCAAACCGGACG	25	25	100.0
P5	TGAGTCCAAACCGGTAA	25	25	100.0
P6	GACTGCGTACGAATTAAT	23	23	100.0
P7	GACTGCGTACGAATTAAC	27	27	100.0
Total		153	152	
Mean			21	99.3

and YT94-128 (number 19), ROC22 (number 18) and YT94-128 (number 19), and YT94-128 (number 19) and ROC22 (number 20). All 4 isolates were collected from Guangxi.

UPGMA clustering analysis based on the SP-SRAP data set also divided the 100 *S. scitamineum* isolates into three clusters, I, II, and III, at the distance index of 0.01 (Figure 4). The first cluster (I) consisted of 25 isolates, of which 22 were from Guangxi, two from Guangdong, and one from Fujian. The second cluster (II) contained 24 isolates, of which 11 were from Guangdong, seven from Fujian, two from Guangxi, two from Jiangxi, one from Yunnan, and one from Sichuan. The third cluster (III) was made up of 51 isolates, of which 25 were from Yunnan, 12 from Guangxi, six from Jiangxi, five from Hainan, two from Sichuan, and one from Fujian. It should be noted that the isolates from Hainan were grouped in the same cluster III, while the isolates from Guangxi and Fujian were distributed into all three clusters (I, II, and III), the

isolates from Guangdong fell into clusters I and II, the isolates from Sichuan fell into clusters I and III, and the isolates from Jiangxi and Yunnan fell into clusters II and III, respectively.

PCA analysis of the 100 *S. scitamineum* isolates was performed based on SP-SRAP data by using the software NTSYS [30]. The results were shown in Figure 5. Amongst the 100 analysed isolates, 97 were assorted into three different groups, named as A, B, and C, except for the isolate number 24 from Guangxi, the isolate number 95 from Hainan, and the isolate number 98 from Sichuan. The group A includes 22 isolates from Guangxi Province. The group B includes 22 isolates, of which eight from Fujian, 11 from Guangdong, one from Guangxi, one from Yunnan, and one from Jiangxi. The group C includes 53 isolates, of which 12 from Guangxi, 25 from Yunnan, seven from Jiangxi, four from Hainan, two from Guangdong, two from Sichuan, and one from Fujian. It is notable that 20 out of 22 isolates in the group A were

Biogeographical Variation and Population Genetic Structure of Sporisorium scitamineum in Mainland China: Insights from ISSR and SP-SRAP Markers

79

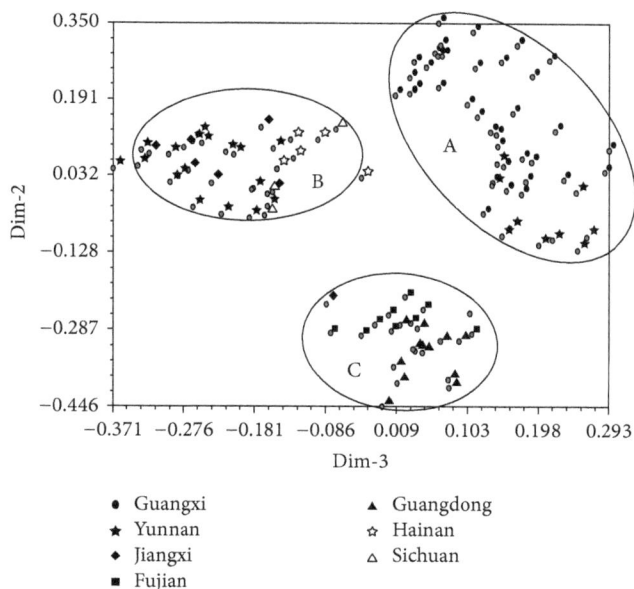

FIGURE 3: Principal components analysis using ISSR data for the 100 S. scitamineum isolates (numbers 1–100).

also included in the cluster I of UPGMA, 16 out of 22 in the group B were also included in the cluster II of UPGMA, and 47 out of 53 in the group C were also included in the cluster III of UPGMA (Figure 4). All the isolates in the group A were collected from Guangxi, the most of the isolates in the group B were collected from Fujian and Guangdong, and the isolates in the group C were composed of those from seven different provinces.

3.4. Genetic Diversity Analysis Based on the Combined ISSR and SP-SRAP Data Set.

By combining the data of ISSR and SP-SRAP, a total of 274 bands were recorded, of which 257 (93.80%) were polymorphic on the 100 S. scitamineum isolates. UPGMA cluster analysis based on these combined data revealed three clusters, named I, II, and III (Figure 6). Of the 100 isolates, 76 (76%) were grouped in cluster I, including all the isolates from Guangxi, Fujian, Guangdong, Hainan, and Sichuan, plus nine isolates from Yunnan and one isolate from Jiangxi. Only four isolates (4%) from Yunnan were grouped in cluster II. The remaining 20 isolates (20%), of which 13 were from Yunnan and seven from Jiangxi, were grouped into the cluster III. Comparable to those observed above, the isolates collected from the same region tend to be grouped into one cluster.

The results of PCA analysis of the 100 S. scitamineum isolates based on the combined ISSR + SP-SRAP data set were shown in Figure 7. It showed that the group A contained 45 isolates, including all the 36 isolates sampled from Guangxi and nine isolates from Yunnan. The group B consisted of 23 isolates, including all the nine isolates sampled from Fujian, all the 13 isolates from Guangdong, and one isolate from Jiangxi. The group C contained 32 isolates, including 17 isolates from Yunnan, seven isolates from Jiangxi, all five isolates from Hainan, and all three isolates from Sichuan.

TABLE 4: Population genetic parameters estimated for different subpopulations of S. scitamineum collected from seven provinces by using the ISSR, SP-SRAP and ISSR + SP-SRAP data sets.

Geographic population[a]	Number of isolates	N_a[b]	N_e[c]	h[d]	I[e]
Guangxi	36	1.86^A	1.48^A	0.28^A	0.42^A
		1.97^B	1.57^B	0.33^B	0.50^B
		1.92^{A+B}	1.53^{A+B}	0.31^{A+B}	0.47^{A+B}
Yunnan	26	1.85^A	1.49^A	0.29^A	0.43^A
		1.96^B	1.61^B	0.35^B	0.52^B
		1.91^{A+B}	1.56^{A+B}	0.32^{A+B}	0.48^{A+B}
Guangdong	13	1.70^A	1.36^A	0.22^A	0.34^A
		1.85^B	1.50^B	0.30^B	0.45^B
		1.78^{A+B}	1.44^{A+B}	0.26^{A+B}	0.40^{A+B}
Fujian	9	1.54^A	1.31^A	0.19^A	0.28^A
		1.75^B	1.46^B	0.27^B	0.40^B
		1.65^{A+B}	1.39^{A+B}	0.23^{A+B}	0.35^{A+B}
Jiangxi	8	1.64^A	1.38^A	0.22^A	0.34^A
		1.84^B	1.52^B	0.30^B	0.46^B
		1.73^{A+B}	1.45^{A+B}	0.26^{A+B}	0.40^{A+B}
Hainan	5	1.43^A	1.29^A	0.17^A	0.25^A
		1.63^B	1.44^B	0.25^B	0.37^B
		1.54^{A+B}	1.37^{A+B}	0.22^{A+B}	0.32^{A+B}
Sichuan	3	1.39^A	1.31^A	0.17^A	0.25^A
		1.58^B	1.46^B	0.26^B	0.37^B
		1.49^{A+B}	1.39^{A+B}	0.22^{A+B}	0.31^{A+B}
Mean	14.3	1.63^A	1.37^A	0.22^A	0.33^A
		1.79^B	1.51^B	0.29^B	0.44^B
		1.72^{A+B}	1.45^{A+B}	0.26^{A+B}	0.39^{A+B}

[A]Genetic diversity of the 100 S. scitamineum isolates based on ISSR markers.
[B]Genetic diversity of the 100 S. scitamineum isolates based on SP-SRAP markers.
[A+B]Genetic diversity of the 100 S. scitamineum isolates based on ISSR and SP-SRAP markers.
[a]The 100 S. scitamineum isolates were devided into 7 subpopulations according to their geographic origins.
[b]Mean observed number of alleles.
[c]Mean effective number of alleles.
[d]Mean of Nei's gene diversity.
[e]Mean of Shannon's Information index.

Compared to the results of UPGMA analysis, all 45 isolates in group A and all 23 isolates in group B were also grouped in the cluster I of UPGMA, while group C contained all four isolates in cluster II and all 20 isolates in cluster III, plus the remaining eight isolates in cluster I. The isolates isolated from the same district in a province tend to group together as revealed by both UPGMA and PCA analyses.

3.5. Population Genetic Analysis Based on the ISSR, SP-SRAP, and ISSR + SP-SRAP Data Set.

Population genetic parameters for the 121 ISSR loci identified among the 100 S. scitamineum isolates were estimated (Table 4). The results showed that the observed (N_a) and effective (N_e) numbers of alleles were higher in the population of Guangxi ($N_a = 1.86$, $N_e = 1.48$) and Yunnan ($N_a = 1.85$, $N_e = 1.49$) compared to other subpopulations. By comparison, the gene diversity index (h) and the Shannon's information index (I) were also relatively higher in the subpopulation of Yunnan ($h = 0.29$,

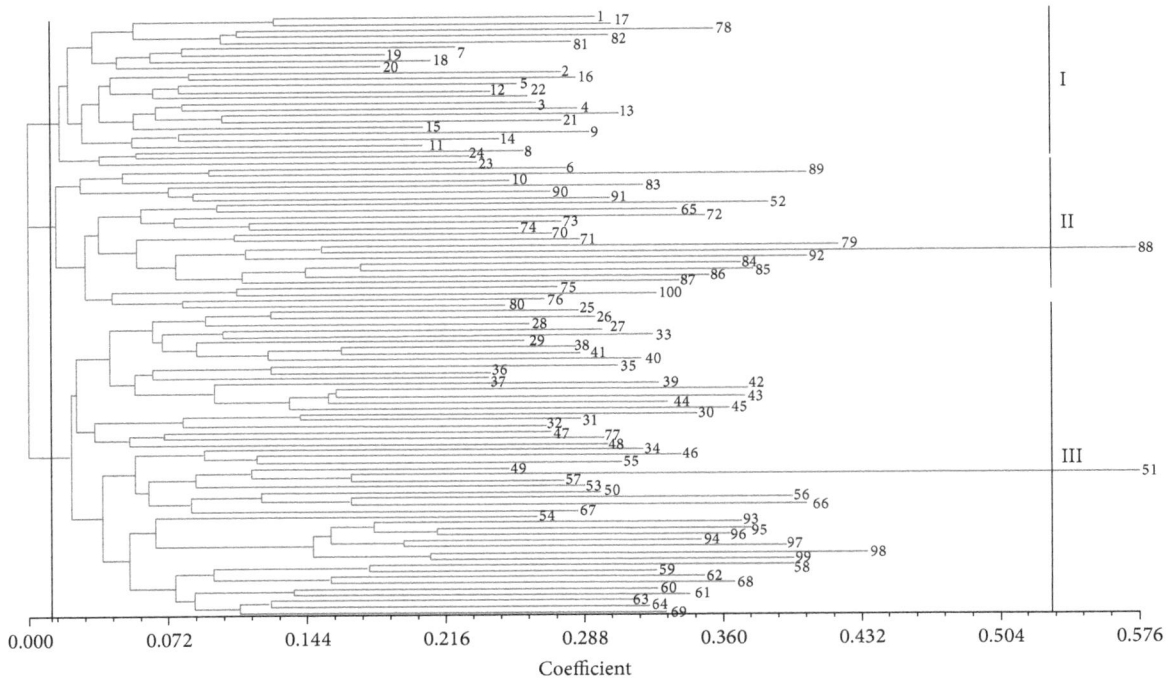

FIGURE 4: Dendrogram of the 100 *S. scitamineum* isolates (numbers 1–100) generated by UPGMA based on SP-SRAP data.

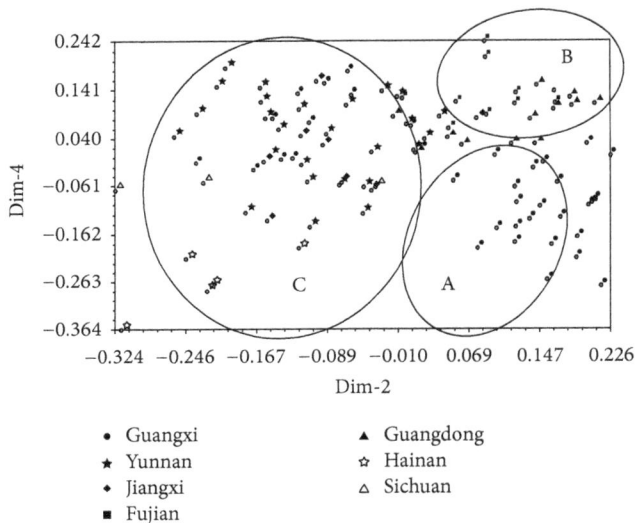

FIGURE 5: Principal components analysis of the 100 *S. scitamineum* isolates (numbers 1–100) based on SP-SRAP data.

indicating that lower rate of gene flow has occurred between populations.

Population genetic parameters for the 153 SP-SRAP loci identified among the 100 isolates was estimated (Table 4). Observed (N_a) and effective (N_e) numbers of alleles were higher in the subpopulations of Guangxi (N_a = 1.97, N_e = 1.57) and Yunnan (N_a = 1.96, N_e = 1.61) compared to other subpopulations. By comparison, the gene diversity (h) and the Shannon's information index (I) values were also relatively higher in the subpopulations of Yunnan (h = 0.35, I = 0.52) than those in the subpopulations of Hainan (h = 0.25, I = 0.37) and Sichuan (h = 0.26, I = 0.37), respectively. The total gene diversity (H_t) and gene diversities between subpopulations (H_s) values in SP-SRAP analysis were estimated to be 0.38 and 0.29, respectively. Gene diversity and differentiation among populations (G_{st}) were estimated to be 0.22, indicating that 22% of the total genetic variation was originated between populations, while the gene flow (Nm) was estimated to be 1.78, indicating that higher rate of gene flow has occurred between populations.

Population genetic parameters for the 274 ISSR + SP-SRAP loci identified among the 100 isolates was estimated (Table 4). Observed (N_a) and effective (N_e) numbers of alleles were higher in the subpopulations of Guangxi (N_a = 1.92, N_e = 1.53) and Yunnan (N_a = 1.91, N_e = 1.56) compared to other subpopulations. The gene diversity (h) and the Shannon's information index (I) values were also relatively higher in the subpopulations of Yunnan (h = 0.32, I = 0.48), but relatively lower in the subpopulations of Hainan (h = 0.22, I = 0.32) and Sichuan (h = 0.22, I = 0.31), respectively. The total gene diversity (H_t) and gene diversities between subpopulations (H_s) values in ISSR + SP-SRAP

I = 0.43) than those in the subpopulations of Hainan (h = 0.17, I = 0.25) and Sichuan (h = 0.17, I = 0.25), respectively. By using ISSR data set, the total gene diversity (H_t) and gene diversities between subpopulations (H_s) were estimated to be 0.34 and 0.22, respectively. Gene diversity and differentiation among populations (G_{st}) were estimated to be 0.35, indicating that 35% of the total genetic variation was originated between populations, while gene flow (Nm) was estimated to be 0.94,

Biogeographical Variation and Population Genetic Structure of Sporisorium scitamineum in Mainland China:
Insights from ISSR and SP-SRAP Markers

81

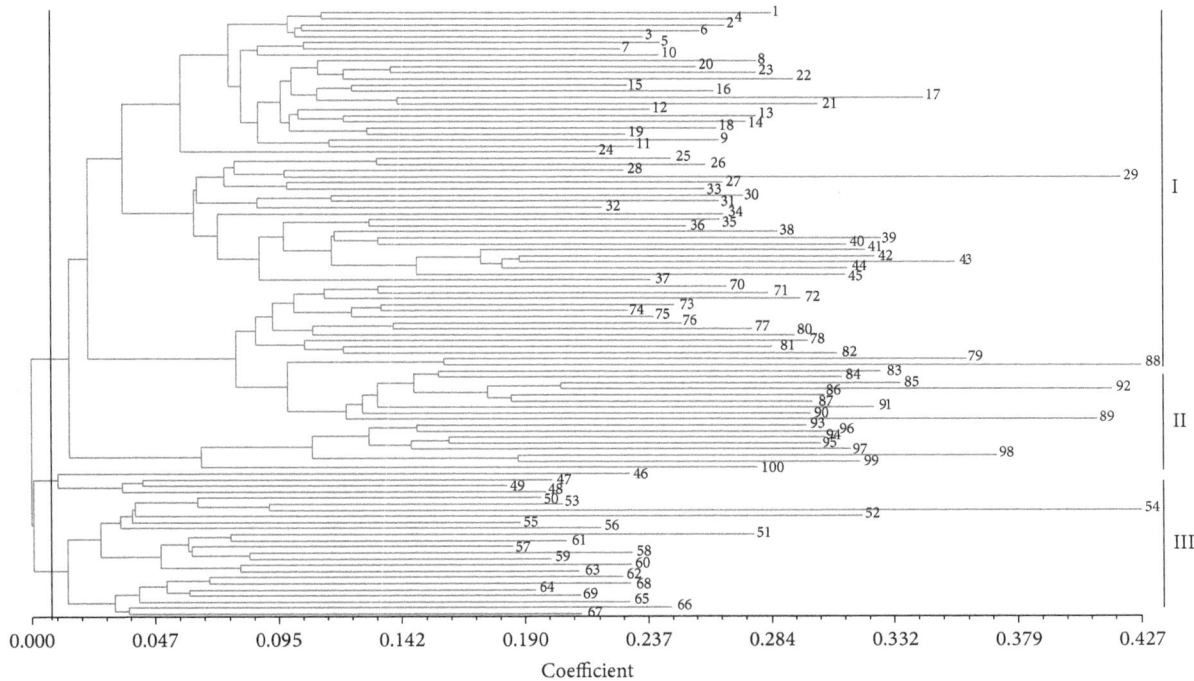

FIGURE 6: Dendrogram of the 100 *S. scitamineum* isolates (numbers 1–100) generated by UPGMA based on the combined ISSR + SP-SRAP data set.

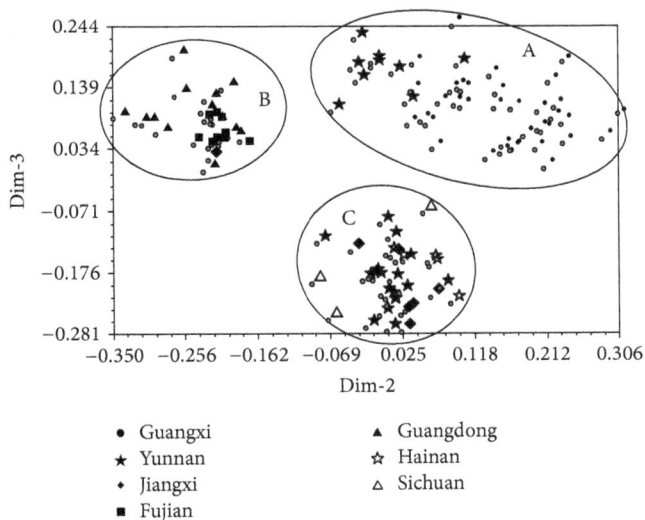

FIGURE 7: Principal components analysis (PCA) of *S. scitamineum* (numbers 1–100) based on the combined ISSR + SP-SRAP data set.

analysis were estimated to be 0.26 and 0.36, respectively. The genetic diversity attributable to differentiation among populations (G_{st}) was estimated to be 0.27, indicating that 27% of the total genetic variation was originated between populations, while the rate of gene flow (N_m) was estimated to be 1.33, relatively lower than that of SP-SRAP (1.78) but higher than that of ISSR (0.94).

3.6. Analysis of ITS Regions for Six Pairs of Isolates with Low Genetic Similarity between Pairs. A total of 12 isolates in six pairs (number 45 and number 98, number 10 and number 89, number 18 and number 85, number 44 and number 47, number 55 and number 92, and number 63 and number 83), having low genetic similarity with the remaining pairs of the 100 isolates were selectedfor further PCR analysis of their rDNA-ITS, including ITS1, ITS2, and 5.8S rDNA, using the primer pair ITS1 and ITS4 [29]. PCR amplification revealed a single 755 bp fragment present in all the 12 *S. scitamineum* isolates. The PCR products from each of these 12 isolates were sequenced. Sequence analysis showed that these isolates shared 94.09% to 96.98% similarity with an average of 96.34%, suggesting that rDNA-ITS regions contain lower sequence variations and may be not suitable for studying the genetic diversity of *S. scitamineum* in Mainland China.

4. Discussion and Conclusions

4.1. Confirmation of S. scitamineum Samples by ITS Sequence Analysis. Because of the limited genetic variability within the ITS region, ITS sequence data can normally define organisms at the genus level and resolve relationships between closely related species [21, 33]. In order to confirm the authenticity of each of the 100 tested *S. scitamineum* isolates, PCR amplifications of the ITS region were performed using the specific primer pair ITS1 and ITS4. As expected a single 755 bp fragment was obtained on each isolates. PCR amplifications with *S. scitamineum*—specific primer pair *b*E4 and *b*E8 yielded, as expected a single 420 bp fragment from all isolates.

These results showed that a single spore isolated from one smut whip can be used to generate a genetically fixed clone representing its original local population. Furthermore, sequence analysis of ITS regions of the six pairs of 12 isolates with lower genetic similarity indexes (from 0.4628 to 0.5627) showed very high sequence conservation between these isolates (sequence similarity ranged from 94.09% to 96.98%, with an average of 96.34%). This result confirms a previous observation by Singh et al. [6], that the sequence mutation rate in the ITS regions is very low among the smut pathogen strains.

4.2. Sugarcane Smut Race Classification by ISSR, SP-SRAP, and ISSR + SP-SRAP.
Several races of *S. scitamineum* are known to exist but the race type is still poorly understood [4, 7, 34–37]. At the 24th International Society of Sugar Cane Technologists Congress held in Brisbane, the members of the Pathology Section agreed that only two races (1 and 2) of the sugarcane smut pathogen exist in the countries investigated, except in Chinese Taiwan, where a third race was identified. Molecular analysis has so far yielded data of little value to the race classification of sugarcane smut. Researchers were unable to separate into races based on polymorphisms detected by AFLP analysis [4, 33]. Using RAPD and SRAP analysis or a combination of the two, Que et al. were unable to separate 23 *S. scitamineum* isolates into clusters based on smut race [7]. It demonstrated that the analysis of DNA from *S. scitamineum* from different geographic areas, by the RAPD, SRAP, and AFLP methods has so far yielded data of little value to the race classification of sugarcane smut.

Similar to the previous study [7], the results of this study showed no evidence that the smut could be divided into two or three races based on ISSR, SP-SRAP, and combined ISSR + SP-SRAP combined analysis. In the present study, smut isolate number 79 was classified as race 1 because it was isolated from sugarcane variety NCo310 which is susceptible to race 1 and resistant to race 2, respectively, based on the use of host differential for smut race identification. Isolates number 40 from YT86-368, number 80, and number 88 from YT89-113, and number 92 from YT96-86 can also be considered to be race 1 because they share a high genetic similarity to the isolate number 79. What should be stressed is that ROC16 is assumed to be smut resistant sugarcane variety in Guangxi Province. However, five smut isolates had been collected from ROC16 in Guangdong, Fujian and Jiangxi Provinces, indicating that its resistance to *S. scitamineum* requires further exploration.

4.3. Biogeographical Variation within the S. scitamineum Population Revealed by ISSR, SP-SRAP, and Combined ISSR + SP-SRAP Analysis.
Ma et al. studied 102 isolates of *S. reilianum* collected from six provinces of China [38]. They found that the genetic diversity of *S. reilianum* was associated essentially with the geographic origins of the pathogen instead of their host genotypes [38]. The similar results were also obtained for *S. commune* [39]. In previous studies, the biogeographical variation within the population of *S. scitamineum* was studied by RAPD, SRAP, SSR, and AFLP methods [4, 7, 9]. Que et al.

[7] found that the analyses of RAPD, SRAP, and combined RAPD-SRAP did not provide any evidence of the coevolution mechanism between *S. scitamineum* and its sugarcane host, while coevolutionary phenomena between pathogens and their hosts have been proposed in *Magnicellulatae* and *Podosphaera* [40, 41]. In the study of Que et al. [7], the analyses of RAPD, SRAP, and combined RAPD-SRAP on 23 isolates of smut pathogen suggested that the molecular evolution of *S. scitamineum* tends to be associated with their geographical origins, not to their host origins. The genetic diversity within a collection of 23 isolates of *S. scitamineum* from Mainland China was relatively high with a genetic dissimilarity index value ranging from 0.071 to 0.825 for RAPD analysis, from 0.045 to 0.947 for SRAP analysis, and from 0.026 to 0.838 for combined RAPD-SRAP analysis [7], while American and African *S. scitamineum* populations were extremely low in genetic diversity and all strains seem to belong to a single lineage according to a study of 142 single-teliospores using SSR markers [9]. In the present study, we used a larger number of strains (100) isolated from all the seven sugarcane cultivated provinces (20 districts) in Mainland China, of which 36 strains were isolated in Guangxi, the largest sugarcane-producing province in Mainland China. The average genetic similarity index estimated for the 100 isolates of *S. scitamineum* was 0.66 by ISSR and 0.62 by SP-SRAP. Our study showed that the genetic diversity among the Chinese smut pathogen population was relatively high and properly estimated by the two adopted methods with the SP-SRAP markers slightly more polymorphic than ISSR markers.

One of the aims of the present study was to estimate the genetic relationship between the subpopulations of *S. scitamineum* from different provinces of Mainland China and from different districts in Guangxi province which accounts for about 62% of the total sugarcane cultivated area in Mainland China. Also included in this study were 26 strains isolated from Yunnan province which is considered to be the plant genetic diversity center [42] because of the ecological diversity found in this province. The UPGMA cluster analyses based on the ISSR and SP-SRAP data sets revealed that the genetic diversity within the isolates from Guangxi, Fujian, and Guangdong was lower than those from Yunnan and Jiangxi (Figures 2 and 4). The isolates of Yunnan can be further divided into three to four subgroups by ISSR or by SP-SRAP analysis (Figures 2 and 4). Among all genetic diversity indexes adopted in this study, the highest values were observed for the isolates from Yunnan and the lowest values were observed for those from Hainan, indicating a higher level of genetic diversity in the subpopulation of Yunnan and a lower one in that of Hainan. This could be explained by the highly variable climatic and geographic factors of Yunnan province contrasting to those of Hainan. Our results in this study support the hypothesis that Yunnan province could be the biological genetic diversity center for *S. scitamineum*.

PCA analysis showed that the strains collected from the same region tend to be grouped together in the resulting plots (Figures 3, 5, and 7). In order to understand the genetic variability of the isolates isolated from one sugarcane genotype but collected from various geographical origins, we

Biogeographical Variation and Population Genetic Structure of Sporisorium scitamineum in Mainland China: Insights from ISSR and SP-SRAP Markers

83

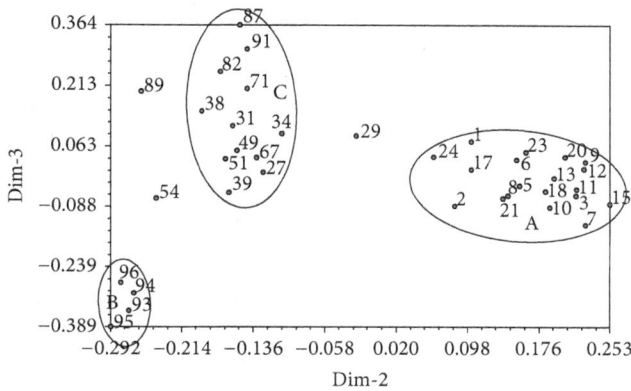

FIGURE 8: Principal components analysis of 38 *S. scitamineum* isolates from ROC22 based on the combined ISSR + SP-SRAP data set. Notes: A: 19 ↑ Guangxi (1, 2, 3, 5, 6, 7, 8, 9, 10, 11, 12, 13, 15, 17, 18, 20, 21, 23, 24); B: 4 ↑ Hainan (93, 94, 95, 96); C: 15 ↑ Guangxi (27, 29, 31, 34), Yunnan (38, 39, 49, 51, 54), Jiangxi (67), Fujian (71), Guangdong (29, 82, 87, 91).

performed a PCA analysis on a total of 100 strains, including the 38 isolates from the sugarcane variety ROC22, 10 from Liucheng03-182, and seven from ROC16 using ISSR and SP-SRAP data sets. The results showed that the 38 isolates from ROC22 were divided into three different groups, named as A, B, and C (Figure 8). Among these 38 isolates, 19 isolates collected from Guangxi were classified into A, four isolates from Hainan into the group B, 15 isolates (four from Guangxi, five from Yunnan, four from Guangdong, one from Fujian, and one from Jiangxi) into the group C. Similar results were also obtained for the strains isolated from the sugarcane varieties Liucheng03-182 and ROC16 (data not shown).

Our results revealed the presence of geographical subdivisions among the 100 *S. scitamineum* isolates, that is, the isolates of the smut pathogen collected from a same region tend to be clustered in the same group. Isolates from a same province were shown to share a high genetic similarity, while the isolates from different provinces were shown to be more genetically distant with lower genetic similarity values among them (Figures 2, 4, and 6; Table 4).

The highest genetic diversity was found within the subpopulation of Yunnan, in contrast to the lower diversity found among the isolates from Guangxi and Hainan. From the above results, we observed that the geographical origins contributed more to the clustering result than the host origins or sugarcane genotypes. This finding of geography-specific population of *S. scitamineum* is in agreement with the results of study on *S. commune* [39]. In recent years, the sugarcane varieties and seed-canes have been exchanged frequently without any quarantine and limitation between different provinces or regions in Mainland China and even introduced from Taiwan into Mainland China in large amount without quarantine for the past ten years, including the introduction of ROC22. The introduction of *S. scitamineum* together with the varieties or seed-canes may have increased the genetic variability of the pathogen into certain regions. Prevention and control of smut may be more difficult if the newly

introduced strain may infect varieties resistant to endemic strains in the region.

4.4. Population Genetic Analysis Advances Our Understanding of the Genetic Structure of S. scitamineum Populations.

Population genetic analysis, which can provide strong evidence about genetic diversity at the level of population or subpopulation, should be helpful for our understanding of the genetic structure of the studied populations [9, 31, 32, 43, 44]. Based on our ISSR, SP-SRAP, and ISSR + SP-SRAP data sets, the total gene diversity (H_t) within the population of 100 isolates of smut pathogen was estimated to be 0.34, 0.22, and 0.26, respectively; the gene diversity between subpopulations (H_s) was estimated to be 0.38, 0.29, and 0.36, respectively; the gene diversity attributable to differentiation among populations (G_{st}) was estimated to be 0.35, 0.22, and 0.27, respectively; the gene flow (Nm) was estimated to be 0.94, 1.78, and 1.33, respectively. In each case, the estimated values of H_t and Nm are higher for SP-SRAP than for ISSR and ISSR + SP-SRAP, and the highest value of H_s was found for ISSR + SP-SRAP data set, while the highest value of G_{st} was found for ISSR data set. Therefore, the combined ISSR + SP-SRAP analysis should be more appropriate for the analysis of gene diversity between subpopulations (H_s) and for the analysis of gene flow (Nm), SP-SRAP works better than ISSR and ISSR + SP-SRAP data set.

The gene flow parameter Nm, which means the product of the effective population number N and the rate of migration among populations, reflects the degree of genetic differentiation within populations. Methods for estimating how much gene flow occurs in natural populations can be divided into two categories; one is direct methods which involve estimating dispersal distances and reproductive success of individuals that disperse by direct observation [45]; the other one is indirect methods which rely on allele frequencies or differences in DNA sequences to estimate levels of gene flow that must have occurred in order to explain observed patterns [46]. In the present study, 100 strains can be assorted into seven subpopulations (Figures 2, 4, and 6), of which the distribution has a certain degree of discontinuity, which means that all these subpopulations should undertake natural genetic differentiation to some extent due to their various habitats. What should also be stressed is that the isolates in these subpopulations did not display any distinct phenotypic properties. Comparing the relative amounts of gene flow taking place among populations is the first step towards predicting the population structure of *S. scitamineum*. According to the value of proportion of gene diversity attributable to differentiation among populations (G_{st}), the values of gene flow parameter Nm were estimated by indirect methods to be 0.94, 1.78, and 1.33, respectively, for ISSR, SP-SRAP, and combined ISSR + SP-SRAP analysis. All three analyses showed that there does exist significant gene flow between the seven *S. scitamineum* subpopulations in Mainland China. In addition to environmental factors that affect patterns of gene flow at the macrogeographic and regional scales, population structuring within ecological zones and habitats can potentially ruin efforts to drive genes into wild *S. scitamineum* populations [9, 32, 43, 44]. As there is always considerable variation within populations and among populations of sugarcane smut fungus, environmental heterogeneity may play an important role in these population variations [9, 47]. It can be inferred from above that the population genetic differentiation of sugarcane smut fungus depends at a large extent on the heterogeneity of habitats and is the result of its long-term adaptation to the ecological environment. Our conclusion is in agreement with the results of previous studies by Raboin et al. [9]and Zheng et al. [47].

Conflict of Interests

The authors declare that there is no conflict of interests regarding the publication of this paper.

Acknowledgments

This work was supported by National Natural Science Foundation of China (31340060), the earmarked fund for the Modern Agriculture Technology of China (CARS-20), the 948 Program on the Introduction of International Advanced Agricultural Science and Technique of Department of Agriculture (2014-S18), Research Funds for Distinguished Young Scientists in Fujian Provincial Department of Education (K80MKT04A), and Research Funds for Distinguished Young Scientists in Fujian Agriculture and Forestry University (xjq201202).

References

[1] R. K. Chen, L. P. Xu, Y. Q. Lin et al., *Modern Sugarcane Genetic Breeding*, China Agriculture Press, Beijing, China, 2011.

[2] J. W. Hoy, C. A. Hollier, D. B. Fontenot, and L. B. Grelen, "Incidence of sugarcane smut in Louisiana and its effects on yield," *Plant Disease*, vol. 70, no. 11, pp. 59–60, 1986.

[3] L. P. Xu and R. K. Chen, "Current status and prospects of smut and smut resistance breeding in sugarcane," *Fujian Journa of Agricultural Science*, vol. 15, no. 2, pp. 26–31, 2000.

[4] K. S. Braithwaite, G. Bakkeren, B. J. Croft, and S. M. Brumbley, "Genetic variation in a worldwide collection of the sugarcane smut fungus *Ustilago scitaminea*," in *Proceedings of the Australian Society of Sugar Cane Technology*, vol. 26, pp. 48–56, 2004.

[5] N. Singh, B. M. Somai, and D. Pillay, "Smut disease assessment by PCR and microscopy in inoculated tissue cultured sugarcane cultivars," *Plant Science*, vol. 167, no. 5, pp. 987–994, 2004.

[6] N. Singh, B. M. Somai, and D. Pillay, "Molecular profiling demonstrates limited diversity amongst geographically separate strains of *Ustilago scitaminea*," *FEMS Microbiology Letters*, vol. 247, no. 1, pp. 7–15, 2005.

[7] Y. X. Que, L. P. Xu, J. W. Lin, R. K. Chen, and M. P. Grisham, "Molecular variation of Sporisorium scitamineum in Mainland China revealed by RAPD and SRAP markers," *Plant Disease*, vol. 96, no. 10, pp. 1519–1525, 2012.

[8] W. J. Lu, W. F. Li, and Y. K. Huang, "Research advances on sugar cane Smut disease occurrence and control," *Sugar Crops China*, no. 3, pp. 64–66, 2008.

[9] L.-M. Raboin, A. Selvi, K. M. Oliveira et al., "Evidence for the dispersal of a unique lineage from Asia to America and Africa in the sugarcane fungal pathogen *Ustilago scitaminea*," *Fungal Genetics and Biology*, vol. 44, no. 1, pp. 64–76, 2007.

[10] L. N. Thokoane and R. S. Rutherford, "cDNA-AFLP differential display of sugarcane (*Saccharum* spp, hybrids) genes induced by challenge with the fungal pathogen *Ustilago scitaminea* (sugarcane smut)," in *Proceedings of the South African Sugar Technologists Association*, vol. 75, pp. 104–107, 2001.

[11] O. Borrás-Hidalgo, B. P. H. J. Thomma, E. Carmona et al., "Identification of sugarcane genes induced in disease-resistant somaclones upon inoculation with *Ustilago scitaminea* or *Bipolaris sacchari*," *Plant Physiology and Biochemistry*, vol. 43, no. 12, pp. 1115–1121, 2005.

[12] M. Lao, A. D. Arencibia, E. R. Carmona et al., "Differential expression analysis by cDNA-AFLP of *Saccharum* spp. after inoculation with the host pathogen *Sporisorium scitamineum*," *Plant Cell Reports*, vol. 27, no. 6, pp. 1103–1111, 2008.

[13] Y.-X. Que, J.-W. Lin, X.-X. Song, L.-P. Xu, and R.-K. Chen, "Differential gene expression in sugarcane in response to challenge by fungal pathogen *Ustilago scitaminea* revealed by cDNA-AFLP," *Journal of Biomedicine & Biotechnology*, vol. 2011, Article ID 160934, 10 pages, 2011.

[14] Y.-X. Que, Z.-X. Yang, L.-P. Xu, and R.-K. Chen, "Isolation and identification of differentially expressed genes in sugarcane infected by *Ustilago scitaminea*," *Acta Agronomica Sinica*, vol. 35, no. 3, pp. 452–458, 2009.

[15] Y. X. Que, L. P. Xu, W. Lin et al., "Application of *E. arundinaceus* cDNA microarray in the study of differentially expressed genes induced by *U. scitaminea*," *Acta Agronomica Sinica*, vol. 35, no. 5, pp. 940–945, 2009.

[16] Q. B. Wu, L. P. Xu, J. L. Guo, Y. C. Su, and Y. X. Que, "Transcriptome profile analysis of sugarcane responses to *Sporisorium scitaminea* infection using Solexa sequencing technology," *BioMed Research International*, vol. 2013, Article ID 298920, 9 pages, 2013.

[17] Y. Que, L. Xu, J. Lin, M. Ruan, M. Zhang, and R. Chen, "Differential protein expression in sugarcane during sugarcane-*Sporisorium scitamineum* interaction revealed by 2-DE and MALDI-TOF-TOF/MS," *Comparative and Functional Genomics*, vol. 2011, Article ID 989016, 10 pages, 2011.

[18] H. H. Albert and S. Schenck, "PCR amplification from a homolog of the bE mating-type gene as a sensitive assay for the presence of *Ustilago scitaminea* DNA," *Plant Disease*, vol. 80, no. 10, pp. 1189–1192, 1996.

[19] Y. C. Su, S. S. Wang, J. L. Guo, B. T. Xue, L. P. Xu, and Y. X. Que, "A TaqMan Real-Time PCR assay for detection and quantification of *Sporisorium scitamineum* in sugarcane," *The Scientific World Journal*, vol. 2013, Article ID 942682, 9 pages, 2013.

[20] Y. X. Que, L. P. Xu, T. T. Zou, and R. K. Chen, "Primary analysis of molecular diversity in populations of the fungus *Ustilago scitaminea* Syd.," *Journal of Agricultural Biotechnology*, vol. 12, no. 6, pp. 685–689, 2004.

[21] M. Stoll, M. Piepenbring, D. Begerow, and F. Oberwinkler, "Molecular phylogeny of *Ustilago* and *Sporisorium* species (*Basidiomycota*, *Ustilaginales*) based on internal transcribed spacer (ITS) sequences," *Canadian Journal of Botany*, vol. 81, no. 9, pp. 976–984, 2003.

[22] M. Stoll, D. Begerow, and F. Oberwinkler, "Molecular phylogeny of *Ustilago*, *Sporisorium*, and related taxa based on combined analyses of rDNA sequences," *Mycological Research*, vol. 109, no. 3, pp. 342–356, 2005.

Biogeographical Variation and Population Genetic Structure of Sporisorium scitamineum in Mainland China:
Insights from ISSR and SP-SRAP Markers

85

[23] D. Liu, X. He, G. Liu, and B. Huang, "Genetic diversity and phylogenetic relationship of Tadehagi in southwest China evaluated by inter-simple sequence repeat (ISSR)," *Genetic Resources and Crop Evolution*, vol. 58, no. 5, pp. 679–688, 2011.

[24] Y. Wang, N. S. Zhuang, H. Q. Gao, and D. Y. Huang, "ISSR analysis for sugarcane germplasm," *Journal of Hunan Agricultural University*, vol. 33, pp. 176–182, 2007.

[25] M. Staats, P. van Baarlen, and J. A. L. van Kan, "Molecular phylogeny of the plant pathogenic genus Botrytis and the evolution of host specificity," *Molecular Biology and Evolution*, vol. 22, no. 2, pp. 333–346, 2005.

[26] M. Staats, P. van Baarlen, and J. A. L. van Kan, "AFLP analysis of genetic diversity in populations of *Botrytis elliptica* and *Botrytis tulipae* from the Netherlands," *European Journal of Plant Pathology*, vol. 117, no. 3, pp. 219–235, 2007.

[27] G. Li and C. F. Quiros, "Sequence-related amplified polymorphism (SRAP), a new marker system based on a simple PCR reaction: its application to mapping and gene tagging in *Brassica*," *Theoretical and Applied Genetics*, vol. 103, no. 2-3, pp. 455–461, 2001.

[28] Z. C. Yang, G. R. Xiong, X. T. Chen, S. Z. Zhang, and B. P. Yang, "Isolation and identification of *Ustilago scitaminea* in Hainan province," *Chinese Journal of Tropical Crops*, vol. 31, no. 5, pp. 828–833, 2010.

[29] T. J. White, T. Bruns, S. Lee, and J. Taylor, "Amplifcation and direct sequencing of fungal ribosomal RNA genes for phylogenetics," in *PCR Protocols, a Guide to Methods and Applications*, New York Academic Press, New York, NY, USA, 1990.

[30] F. J. Rohlf, *NTSYS-pc: Numerical Taxonomy and Multivariate Analysis System Version 2. 0*, Department of Ecology and Evolution, State University of New York, 1998.

[31] J. M. McDermott and B. A. McDonald, "Gene flow in plant pathosystems," *Annual Review of Phytopathology*, vol. 31, pp. 353–373, 1993.

[32] R. M. Hong, Y. S. Wang, and J. H. Yang, "Analysis of phenotypic and genetic variation in clonal population of *Leymus chinensis* and others," *Journal of Northeast Agricultural University*, vol. 32, no. 4, pp. 376–395, 2001.

[33] G. Bakkeren, J. W. Kronstad, and G. André Lévesque, "Comparison of AFLP fingerprints and ITS sequences as phylogenetic markers in *Ustilaginomycetes*," *Mycologia*, vol. 92, no. 3, pp. 510–521, 2000.

[34] B. B. Mundkur, "Taxonomy of the sugar-cane smuts," *Kew Bulletin*, vol. 10, pp. 525–533, 1939.

[35] L. S. Leu and W. S. Teng, "Pathogenic strains of *Ustilago scitaminea* Sydow," *Sugarcane Pathologists' Newsletter*, no. 8, pp. 12–13, 1972.

[36] K. K. Wu, D. J. Heinz, and D. M. Hogarth, "Association and heritability of sugarcane smut resistance to races A and B in Hawaii," *Theoretical and Applied Genetics*, vol. 75, no. 5, pp. 754–760, 1988.

[37] S. A. Ferreira and J. C. Comstock, "Smut," in *Diseases of Sugarcane-Major Disease*, C. Ricaud, B. T. Egan, A. G. Gillaspie, and C. G. Hunghes, Eds., pp. 211–229, Elsevier, Amsterdam, The Netherlands, 1989.

[38] J. T. Ma, D. X. Wang, X. J. Gong, and J. Gao, "Analysis on genetic diversity of *Sporisorium reilianum* from 6 provinces and regions in China," *Journal of Maize Sciences*, vol. 16, no. 6, pp. 139–143, 2008.

[39] T. Y. James, J. M. Moncalvo, S. Li, and R. Vilgalys, "Polymorphism at the ribosomal DNA spacers and its relation to breeding structure of the widespread mushroom *Schizophyllum commune*," *Genetics*, vol. 157, no. 1, pp. 149–161, 2001.

[40] T. Hirata, J. H. Cunnington, U. Paksiri et al., "Evolutionary analysis of subsection *Magnicellulatae* of *Podosphaera* section *Sphaerotheca* (*Erysiphales*) based on the rDNa internal transcribed spacer sequences with special reference to host plants," *Canadian Journal of Botany*, vol. 78, no. 12, pp. 1521–1530, 2000.

[41] S. Takamatsu, S. Niinomi, M. Harada, and M. Havrylenko, "Molecular phylogenetic analyses reveal a close evolutionary relationship between *Podosphaera* (*Erysiphales*: *Erysiphaceae*) and its rosaceous hosts," *Persoonia*, vol. 24, pp. 38–48, 2010.

[42] H. Zhang, J. Sun, M. Wang et al., "Genetic structure and phylogeography of rice landraces in Yunnan, China, revealed by SSR," *Genome*, vol. 50, no. 1, pp. 72–83, 2007.

[43] S. I. Wright and B. S. Gaut, "Molecular population genetics and the search for adaptive evolution in plants," *Molecular Biology and Evolution*, vol. 22, no. 3, pp. 506–519, 2005.

[44] K. R. Merrill, S. E. Meyer, and C. E. Coleman, "Population genetic analysis of *Bromus tectorum* (*Poaceae*) indicates recent range expansion may be facilitated by specialist genotypes," *American Journal of Botany*, vol. 99, no. 3, pp. 529–537, 2012.

[45] M. C. Whitlock and D. E. Mccauley, "Indirect measures of gene flow and migration: $F(ST) \neq 1/(4Nm + 1)$," *Heredity*, vol. 82, no. 2, pp. 117–125, 1999.

[46] S. Wright, "Evolution in Mendelian populations," *Genetics*, vol. 16, no. 2, pp. 97–159, 1931.

[47] H. M. Zheng, T. M. Hu, and Q. Z. Wang, "Research of genetic diversity in seven Kobresia by AFLP Tibet plateau of China," *Scientia Agricultura Sinica*, vol. 41, no. 9, pp. 2820–2825, 2008.

10

YgaE Regulates Out Membrane Proteins in *Salmonella enterica* Serovar Typhi under Hyperosmotic Stress

Min Wang,[1] Ping Feng,[1] Xun Chen,[2] Haifang Zhang,[3] Bin Ni,[3] Xiaofang Xie,[1] and Hong Du[1]

[1] *Clinical Laboratory, The Second Affiliated Hospital of Soochow University, Suzhou 215004, China*
[2] *Clinical Laboratory Center, Xiyuan Hospital, China Academy of Chinese Medical Sciences, Beijing 100091, China*
[3] *Department of Biochemistry and Molecular Biology, School of Medical Technology, Jiangsu University, Zhenjiang 212013, China*

Correspondence should be addressed to Hong Du; hong_du@126.com

Academic Editors: R. Tofalo and J. Yoon

Salmonella enterica serovar Typhi (*S.* Typhi) is a human-specific pathogen that causes typhoid fever. In this study, we constructed Δ*ygaE* mutant and a microarray was performed to investigate the role of *ygaE* in regulation of gene expression changes in response to hyperosmotic stress in *S.* Typhi. qRT-PCR was performed to validate the microarray results. Our data indicated that *ygaE* was the repressor of *gab* operon in *S.* Typhi as in *Escherichia coli* (*E. coli*), though the sequence of *ygaE* is totally different from *gabC* (formerly *ygaE*) in *E. coli*. OmpF, OmpC, and OmpA are the most abundant out membrane proteins in *S.* Typhi. Here we report that YgaE is a repressor of both OmpF and OmpC at the early stage of hyperosmotic stress. Two-dimensional electrophoresis was applied to analyze proteomics of total proteins in wild-type strain and Δ*ygaE* strain and we found that YgaE represses the expression of OmpA at the late stage of hyperosmotic stress. Altogether, our results implied that YgaE regulates out membrane proteins in a time-dependent manner under hyperosmotic stress in *S.* Typhi.

1. Introduction

S. Typhi is a human-specific pathogen, which produces typhoid fever. Once ingested through contaminated water or food, *S.* Typhi invades intestinal epithelial cells and can enter the host bloodstream and disseminate to deep organs. Lacking of standard water supply and sanitation, typhoid remains a major health problem in developing world [1–3]. In contaminated water or food the external osmolarity is in the order of 50 mM NaCl; however, in the lumen of the host intestine *S.* Typhi cells are exposed to a significant increase in osmolarity: 300 mM NaCl in the small intestine [4]. Considering the significant morbidity and mortality associated with this disease [5], it is important to understand the gene regulation mechanisms in *S.* Typhi in response to hyperosmotic environments.

The outer membrane (OM) of Gram-negative bacteria constitutes the first permeability barrier that protects the cells against environmental stresses including chemical and biological attacks [6]. Simultaneously, it allows the selective uptake of essential nutrients and the secretion of metabolic waste products. The OM is a sophisticated organization of lipid and protein components. The outer membrane proteins (OMPs), called porins, are characterized by a β-barrel structure and form water-filled channels for the passage of a large variety of hydrophilic molecules [7–9].

In *E. coli*, the two general porins OmpF and OmpC are among the most abundant OMPs (about 10^5 copies per cell) and serve as general pathways for the influx of small molecules (e.g., molecular weight under 600). They consist of three 16-stranded β-barrels, each of which forms a channel that is restricted in the middle due to the inward folding of a loop (loop L3) [10]. The expression of *E. coli* porins has been extensively studied. The OmpF-OmpC balance is highly regulated by different genetic control systems. Changes in osmolarity profoundly affect expression of OmpF and OmpC. OmpC is preferentially expressed in high osmolarity, whereas OmpF expression is favored in low osmolarity [11]. Other factors, including local anesthetics [12], pH [13], and nutrition limitation [14] also influence *ompF* and *ompC* transcription in an EnvZ/OmpR-dependent manner. Noteworthy, growth conditions where nutrient levels are high, such as in mammal intestinal tracts, favor the expression of OmpC, which has a

TABLE 1: Strains and plasmids used in this study.

Strains or plasmid	Relevant characteristics	Source
Strains		
Salmonella enterica serovar Typhi GIFU10007	Wild-type strain; z66$^+$	Gifu University [22]
Δ*ygaE*	GIFU10007; z66$^+$	This study
Δ*ygaE* (pBAD)	Δ*ygaE* harboring pBAD plasmid;	This study
Δ*ygaE* (pBAD*ygaE*)	Δ*ygaE* harboring pBAD-*ygaE* recombinant plasmid;	This study
E. coli SY372λpir	*E. coli* host strain of suicide plasmid	Gifu University [23]
Plasmids		
pGMB151	Suicide plasmid; sacB; Amp^r	Gifu University [23]
pBAD/gIII	Expression vector; Amp^r	Invitrogen
pBAD*ygaE*	pBAD/gIII containing *ygaE* gene	This study

TABLE 2: Primers used in this study.

Name	Sequence
ygaE-F1A(+BamH I)	5′-TTGGATCCGGTACTGTCCCCATTATGT
ygaE-F1B(+Sal I)	5′-CCGTCGACATAGCCTTTTTGATTCACC
ygaE-F2A(+Sal I)	5′-TAGTCGACGAGATGCTGGAAGATAAAC
ygaE-F2B(+BamH I)	5′-CTGGATCCTCTGCTACACTTTTCTTTG
ygaE-PA(+Nco I)	5′-AACCATGGAGATGACCGCCCTTTCCCAA
ygaE-PB(+Sal I)	5′-AAGTCGACCTACATTTTTCCTGCCAT
P-*ompF*-A	5′-GGA ATACCGTACTAA AGCA
P-*ompF*-B	5′-GATACTGGATACCGA AAGA
P-*ompC*-A	5′-ATCAGA ACAACACCGCTAA
P-*ompC*-A	5′-GTTGCTGATGTCCTTACC

smaller channel than OmpF, thus limiting the influx of large and charged molecules such as bile salts and antibiotics. Conversely, OmpF will be the major porin under ex vivo growth conditions with nutritional deficiency, as its larger pore will allow efficient influx of nutrients [15]. In contrast, expression of OmpC in S. Typhi is not influenced by osmolarity, while OmpF is regulated as in *E. coli* [16].

OmpA is another abundant OMP. It is monomeric, and it is unusual in that it can exist in two different conformations [17]. A minor form of the protein, with an unknown number of transmembrane strands, can function as a porin, but the major, nonporin form has only eight transmembrane strands, and the periplasmic domain of this form performs a largely structural role [18]. The function of OmpA is thought to contribute to the structural integrity of the outer membrane along with murein lipoprotein [19] and peptidoglycan associated lipoprotein [20].

In *E. coli*, *gabC* (formerly *ygaE*) was reported to belong to the *gabDTPC* operon, and *gabC* is the repressor of the operon [21]. The function of *gab* operon is mainly revolved in γ-aminobutyrate (GABA) catabolism but does not contribute to the catabolism of any other nitrogen source [21]. However, the function of *ygaE* in S. Typhi has not been extensively studied. In this work, we showed that *ygaE* regulates out membrane proteins OmpF/OmpC at the early stage of hyperosmotic

stress and OmpA at the late stage of hyperosmotic stress in S. Typhi.

2. Materials and Methods

2.1. Bacterial Strains and Conditions of Culture. The bacterial strains and plasmids used in this study are listed in Table 1. Bacteria were grown in LB broth at 37°C with shaking (250 rpm). As for antibiotic sensitivity assay, Muller Hinton agar was used. For low environmental osmolarity, the growth medium contained a final concentration of 50 mM NaCl. For hyperosmotic environment, Nacl was added into the medium at a final concentration of 300 mM. The complemented strains were induced by L-arabinose (0.2% wt/vol). When appropriate, ampicillin was added to the medium at a final concentration of 100 μg/mL.

2.2. Construction of the ygaE Deletion Mutant Strain. The *ygaE* deletion mutant (Δ*ygaE*) was prepared by homologous recombination according to a previously described method [24] with a *ygaE*-deletion suicide plasmid lacking 327 bp of the *ygaE* gene. The specific primers used for deletion of *ygaE* were listed in Table 2. The mutant strain was selected by PCR as described previously [23], verified by sequencing, and designated Δ*ygaE*.

2.3. Complement of ygaE in the ΔygaE Mutant Strain. Specific primers *ygaE*-PA and *ygaE*-PB (Table 2) were designed to amplify a *ygaE* promoterless DNA fragment with *pfu* DNA polymerase (Takara). Nco I and Sal I sites were added to the 5′-termini of primers *ygaE*-PA and *ygaE*-PB, respectively. The amplicon was digested by Nco I and Sal I and inserted into the expression vector pBAD/gIII (Invitrogen), which was predigested with the same restriction enzymes, to form the recombinant plasmid pBAD*ygaE*. The positive plasmid was verified by sequence analysis. Δ*ygaE* was transformed with pBAD*ygaE* and designated as Δ*ygaE* (pBAD*ygaE*). As control, Δ*ygaE* was transformed with pBAD and designated as Δ*ygaE* (pBAD). Expression of *ygaE* in Δ*ygaE* (pBAD*ygaE*) was induced by L-arabinose (0.2% wt/vol).

2.4. RNA Extraction and Transcriptional Profiling by Genomic DNA. Wild-type and Δ*ygaE* strains were cultured overnight at 37°C with shaking (250 rpm) in LB broth (with final concentration of 50 mM). After dilution into fresh medium, cultures were incubated to exponential growth (OD 0.5 at 600 nm). To induce hyperosmotic stress, NaCl was added to a final concentration of 300 mM and bacteria were incubated with shaking for a further 30 min at 37°C. Bacteria were collected by centrifugation and total RNA was extracted using an RNeasy kit (minicolumn, Qiagen, Germany) according to the manufacturer's instructions. The quality and quantity of the extracted RNA were determined by agarose gel electrophoresis and analysis with a ND-1000 spectrophotometer (NanoDrop Technologies). Extracted RNA was treated with 1 U of RNase-free DNase I (TaKaRa) at 37°C for 10 min to remove traces of DNA and then incubated at 85°C for 15 min to inactivate the DNase. cDNA probes were synthesized using 20 μg of RNA. A genomic DNA microarray designed for *S.* Typhi was used in this study and fluorescence labeling of cDNA probes, hybridization, and microarray scanning were performed as described previously [25].

2.5. Data Analysis. GENEPIX PRO 6.0 (Molecular Devices) was used for signal quantification. The densitometric values of the spots with DNA sequences representing open reading frames (ORFs) were normalized to the average overall intensity of the slide in global normalization mode. Data were exported into an Excel (Microsoft Corporation) spreadsheet for subsequent analysis as described previously [25] with minor modifications. In brief, the two-channel fluorescent intensity ratios were calculated for each individual spot on each slide; the average intensity ratio of the same gene from different slides was taken as the mean change in gene expression level. This was expressed as log2 (ratio) and entered as one data point in the gene expression profile plot view. Only genes that displayed at least eight valid values in 12 replicate analyses were subject to further analysis.

2.6. Quantitative Real-Time RT-PCR (qRT-PCR) Assay. Total RNA extracted after 30 min of hyperosmotic stress as above was subjected to qRT-PCR as described previously [26]. The PCR primers used for qRT-PCR are listed in Table 2. Each experiment was performed with three RNA samples

from three independent experiments. Student's *t*-test was used for the statistical analysis. Differences were considered statistically significant when P was <0.05 in all cases.

2.7. Measurement of Bacterial Growth. Wild-type and *ygaE* mutant were overnight cultured; then 200 μL aliquots of the culture were diluted to 20 mL fresh LB medium (with NaCl concentration of 300 mM) and incubated at 37°C with shaking (250 rpm). The growth was measured every two hours using a BioPhotometer (Eppendorf). The measurement was performed three times. Student's *t*-test was used for the statistical analysis. Differences were considered statistically significant when P was <0.05.

2.8. Measurement of Antibiotic Susceptibility. The antibiotic susceptibility testing was done by using the modified Kirby-Bauer disk diffusion method on Muller Hinton agar (OXOID) with a final NaCl concentration of 300 mM. The antibiotic disks which were used in this study were cefotaxime (CTX), ampicillin (AMP), piperacillin (PRL), ceftazidime (CAZ), compound sulfamethoxazole (SXT), and chloramphenicol (C). The zone size around each antimicrobial disk was measured. The experiment was performed three times. Student's *t*-test was used for the statistical analysis. Differences were considered statistically significant when P was <0.05.

2.9. Protein Extraction. Wild-type and Δ*ygaE* strains were cultured overnight at 37°C with shaking (250 rpm) in LB broth (with final concentration of 50 mM). After dilution into fresh medium, cultures were incubated to exponential growth (OD 0.5 at 600 nm). To induce hyperosmotic stress, NaCl was added to a final concentration of 300 mM and bacteria were incubated with shaking for a further 120 min at 37°C. Bacteria were collected by centrifugation. The cell pellets were washed twice with ice-cold PBS, resuspended in PBS, and sonicated for 10 sec with a Sonoplus sonicator (Bandelin electronic, Germany). The cells were collected by centrifugation at 5,000 g for 20 min. The resulting cell pellet was resuspended in sample lysis solution, which was composed of 7 M urea, 2 M thiourea containing 4% (w/v) 3-[(3-cholamidopropyl) dimethylammonio] -1-propanesulfonate (CHAPS), 1% (w/v) dithiothreitol (DTT) 2% (v/v) pharmalyte, and 1 mM benzamidine. Proteins were extracted for 1 h at room temperature with vortexing. After centrifugation at 15,000 g for 1h at 15°C, the insoluble material was discarded, and the soluble fraction was harvested and used for 2-DE.

2.10. Two-Dimensional Electrophoresis (2-DE). The total proteins were dissolved in IPG rehydration/sample buffer (8 M urea, 2% CHAPS, 50 mM DTT, 0.2% Bio-Lyte 4/7 ampholyte, 0.001% Bromophenol Blue; Bio-Rad) and centrifuged at 12,000 g for 15 min at room temperature to remove nondissolved materials. The protein content was determined using the PlusOne 2D Quant Kit (Amersham Pharmacia Biotech). A 7 cm Immobiline DryStrip (IPG, Immobilized pH Gradient, pH range 4–7; Bio-Rad) was rehydrated at 50 V for 12 h, in IPG rehydration/sample buffer containing 150 mg of the protein sample in a total volume of 125 mL. Isoelectric

TABLE 3: Gene expression changes in $\Delta ygaE$ under hyperosmotic stress discussed in this study.

Gene name	Description of gene product	$\log_2(\Delta ygaE/\mathrm{WT})$
gab operon genes		
ygaF	Putative GAB DTP gene cluster repressor	1.68
gabD	Succinate-semialdehyde dehydrogenase	1.44
gabT	4-Aminobutyrate aminotransferase	1.65
ygaE	Putative transcriptional regulator	−3.62
Outer membrane protein genes		
ompC	Out membrane protein C	1.76
ompF	Outer membrane protein F precursor	1.37

focusing was performed using a Bio-Rad PROTEAN IEF cell (Bio-Rad) and focusing was conducted by stepwise increase of the voltage as follows: 250 V for 0.5 h, 500 V for 0.5 h, 4000 V for 3 h, and 4000 V until 25,000 Vh. The temperature was maintained at 20°C. After IEF separation, each IPG strip was washed in 3 mL of equilibration buffer 1 (75 mMTris—HCl [pH 8.8], 6 Murea, 2% SDS, 29.3% [v/v] glycerol, 1% DTT) for 15 min and in 3 mL of equilibration buffer 2 (75 mM Tris—HCl [pH 8.8], 6 M urea, 2% SDS, 29.3% [v/v] glycerol, 2.5% iodoacetamide) for an additional 15 min·IPG strips were then placed over a 12% resolving polyacrylamide gel and electrophoresis was performed in two steps at 10°C: 15 mA/gel for 30 min and 30 mA/gel until the tracking dye reached the bottom of the gels. All gels were stained with colloidal Coomassie Brilliant Blue G-250 (CBB). Gel evaluation and data analysis were carried out using the PDQuest v 7.3 program (Bio-Rad). Three replicates were run for the sample.

2.11. Mass Spectrometry Analysis of Protein Spots and Database Searches. Spots unique to both strains were excised from the 2-DE gels and sent to Shanghai GeneCore BioTechnologies Co., Ltd for tryptic in-gel digestion, MALDI-TOF-MS, and MALDI-TOF/TOF-MS Data from MALDI-TOF-MS and MALDI-TOF/TOF-MS acquisitions were used in a combined search against the NCBInr protein database using MASCOT (Matrix Science) with the parameter sets of trypsin digestion, one max missed cleavages, variable modification of oxidation (M), and peptide mass tolerance for monoisotopic data of 100 ppm. Originally, the MASCOT server was used against the NCBI for peptide mass fingerprinting (PMF). The criteria used to accept protein identifications were based on PMF data, including the extent of sequence coverage, number of peptides matched, and score of probability. Protein identification was assigned when the following criteria were met: at least four matching peptides and sequence coverage greater than 15% [27, 28]. The identification of protein spots with a lower Mascot Score required further confirmation by MALDI-TOF/TOF-MS.

3. Results and Discussion

3.1. YgaE Represses the Expression of gab Operon under Hyperosmotic Stress. In our previous work, we investigated the global transcriptional profiles of S. Typhi $\Delta rpoE$, $\Delta rpoS$, and $\Delta rpoE/\Delta rpoS$ strains after 30 min of hyperosmotic stress

by *Salmonella* genomic DNA microarray. The results of microarray indicated that the expression level of *ygaE* is dramatically reduced in $\Delta rpoE/\Delta rpoS$ strain [29], while no apparent downregulation is observed in either $\Delta rpoE$ or $\Delta rpoS$ strain (data not shown). We speculated that *ygaE*, coregulated by RpoE and RpoS, is required for survival under extreme stresses of S. Typhi. To investigate the role of *ygaE* in the regulation of gene expression changes in response to hyperosmotic stress in S. Typhi, the *ygaE* mutant was constructed by homologous recombination mediated by suicide plasmid. Then, a genomic DNA microarray was performed to analyze the global transcriptional profiles of wild-type and $\Delta ygaE$ strains after 30 min of exposure to hyperosmotic stress.

The microarray results exhibited that, compared to wild-type strain, the expression, of *ygaF*, *gabD*, and *gabT* were obviously upregulated in $\Delta ygaE$ strain (Table 3) after exposure to hyperosmotic stress 30 min, which indicated that *ygaE* is the repressor of these genes at the early stage of hyperosmotic stress.

In *E. coli*, it was reported that there is an operon structure for the *gab* genes and that four genes form the *gabDTPC* operon [21]. The evidence that the first three genes are members of the *gab* operon is unambiguous [21]. Though the evidence that *gabC* (*ygaE*) is also a member of this operon is reasonably convincing, there still exist disputes. Firstly, it is unusual for a repressor to be encoded within the operon that it regulates. Next, a four-gene *gab* operon transcript was failed to be found [21]. *ygaF*, the gene preceding *gabD*, is not included in the *gab* operon in *E.coli* for several strong evidence [21].

In *E. coli*, *gabT* codes for a GABA transaminase that generates succinic semialdehyde. *gabD* specifies an NADP-dependent succinic semialdehyde dehydrogenase, which oxidizes succinic semialdehyde to succinate [30]. GabC does not obviously respond to a specific inducer. GabC is in the FadR subfamily of the GntR family of transcriptional regulators [31]. In *S.* Typhi, *gabD* encodes for a succinate-semialdehyde dehydrogenase, *gabT* encodes for a 4-aminobutyrate aminotransferase, and *ygaF* is a putative GAB DTP gene cluster repressor (Table 3). We compared the sequences of *gab* operon of *S.* Typhi to that of *E. coli* and found they are about 80% homologous. However, despite the same regulation pattern to *gab* operon, the sequence of *ygaE* in *S.* Typhi is totally different from *gabC* in *E. coli,* which also indicates that *ygaE* in *S.* Typhi may play other roles that is not found

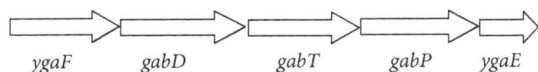

FIGURE 1: The *gab* operon structure in *S.* Typhi. The arrowhead represents the length of the gene; the arrowhead of *ygaE* corresponds to 0.678 kb.

in *E. coli*. The gene organization of *gab* operon in *S.* Typhi was shown in Figure 1. Our microarray results suggested that YgaE can response to osmotic pressure in early stage to repress the expression of *gab* operon. However, the concrete regulation mechanism, whether *ygaF* is included in the *gab* operon and the functions of *gab* operon in *S.* Typhi, still needs further study.

The expression of *gabP* was failed to be detected both in wild-type and Δ*ygaE* strains due to the lack of *gabP* probe on the microarray used in this study.

3.2. YgaE Represses the Expression of ompF/ompC at the Early Stage of Hyperosmotic Stress.

To conquer the often hostile environments they face, the bacteria have evolved a sophisticated cell envelope. The cell envelope of bacteria not only protects them from hazards but also provides them with channels for nutrients from the outside and wastes from the inside. In the envelope, there are three major compartments: the out membrane (OM), the periplasm, and the inner membrane (IM). The OM is a distinguishing feature of Gram-negative bacteria; Gram-positive bacteria lack this organelle [18]. The proteins of OM can be divided into two classes: lipoproteins and β-barrel proteins; the latter is the so-called out membrane proteins (OMPs). OmpF and OmpC are two abundant OMPs, which together are present at approximately 10^5 copies per cell and they serve as general pathways for the influx of small molecules (e.g., molecular weight under 600) [32]. In *E. coli*, *ompC* is preferentially expressed under conditions of high osmolarity [8]. However, *ompC* is regulated differently in *S.* Typhi, in which *ompC* is expressed constitutively under conditions of high and low osmolarity [33, 34], while *ompF* is preferentially expressed under low osmolarity as in *E. coli* [16].

Interestingly, our microarray results indicated that, compared to wild-type strain, the expression of *ompC* and *ompF* are obviously upregulated in Δ*ygaE* mutant strain under hyperosmotic stress and the expression of *ompC* is slightly more abundant than *ompF* in Δ*ygaE* strain (Table 3). The results of RT-PCR validated it; the expressions of *ompC* and *ompF* were increased fourfold, threefold, respectively, in Δ*ygaE* strain (Figures 2(a) and 2(b)), compared to wild-type strain after exposure to hyperosmotic stress 30 min. The completion of *ygaE* in Δ*ygaE* strain repressed the expression of *ompC* and *ompF* to wild type level (Figures 2(a) and 2(b)). The expression of the two genes in the strain which contained pBAD as control was similar to Δ*ygaE* strain (Figures 2(a) and 2(b)). These results suggested that *ygaE* is a repressor of *ompC* and *ompF*. Apparently, it is beneficial to decrease the influx channels when the osmotic stress is high in the environment, which will help the bacteria survival.

In *S.* Typhi, OmpC is always more abundant than OmpF, regardless of the growth conditions [16]. OmpC and OmpF are regulated by the OmpR and EnvZ proteins in *S.* Typhi, as in *E. coli* [16]. On the other hand, deletion of either *ompC* or *ompF* had no effect on expression of the gene coding for the other major porin: osmoregulation of OmpF synthesis was independent of OmpC expression; likewise, OmpC was still highly expressed in a Δ*ompF* background [16]. There appear to be unknown factors in *S.* Typhi that, together with the EnvZ and OmpR regulatory proteins, determine the particular behavior of OmpC expression [16]. Here we report that YgaE is a repressor of *ompC* in *S.* Typhi; it can be partially explained why the expression of *ompC* is not up-regulated under hyperosmotic stress. As for YgaE also repressing the expression of *ompF* in *S.* Typhi, we speculated that, in the evolution, the bacteria prefer to minus the influx pathways as more as possible to ensure the stability of the inner environment under hyperosmotic condition. The regulation of YgaE to *ompC* seems more obvious than *ompF*, which may due to the more abundant expression of *ompC* than *ompF*. However, whether the regulation of YgaE to *ompC* and *ompF* is direct and the concrete regulation mechanism still need further experiments to explore.

3.3. YgaE Does Not Affluence Growth and Antibiotic Susceptibility under Hyperosmotic Stress.

Porin proteins control the permeability of polar solutes across the outer membrane of *E. coli* [35]. Optimal nutrient access is favored by larger porin channels as in OmpF protein [36]. But high outer membrane permeability is a liability in less favorable circumstances, and access of toxic agents or detergents needs to be minimized through environmental control of outer membrane porosity and the increased proportion of smaller OmpC channels in the outer membrane [37]. OmpF and OmpC of *E. coli* affect antibiotic transport and strain susceptibility [38–40]. Recent simulations pinpointed the specific interactions between antibiotics and key residues in the porin channels [41, 42].

The repression of YgaE to *ompC* and *ompF* means less pathways for nutrition and antibiotics. To investigate whether the repression influences the nutrition influx under hyperosmotic stress in *S.* Typhi, we measured the growth of wild-type strain and Δ*ygaE* mutant in LB medium with a final NaCl concentration of 300 mM. Our results showed that the overall growth of both was similar, though the growth curve of Δ*ygaE* mutant seemed to be slightly higher than wild-type strain during the first ten hours (Figure 3); the differences were of no statistic meaning ($P > 0.05$). This result suggested that YgaE does not influence the growth of *S.* Typhi by the regulation of *ompC* and *ompF*. Next, in order to investigate whether YgaE affects the influx of antibiotics under hyperosmotic stress in *S.* Typhi, we examined the antibiotic susceptibility of wild-type and Δ*ygaE* strain to cefotaxime (CTX), ampicillin(AMP), piperacillin(PRL), ceftazidime(CAZ), compound sulfamethoxazole (SXT), and chloroamphenicol (C) by modified Kirby-Bauer disk diffusion method on Muller Hinton agar (OXOID) with a final NaCl concentration of 300 mM. The results displayed that the susceptibility of both strains to these antibiotics had no obvious differences (Figure 4).

(a)

(b)

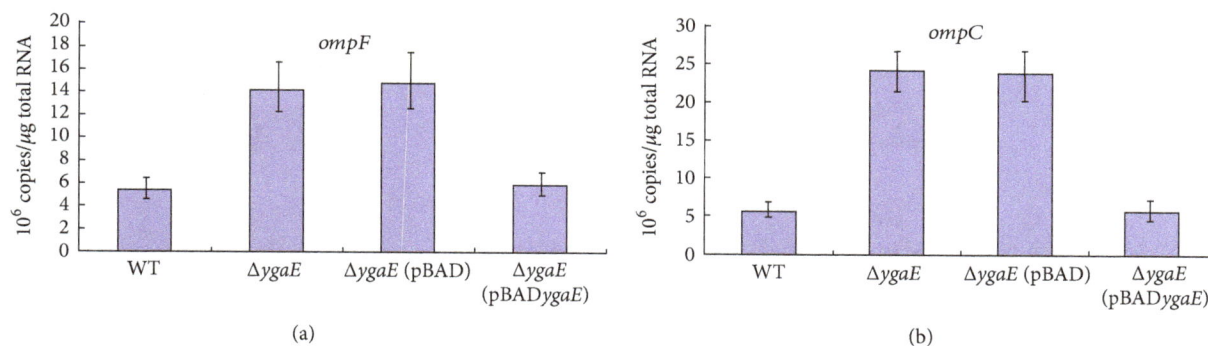

FIGURE 2: qRT-PCR was performed to detect expression of out membrane proteins in wild-type strain(WT), $\Delta ygaE$, $\Delta ygaE$(pBAD), and $\Delta ygaE$(pBADygaE) strains in hyperosmotic LB medium. Data are the mean ± SD from three independent experiments. (a) Expression of *ompF* in the $\Delta ygaE$ strain was increased threefold, compared to the wild-type strain. The complementation of *ygaE* to $\Delta ygaE$ strain repressed the expression of *ompF* to wild-type strain level. The expression of *ompF* in $\Delta ygaE$(pBAD) was similar to $\Delta ygaE$ strain. (b) Expression of *ompC* in the $\Delta ygaE$ strain was increased fourfold, compared to the wild-type strain. The complementation of *ygaE* to $\Delta ygaE$ strain repressed the expression of *ompC* to wild-type strain level. The expression of *ompC* in $\Delta ygaE$(pBAD) was similar to $\Delta ygaE$ strain.

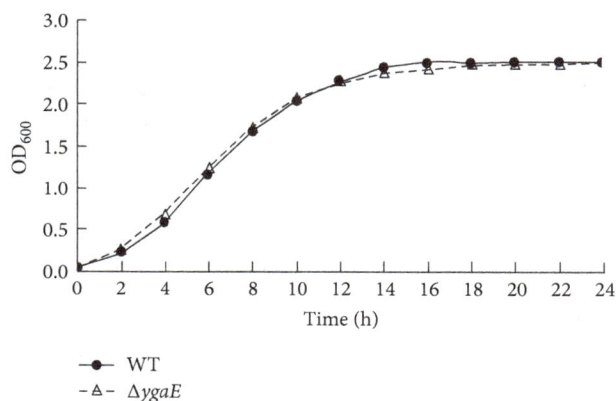

FIGURE 3: Growth of wild type (WT) and $\Delta ygaE$ in LB broth under hyperosmotic stress was monitored at OD_{600}. Each datapoint represents the mean of three independent measurements.

FIGURE 4: Antibiotic susceptibility of wild type and $\Delta ygaE$ under hyperosmotic stress measured by modified Kirby-Bauer disk diffusion method on Muller Hinton agar with a final NaCl concentration of 300 mM. The antibiotic disks which were used in this study were cefotaxime (CTX), ampicillin (AMP), piperacillin (PRL), ceftazidime (CAZ), compound sulfamethoxazole (SXT), and chloramphenicol (C). The zone size around each antimicrobial disk was measured. The experiment was performed three times. Student's *t*-test was used for the statistical analysis. Differences were considered statistically significant when *P* was < 0.05.

One explanation for these phenomena is the repression of YgaE to *ompC* and *ompF* occurs only in the very early stage of hyperosmotic stress as an emergency approach to protect the bacteria. As time goes by, other mechanisms are involved in the process of handling the hyperosmotic stress, the repression of YgaE to *ompC* and *ompF* relieves. Another possibility is that the expression changes of *ompC* and *ompF* on transcription level do not lead to obvious decrease in OmpC and OmpF amount, which causes the unchanging phenotypes. All these speculations still needs more experiments to clarify.

3.4. YgaE Represses the Expression of OmpA at the Late Stage of Hyperosmotic Stress. For revealing proteins probably regulated by YgaE in *S.* Typhi, a comparative proteomics approach was used to distinguish between the two-dimensional electrophoresis profiles of bacterial total proteins in wild-type strain and $\Delta ygaE$ strain. The total proteins of the two strains were obtained after 120 min stress growing in hyperosmotic LB culture and were analyzed in the pH range of 4–7. Protein

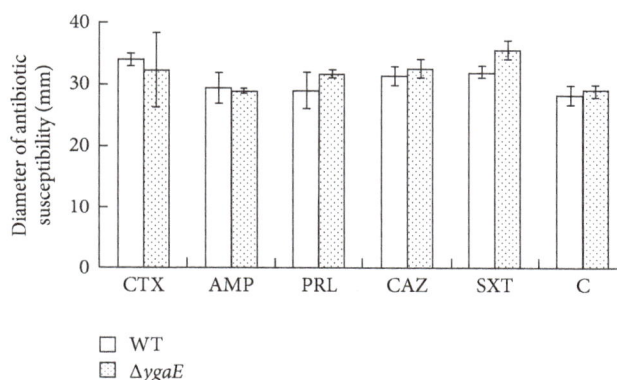

spots that were unique to each strain were chosen for mass spectroscopy (MS) analysis. The MS analysis revealed that one of the unique proteins of $\Delta ygaE$ strain was identified to be OmpA (Figure 5), which indicated that YgaE represses the expression of OmpA at the late stage of hyperosmotic stress in *S.* Typhi.

OmpA is a key regulator of bacterial osmotic homeostasis modulating the permeability and integrity of the outer membrane in *E. coli* [43]. The predicted sequences of *S.* Typhi and *E. coli* OmpA proteins are nearly (>90%) identical [44]. In *S.* Typhi, OmpA is crucial for maintaining envelope integrity and preventing hemolysis through MV secretion [44]. OmpA can exist in two different conformations; a small part of this

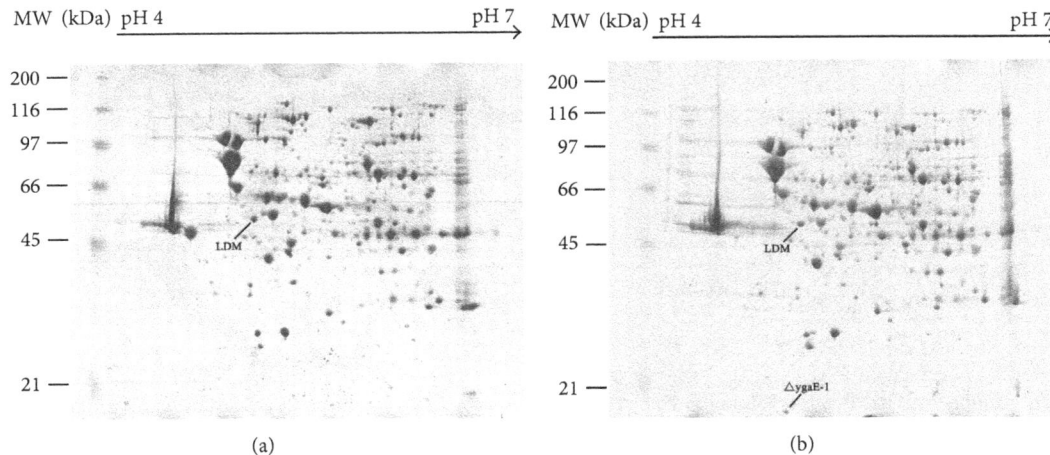

(a)   (b)

FIGURE 5: Two-dimensional electrophoresis was performed to compare the profiles of bacterial total proteins in wild-type strain and $\Delta ygaE$ strains after 120 min hyperosmotic shock. Protein spots that were unique to each strain were chosen for mass spectroscopy (MS) analysis. $\Delta ygaE$-1 spot, which was discussed in this study, was unique to $\Delta ygaE$ strain and determined by MS to be the out membrane protein OmpA.

protein functions as a porin and the major part functions as an important structural protein [18]. Our microarray results revealed that the expression of *ompA* was similar in wild-type strain and $\Delta ygaE$ strain at the early stage of hyperosmotic stress. However, the 2-DE results showed that YgaE is a repressor of OmpA at the late stage of hyperosmotic stress. Oppositely, the expressions of *ompC* and *ompF* were obviously repressed by YgaE at the early stage and no apparent regulation of OmpC and OmpF by YgaE was observed in the 2-DE results at the late stage of hyperosmotic stress. These results together suggested that YgaE regulates out membrane proteins in a time-dependent manner under hyperosmotic stress in *S.* Typhi. The meaning of this regulation model lies in the fact that once the bacteria suddenly transfer to hyperosmotic environment, YgaE responses immediately to minus the influx channels to maintain the stability inside. Gradually, the bacteria adjust themselves and adapt to the environment, YgaE no longer tightly represses the expression of the two major porins OmpC and OmpF but transfers to repress the relatively less important porin OmpA, which also contributes to inner stable state of bacteria.

4. Conclusion

In the lumen of the host intestine, *S.* Typhi cells are exposed to a significant increase in osmolarity. The bacterial responses to hyperosmotic stress are complex. Our previous work found that RpoE and RpoS are two important sigma factors response to hyperosmotic stress and there are compensation and crosstalk between them [29]. YgaE is coregulated by RpoE and RpoS under hyperosmotic stress [29], which indicated its important role in *S.* Typhi under hyperosmotic stress. In this study, we firstly report that other than a repressor of *gab* operon in *S.* Typhi, YgaE also represses the expression of out membrane proteins. The repression of OMPs by YgaE is executed in a time-dependent manner: OmpC and OmpF are repressed in the early stage and OmpA is repressed in

the late stage. We speculate that the reason of regulation pattern transformation of YgaE may be due to the expression variation of RpoE and RpoS: in the early stage of hyperosmotic stress, the decrease of RpoE and increase of RpoS [29] stimulate the expression of YgaE, and the accumulated YgaE represses the expression of OmpF and OmpC. In the late stage of hyperosmotic stress, RpoE and RpoS reached a balanced level [29]; once YgaE senses the balance, it transfers to repress the expression of OmpA.

This study provides new insight into the regulation of out membrane proteins under hyperosmotic stress in *S.* Typhi, which will help us to better understand the adaptation of *S.* Typhi to hyperosmotic shock once invading the host.

Conflict of Interests

The authors declare that there is no conflict of interests regarding the publication of this paper.

Acknowledgments

This study was supported by the Natural Science Foundation of Jiangsu Province, China (no. BK2011301), the startup fund of Soochow University (no. SDY2012B28), the startup fund of the Second Affiliated Hospital of Soochow University (no. SDFEYGJ1301), and Science and Technology Program of Suzhou (SYS201236). Min Wang and Ping Feng are co-first authors.

References

[1] N. F. Crum-Cianflone, "Salmonellosis and the gastrointestinal tract: more than just peanut butter," *Current Gastroenterology Reports*, vol. 10, no. 4, pp. 424–431, 2008.

[2] R. Bhunia, Y. Hutin, R. Ramakrishnan, N. Pal, T. Sen, and M. Murhekar, "A typhoid fever outbreak in a slum of South Dumdum municipality, West Bengal, India, 2007: evidence for

foodborne and waterborne transmission," *BMC Public Health*, vol. 9, article 115, 2009.

[3] J. A. Crump and E. D. Mintz, "Global trends in typhoid and paratyphoid fever," *Clinical Infectious Diseases*, vol. 50, no. 2, pp. 241–246, 2010.

[4] X. Huang, H. Xu, X. Sun, K. Ohkusu, Y. Kawamura, and T. Ezaki, "Genome-wide scan of the gene expression kinetics of *Salmonella enterica* serovar Typhi during hyperosmotic stress," *International Journal of Molecular Sciences*, vol. 8, no. 2, pp. 116–135, 2007.

[5] S. E. Majowicz, J. Musto, E. Scallan et al., "The global burden of nontyphoidal *Salmonella* gastroenteritis," *Clinical Infectious Diseases*, vol. 50, no. 6, pp. 882–889, 2010.

[6] H. Nikaido, "Molecular basis of bacterial outer membrane permeability revisited," *Microbiology and Molecular Biology Reviews*, vol. 67, no. 4, pp. 593–656, 2003.

[7] A. H. Delcour, "Outer membrane permeability and antibiotic resistance," *Biochimica et Biophysica Acta*, vol. 1794, no. 5, pp. 808–816, 2009.

[8] J. W. Fairman, N. Noinaj, and S. K. Buchanan, "The structural biology of β-barrel membrane proteins: a summary of recent reports," *Current Opinion in Structural Biology*, vol. 21, no. 4, pp. 523–531, 2011.

[9] V. M. Aguilella, M. Queralt-Martín, M. Aguilella-Arzo, and A. Alcaraz, "Insights on the permeability of wide protein channels: measurement and interpretation of ion selectivity," *Integrative Biology*, vol. 3, no. 3, pp. 159–172, 2011.

[10] S. W. Cowan, T. Schirmer, G. Rummel et al., "Crystal structures explain functional properties of two *E. coli* porins," *Nature*, vol. 358, no. 6389, pp. 727–733, 1992.

[11] L. A. Pratt, W. Hsing, K. E. Gibson, and T. J. Silhavy, "From acids to *osmZ*: multiple factors influence synthesis of the OmpF and OmpC porins in *Escherichia coli*," *Molecular Microbiology*, vol. 20, no. 5, pp. 911–917, 1996.

[12] M. Villarejo and C. C. Case, "*envZ* mediates transcriptional control by local anesthetics but is not required for osmoregulation in *Escherichia coli*," *Journal of Bacteriology*, vol. 159, no. 3, pp. 883–887, 1984.

[13] M. Heyde and R. Portalier, "Regulation of major outer membrane porin proteins of *Escherichia coli* K 12 by pH," *MGG Molecular & General Genetics*, vol. 208, no. 3, pp. 511–517, 1987.

[14] X. Liu and T. Ferenci, "Regulation of porin-mediated outer membrane permeability by nutrient limitation in *Escherichia coli*," *Journal of Bacteriology*, vol. 180, no. 15, pp. 3917–3922, 1998.

[15] M. Masi and J. M. Pagès, "Structure, function and regulation of outer membrane proteins involved in drug transport in enterobactericeae: the OmpF/C—TolC case," *The Open Microbiology Journal*, vol. 7, pp. 22–33, 2013.

[16] I. Martínez-Flores, R. Cano, V. H. Bustamante, E. Calva, and J. L. Puente, "The *ompB* operon partially determines differential expression of OmpC in *Salmonella typhi* and *Escherichia coli*," *Journal of Bacteriology*, vol. 181, no. 2, pp. 556–562, 1999.

[17] A. Arora, D. Rinehart, G. Szabo, and L. K. Tamm, "Refolded outer membrane protein A of *Escherichia coli* forms ion channels with two conductance states in planar lipid bilayers," *Journal of Biological Chemistry*, vol. 275, no. 3, pp. 1594–1600, 2000.

[18] T. J. Silhavy, D. Kahne, and S. Walker, "The bacterial cell envelope," *Cold Spring Harbor Perspectives in Biology*, vol. 2, no. 5, 2010.

[19] V. Braun and V. Bosch, "Sequence of the murein-lipoprotein and the attachment site of the lipid," *European Journal of Biochemistry*, vol. 28, no. 1, pp. 51–69, 1972.

[20] J.-C. Lazzaroni and R. Portalier, "The *excC* gene of *Escherichia coli* K-12 required for cell envelope integrity encodes the peptidoglycan-associated lipoprotein (PAL)," *Molecular Microbiology*, vol. 6, no. 6, pp. 735–742, 1992.

[21] B. L. Schneider, S. Ruback, A. K. Kiupakis, H. Kasbarian, C. Pybus, and L. Reitzer, "The *Escherichia coli gabDTPC* operon: specific γ-aminobutyrate catabolism and nonspecific induction," *Journal of Bacteriology*, vol. 184, no. 24, pp. 6976–6986, 2002.

[22] S. Kohbata, H. Yokoyama, and E. Yabuuchi, "Cytopathogenic effect of Salmonella typhi GIFU, 10007 on M cells of murine ileal Peyer's patches in ligated ileal loops: an ultrastructural study," *Microbiology and Immunology*, vol. 30, no. 12, pp. 1225–1237, 1986.

[23] X. Huang, L. V. Phung, S. Dejsirilert et al., "Cloning and characterization of the gene encoding the z66 antigen of *Salmonella enterica* serovar Typhi," *FEMS Microbiology Letters*, vol. 234, no. 2, pp. 239–246, 2004.

[24] H. Du, X. Sheng, H. Zhang et al., "RpoE may promote flagellar gene expression in *Salmonella enterica* serovar Typhi under hyperosmotic stress," *Current Microbiology*, vol. 62, no. 2, pp. 492–500, 2011.

[25] X. Sheng, X. Huang, L. Mao et al., "Preparation of *Salmonella enterica* serovar typhi genomic DNA microarrays for gene expression profiling analysis," *Progress in Biochemistry and Biophysics*, vol. 36, no. 2, pp. 306–312, 2009.

[26] S. Xu, H. Zhang, X. Sheng, H. Xu, and X. Huang, "Transcriptional expression of *fljB:z66*, a flagellin gene located on a novel linear plasmid of *Salmonella enterica* serovar Typhi under environmental stresses," *New Microbiologica*, vol. 31, no. 2, pp. 241–247, 2008.

[27] Y. Wang, L. Yang, H. Xu, Q. Li, Z. Ma, and C. Chu, "Differential proteomic analysis of proteins in wheat spikes induced by Fusarium graminearum," *Proteomics*, vol. 5, no. 17, pp. 4496–4503, 2005.

[28] H. Xu, L. Yang, P. Xu, Y. Tao, and Z. Ma, "cTrans: generating polypeptide databases from cDNA sequences," *Proteomics*, vol. 7, no. 2, pp. 177–179, 2007.

[29] H. Du, M. Wang, Z. Luo et al., "Coregulation of gene expression by sigma factors RpoE and RpoS in *Salmonella enterica* serovar Typhi during hyperosmotic stress," *Current Microbiology*, vol. 62, no. 5, pp. 1483–1489, 2011.

[30] E. Niegemann, A. Schulz, and K. Bartsch, "Molecular organization of the *Escherichia coli* gab cluster: nucleotide sequence of the structural genes *gabD* and *gabP* and expression of the GABA permease gene," *Archives of Microbiology*, vol. 160, no. 6, pp. 454–460, 1993.

[31] S. Rigali, A. Derouaux, F. Giannotta, and J. Dusart, "Subdivision of the helix-turn-helix GntR family of bacterial regulators in the FadR, HutC, MocR, and YtrA subfamilies," *Journal of Biological Chemistry*, vol. 277, no. 15, pp. 12507–12515, 2002.

[32] S. W. Cowan, T. Schirmer, G. Rummel et al., "Crystal structures explain functional properties of two *E. coli* porins," *Nature*, vol. 358, no. 6389, pp. 727–733, 1992.

[33] D. Pickard, J. Li, M. Roberts et al., "Characterization of defined *ompR* mutants of *Salmonella typhi*: ompR is involved in the regulation of Vi polysaccharide expression," *Infection and Immunity*, vol. 62, no. 9, pp. 3984–3993, 1994.

[34] J. L. Puente, A. Verdugo-Rodriguez, and E. Calva, "Expression of *Salmonella typhi* and *Escherichia coliOmpC* is influenced differently by medium osmolarity; dependence on *Escherichia coliOmpR*," *Molecular Microbiology*, vol. 5, no. 5, pp. 1205–1210, 1991.

[35] H. Nikaido and T. Nakae, "The outer membrane of gram-negative bacteria," *Advances in Microbial Physiology*, vol. 20, pp. 163–250, 1980.

[36] H. Nikaido and M. Vaara, "Molecular basis of bacterial outer membrane permeability," *Microbiological Reviews*, vol. 49, no. 1, pp. 1–32, 1985.

[37] X. Liu and T. Ferenci, "Regulation of porin-mediated outer membrane permeability by nutrient limitation in *Escherichia coli*," *Journal of Bacteriology*, vol. 180, no. 15, pp. 3917–3922, 1998.

[38] D. Jeanteur, T. Schirmer, D. Fourel et al., "Structural and functional alterations of a colicin-resistant mutant of OmpF porin from *Escherichia coli*," *Proceedings of the National Academy of Sciences of the United States of America*, vol. 91, no. 22, pp. 10675–10679, 1994.

[39] V. Simonet, M. Malléa, and J.-M. Pagès, "Substitutions in the eyelet region disrupt cefepime diffusion through the *Escherichia coli* OmpF channel," *Antimicrobial Agents and Chemotherapy*, vol. 44, no. 2, pp. 311–315, 2000.

[40] P. S. Phale, A. Philippsen, C. Widmer, V. P. Phale, J. P. Rosenbusch, and T. Schirmer, "Role of charged residues at the OmpF porin channel constriction probed by mutagenesis and simulation," *Biochemistry*, vol. 40, no. 21, pp. 6319–6325, 2001.

[41] A. Kumar, E. Hajjar, P. Ruggerone, and M. Ceccarelli, "Molecular simulations reveal the mechanism and the determinants for ampicillin translocation through OmpF," *Journal of Physical Chemistry B*, vol. 114, no. 29, pp. 9608–9616, 2010.

[42] A. Kumar, E. Hajjar, P. Ruggerone, and M. Ceccarelli, "Structural and dynamical properties of the porins OmpF and OmpC: insights from molecular simulations," *Journal of Physics Condensed Matter*, vol. 22, no. 45, Article ID 454125, 2010.

[43] Y. Wang, "The function of OmpA in *Escherichia coli*," *Biochemical and Biophysical Research Communications*, vol. 292, no. 2, pp. 396–401, 2002.

[44] J. A. Fuentes, N. Villagra, M. Castillo-Ruiz, and G. C. Mora, "The *Salmonella typhihlyE* gene plays a role in invasion of cultured epithelial cells and its functional transfer to S. Typhim'-urium promotes deep organ infection in mice," *Research in Microbiology*, vol. 159, no. 4, pp. 279–287, 2008.

11

Phytase Production by *Aspergillus niger* CFR 335 and *Aspergillus ficuum* SGA 01 through Submerged and Solid-State Fermentation

Gunashree B. Shivanna and Govindarajulu Venkateswaran

Department of Food Microbiology, Central Food Technological Research Institute, Mysore, Karnataka-570 020, India

Correspondence should be addressed to Govindarajulu Venkateswaran; venkatcftri@gmail.com

Academic Editors: L. Ramirez and G. Vaughan

Fermentation is one of the industrially important processes for the development of microbial metabolites that has immense applications in various fields. This has prompted to employ fermentation as a major technique in the production of phytase from microbial source. In this study, a comparison was made between submerged (SmF) and solid-state fermentations (SSF) for the production of phytase from *Aspergillus niger* CFR 335 and *Aspergillus ficuum* SGA 01. It was found that both the fungi were capable of producing maximum phytase on 5th day of incubation in both submerged and solid-state fermentation media. *Aspergillus niger* CFR 335 and *A. ficuum* produced a maximum of 60.6 U/gds and 38 U/gds of the enzyme, respectively, in wheat bran solid substrate medium. Enhancement in the enzyme level (76 and 50.7 U/gds) was found when grown in a combined solid substrate medium comprising wheat bran, rice bran, and groundnut cake in the ratio of 2:1:1. A maximum of 9.6 and 8.2 U/mL of enzyme activity was observed in SmF by *A. niger* CFR 335 and *A. ficuum*, respectively, when grown in potato dextrose broth.

1. Introduction

Phytases (EC 3.1.3.8) and phosphatases have a big share in the market due to their widespread application as a feed supplement [1, 2]. Most cereals and legumes are rich in protein and fat but they have antinutritional factors like phytic acid (*myo*-inositol hexakisphosphate) which discourage their use in food. Phytic acid chelates various metals and proteins, thereby diminishing the bioavailability of proteins and nutritionally important minerals such as Ca^{2+}, Mg^{2+}, and P, Zn^{2+} Fe^{2+} [3]. Phytase (*myo*-inositol hexakisphosphate phosphohydrolase, E.C.3.1.3.8) catalyzes the hydrolysis of phytic acid and its salts (phytates) that generally yield inositol, inositol monophosphate, and inorganic phosphate [4]. The enzymatic degradation of phytic acid will not produce toxic by-products; hence, it is environment friendly [5]. In view of increasing demand for phytase, it is essential to produce phytase in a cost-effective manner using microorganisms.

The production of phytase from fungi has been achieved using three different cultivation methods, namely, solid-state [6], semisolid [7], and submerged fermentation [8, 9]. About 5,000 years ago fungi were cultivated in SSF for the production of food, the oldest known fermentation of rice by *A. oryzae* used to initiate the koji process. Solid-state fermentation (SSF) system has generated a great deal of interest in recent years because it offers several economical and practical advantages including high product concentration, improved product recovery, simple cultivation equipment, and lower plant operational cost [10, 11].

A detailed study on the effect of various cultural conditions for the production of phytase by *Aspergillus niger* CFR 335 in submerged and solid-state fermentation has been carried out [12]. The present study aims at a comparison between the production of phytase enzyme in solid-state and submerged fermentation medium by two fungal strains, *Aspergillus niger* CFR 335 and *Aspergillus ficuum* SGA 01.

2. Experimental Procedures

2.1. Strains, Media, and Growth Conditions. *Aspergillus niger* CFR 335 and *Aspergillus ficuum* SGA 01 were cultivated in complete medium (g/L: glucose 10, yeast extract 2.5, malt extract 5, agar 20, pH 5.5 ± 0.2) slants for 3-4 days at 30°C and fully grown slants were stored at 4°C for further use. All the experiments on submerged fermentations were carried out using potato dextrose broth, two different solid substrates, wheat bran, and a combination of wheat bran, rice bran, and groundnut cake (2 : 1 : 1). All the media ingredients were obtained from Hi-Media Chemicals, India. Reagent chemicals were of analytical grade procured from e-Merck, Hi-Media, and Qualigen Chemicals, India, Ltd. and sodium phytate used as a substrate for phytase, standard phytase, and bovine serum albumin were procured from Sigma Chemicals, USA. Fresh wheat bran used in solid-state fermentation was obtained from the Department of Flour Milling Baking & Confectionary Technology, CFTRI, Mysore, India. Rice bran and groundnut cake were procured freshly from a local market.

2.2. Studies on the Inoculum Size, Age, and Moisture Level. Effect of different inoculum size (0.25, 0.5, 0.75, 1, 1.5 mL, and 2.0 mL/50 g of solid medium and 100 mL of submerged media) containing 2×10^6 spores/mL on phytase production by *Aspergillus niger* CFR 335 and *Aspergillus ficuum* SGA 01 was studied. In another experiment, the effect of inoculum age on phytase production by *A. niger* CFR 335 and *A. ficuum* SGA 01 in both solid-state and submerged fermentation media was studied using 1–10-day-old inoculum. The effect of moisture level on solid-state fermentation was also studied by varying the moisture from 10 to 80% using sterile distilled water and optimum moisture for maximum phytase production was determined. All the inoculated flasks were incubated at 30°C for 10 days and periodically tested for phytase production.

2.3. Studies on Fermentation Temperature, pH, and Time. Effect of fermentation temperature on maximum phytase production was studied by incubating the fungi at temperatures varying from 5 to 60°C at an interval of 5°C. The effect of media pH on maximum phytase production was also studied by setting the media pH between 2.0 and 7.5 at an interval of 0.5. The effect of fermentation time on phytase production was studied by incubating both *A. niger* CFR 335 and *A. ficuum* SGA 01 in submerged (potato dextrose broth) and solid-state (wheat bran) cultivation medium for 10 days and periodically testing the enzyme activity.

2.4. Preparation of Solid-State Fermentation (SSF) Medium. In solid-state fermentation studies, two different media including wheat bran (100%) and a mixed medium with combination of wheat bran, rice bran, and groundnut cake (2 : 1 : 1) were used. Solid substrate media were prepared according to the method of Gunashree and Venkateswaran [12]. After cooling to room temperature, sterile solid media were inoculated with 1 mL and 1.5 mL suspensions of *A. niger*

CFR 335 and *A. ficuum* SGA 01, respectively, containing 2×10^6 spores/mL. The media were mixed thoroughly using a sterile glass rod and then incubated for 8 days at 30°C in a static inclined position.

2.5. Preparation of Submerged Fermentation (SmF) Medium. Aliquots of 100 mL potato dextrose broth (PDB) were taken in 500 mL Erlenmeyer flask and autoclaved for 20 minutes at 121°C and 15 lbs pressure. The media were cooled to room temperature and inoculated with 0.5 and 1 mL suspensions of *A. niger* and *A. ficuum*, respectively, containing 2×10^6 spores/mL. The flasks were incubated for 10 days at 30°C on an orbital shaker at 200 rpm.

2.6. Extraction of Crude Enzyme from SSF. Crude enzyme extraction was carried out by the method of Gunashree and Venkateswaran [12] by soaking moldy bran in 1 : 5 w/v 0.2 M acetate buffer at pH 4.5. The flasks were kept in a rotary shaker for 20 minutes at 200 rpm after thorough mixing of the bran with distilled water. The solids were separated from the aqueous solution by filtering through clean muslin cloth. The aqueous solution was centrifuged at 8944 g for 20 minutes at 4°C in a refrigerated centrifuge. The aqueous supernatant was collected and used for further investigation.

2.7. Extraction of Crude Enzyme in SmF. Extracellular crude phytase from submerged media was extracted by the method of Gunashree and Venkateswaran [12], which entailed initial filtration through Whatman no. 1 filter paper and centrifugation at 8944 g for 10–15 minutes. The culture broth containing phytase was stored at 4°C and used as crude enzyme preparation for further studies.

2.8. Enzyme Assay. Crude enzyme extracted from both solid-state and submerged fermentation media was quantitatively assayed for phytase enzyme as described by the method of Heinonen and Lahti [13]. A standard graph was plotted using potassium dihydrogen phosphate with working concentration ranging from 30 to 360 μM. Protein quantifications were made by the method of Bradford [14] and compared with the standard prepared using bovine serum albumin.

2.9. Statistical Analysis. Data are presented as standard error means (±SEM). Comparisons between solid substrate and submerged fermentation between *Aspergillus niger* CFR 335 and *Aspergillus ficuum* SGA 01 were made with analysis of variance [15]. *P* values were considered significant at $P < 0.05$. All statistical tests were carried out using demo version of Graph Pad Prism software.

3. Results and Discussion

3.1. Inoculum Age. Effect of inoculum age, size, and moisture level for maximum phytase production by *Aspergillus niger* CFR 335 and *Aspergillus ficuum* SGA 01 through solid-state and submerged fermentation was studied. Age of inoculum used in fermentation media has an impact on growth of

Phytase Production by Aspergillus niger CFR 335 and Aspergillus ficuum SGA 01 through Submerged and
Solid-State Fermentation

97

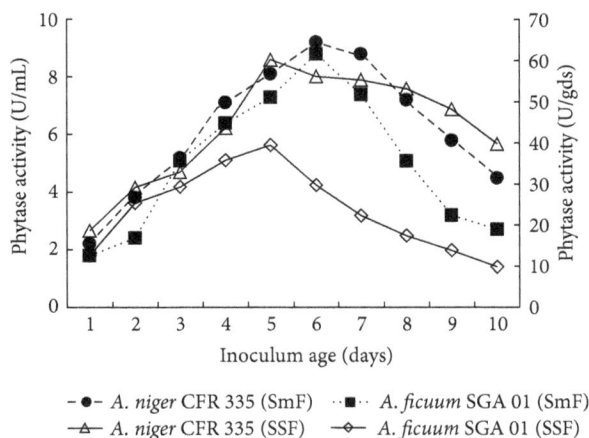

FIGURE 1: Effect of inoculum age on phytase production by *Aspergillus niger* CFR 335 and *Aspergillus ficuum* SGA 01 in submerged and solid-state fermentation.

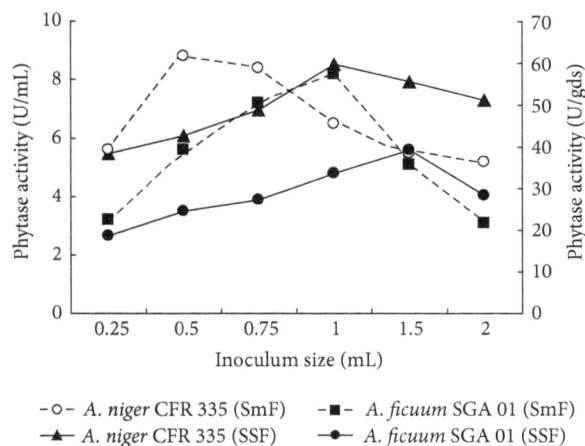

FIGURE 2: Effect of inoculum size on phytase production by *Aspergillus niger* CFR 335 and *Aspergillus ficuum* SGA 01 in submerged and solid-state fermentation.

the organism as well as the amount of metabolite that is produced [6]. Studies on the effect of inoculum age showed that the enzyme production rate increased gradually with increase in inoculum age up to six days and declined when older inocula were used. It was found that six-day old culture resulted in maximum phytase production by both *A. niger* CFR 335 and *A. ficuum* SGA 01 under both fermentations (Figure 1). The enzyme activity was 2.2 and 1.8 U/mL when one-day inocula of *A. niger* CFR 335 and *A. ficuum* SGA 01 were used for submerged fermentation with a gradual increase up to 9.2 and 8.8 U/mL with six-day old inocula. In solid-state fermentation, phytase activity was 12.6 and 18.6 U/gds with one-day old inocula and increased up to 60.2 and 39.4 U/gds with six-day old inocula of *A. niger* CFR 335 and *A. ficuum* SGA 01, respectively. About 51 and 69% reductions in phytase activity of *A. niger* CFR 335 and *A. ficuum* SGA 01 in submerged fermentation were observed and a decline of 34 and 75% was observed in solid-state fermentation when ten-day old inocula were used. This decline in enzyme activity with older inocula may be due to reduced metabolic rate. Studies have been extensively carried out on the effect of culture conditions, particularly inoculum age, media composition, and duration of SSF on phytase production by *A. niger* [16]. Ebune et al. [6] have shown 2- and 5-day-old homogenized pellet as inocula for producing the least and the highest amount of enzyme, respectively, whereas in the present investigation, spore suspension was used.

3.2. Inoculum Size. Inoculum size also plays an important role in the extent of growth and metabolite production by fungi. The results on different inoculum size indicated that there was a gradual increase in the enzyme synthesis with a maximum activity of 8.8 and 8.2 U/mL with 0.5 and 1 mL of *A. niger* CFR 335 and *A. ficuum* SGA 01 spore suspensions (2×10^6 spores/mL) used for submerged fermentation, respectively. In solid-state fermentation of *A. niger* CFR 335 and *A. ficuum* SGA 01, 1 and 1.5 mL of spore suspensions were optimum for maximum enzyme activities of 59.8 and

39.2 U/gds, respectively, (Figure 2). Minimum enzyme activities of 5.6 and 3.2 U/mL for submerged fermentation and 38.3 and 18.6 U/gds for solid-state fermentation were observed when 0.25 mL of *A. niger* CFR 335 and *A. ficuum* SGA 01 inocula was used, respectively. There was decline of 41 and 62.2% in the submerged and 14.4 and 27.6% in the solid-state fermentation of *A. niger* CFR 335 and *A. ficuum* SGA 01, respectively, when 2 mL inocula were used. This may be due to higher fungal growth that leads to increased competition for nutrients and their fast exhaustion thereby can be retained enzyme production [6]. A similar study on the influence of inoculum level on phytase production has also been carried out earlier [17].

3.3. Studies on Moisture Level. Moisture content is one of the most critical factors for microbial growth and enzyme production in solid-state fermentation [18]. The optimum amount of water varies and must be determined for each microbial system [19]. The present study showed that there was linearity between the enzyme production and moisture levels of solid substrate media up to 60%; further increase in the moisture resulted in reduced enzyme production. The fungi failed to grow in lower moisture levels, while in higher levels, both the fungi grew vegetatively resulting in reduced enzyme yield. (Figure 3). There was 90 and 82% decline in the enzyme activity of *A. niger* CFR 335 and *A. ficuum* SGA 01, respectively, with moisture level beyond 60%. This may be attributed to reduced aeration in the substrate and also due to reduced decomposition rate of total organic matter at the lowest and highest moisture contents [17]. Canola meal has been used for phytase production by *Aspergillus ficuum* with optimum moisture of 64% [6] and by *A. carbonarius* strain with optimum moisture ranging from 53 to 60% [17]. A maximum phytase production was also reported at 60% moisture level [20], which is identical to the present findings.

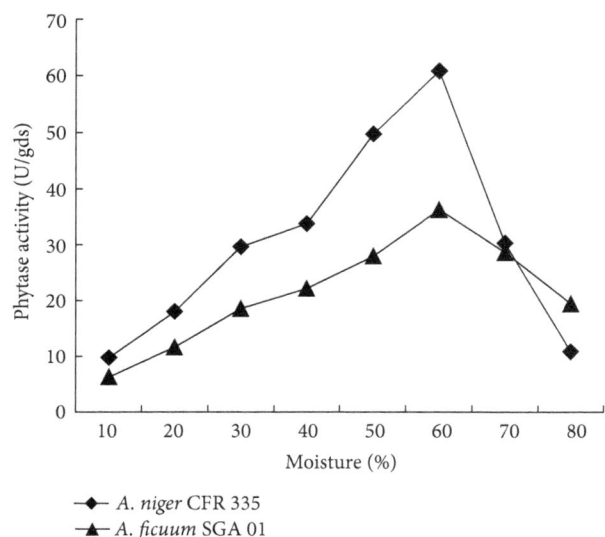

FIGURE 3: Effect of moisture level on phytase production by *Aspergillus niger* CFR 335 and *Aspergillus ficuum* SGA 01 in solid-state fermentation.

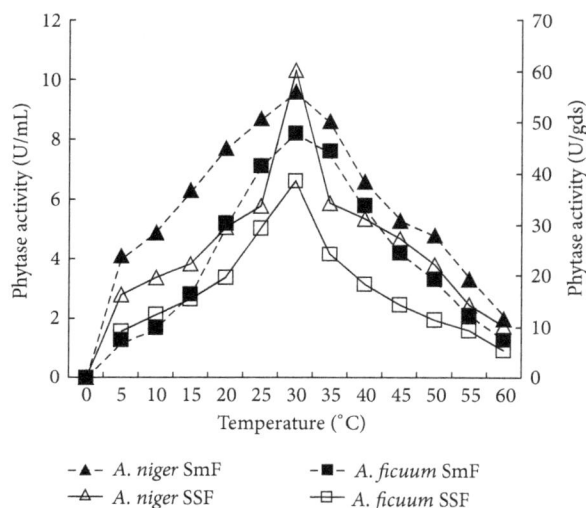

FIGURE 4: Effect of temperature on phytase production by *Aspergillus niger* CFR 335 and *Aspergillus ficuum* SGA 01 in submerged and solid-state fermentation.

3.4. Effect of Temperature and pH. Physical parameters like temperature and pH play a vital role in the growth, production, and stability of any microbial metabolite. Results showed linearity between phytase production, and fermentation temperature up to 30°C for both the fungi (Figure 4). There was gradual decline of >80 and 90% in the enzyme production at 60°C under submerged fermentation of *A. niger* CFR 335 and *A. ficuum* SGA 01, respectively. About 70% reduction in enzyme activity was obtained in both the fungi through solid-state fermentation at 60°C. Similarly, there was a linear trend between phytase production and pH up to 4.5 for both the fungi and the enzyme yield was reduced with increase in pH (Figure 5). There was >90% reduction in the enzyme

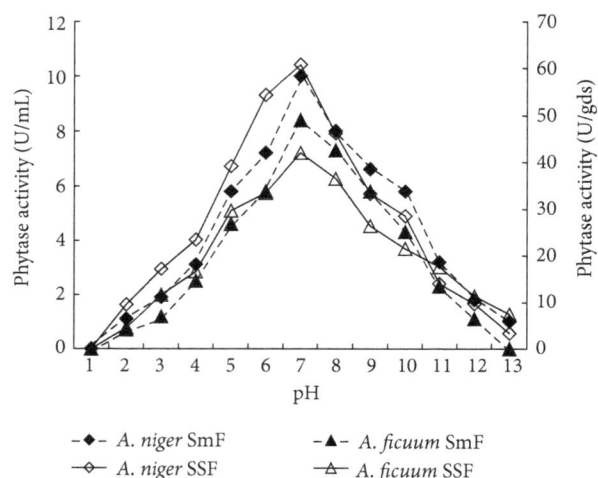

FIGURE 5: Effect of pH on phytase production by *Aspergillus niger* CFR 335 and *Aspergillus ficuum* SGA 01 in submerged and solid-state fermentation.

production at pH beyond 4.5. In the present investigation, there was a linear trend between phytase production and pH up to 4.5 for both the fungi and the enzyme yield was reduced with increase in pH. This is in support of earlier reports where they have shown that pH ranging from 4.5 to 6.0 is optimum for filamentous fungi [21–23].

3.5. Fermentation Time. The effect of fermentation period on phytase production by *A. niger* CFR 335 and *A. ficuum* SGA 01 in submerged and solid-state fermentation was tested by incubating the fungi for 10 days with periodically testing the enzyme activity. Enzyme activity was found to be increased exponentially with increase in the incubation period. The two fungi grew luxuriantly with abundant conidia in solid substrate medium at an early period of 40–48 h of incubation. Activities of 61 U/gds and 38 U/gds were observed in *A. niger* CFR 335 and *A. ficuum* SGA 01, respectively, on 5th day of solid-state fermentation at 30°C. With subsequent cultivation period, there were 73 and 71% reductions in solid-state fermentation and 62 and 71% reductions, in submerged fermentation of *A. niger* CFR 335 and *A. ficuum* SGA 01 respectively (Figure 6). A similar growth study with respect to phytase production was carried out [23] and found luxuriant growth within 48 h of incubation.

3.6. Solid Substrate Media. The enzyme activities of *A. niger* CFR 335 and *A. ficuum* SGA 01 varied when grown in wheat bran (WB) and mixed solid substrate media (MB) comprising 2 : 1 : 1 proportion of wheat bran, rice bran, and groundnut cake. An activity of 76 U/gds and 51 U/gds was observed in *A. niger* CFR 335 and *A. ficuum* SGA 01, respectively, when grown in mixed solid substrate medium (Figure 7). Enhancement of 24% enzyme activity was observed in mixed medium than in mere wheat bran medium. This increase in the enzyme yield may be due to the presence of nutritionally rich rice bran and groundnut cake. Rice bran contains about

Phytase Production by Aspergillus niger CFR 335 and Aspergillus ficuum SGA 01 through Submerged and
Solid-State Fermentation

99

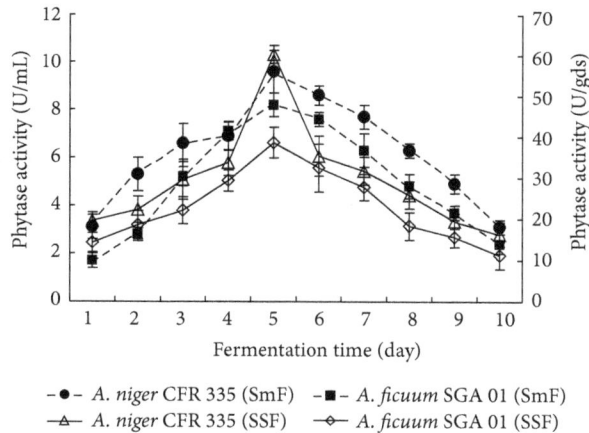

FIGURE 6: Effect of fermentation time on phytase production by *Aspergillus niger* CFR 335 and *Aspergillus ficuum* SGA 01 in submerged and solid-state fermentation.

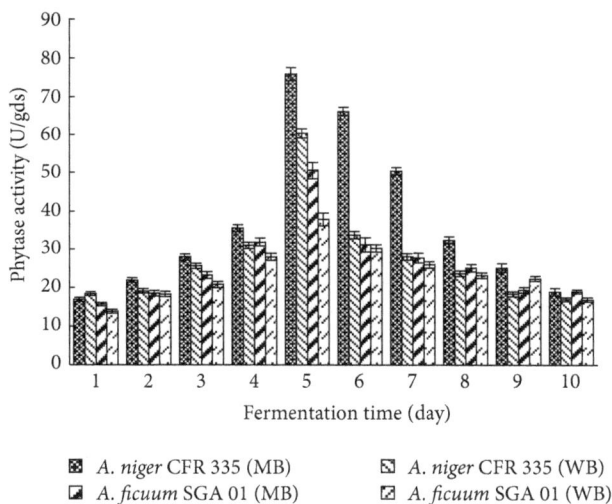

FIGURE 7: Effect of various solid-substrate media on phytase production by *Aspergillus niger* CFR 335 and *Aspergillus ficuum* SGA 01 in submerged and solid-state fermentation.

12 to 13% oil and a high level of dietary fibers such as beta-glucan, pectin, and gum. In addition, it also contains 4-hydroxy-3-methoxycinnamic acid (ferulic acid) which may also be a component of the structure of nonlignified cell walls [24]. Groundnut cake is a rich source of protein and monounsaturated fats [25]. A large number of complex high molecular weight polysaccharides such as cellulose, hemicellulose, lignin, and starch are available in wheat bran and serve as additional carbon sources and also increased total phosphorus content prevents phosphate limitation in the culture medium [23]. Similar findings have been shown earlier [26–28].

4. Conclusion

It is concluded that *Aspergillus niger* CFR 335 and *Aspergillus ficuum* SGA 01 were capable of accumulating phytase after the onset of sexual growth both in submerged and solid-state fermentation media. However, enzyme accumulation was higher in solid-state fermentation than in submerged fermentation medium, because of freely available aqueous content in submerged medium which only supports the vegetative growth of the fungi. This is in contrast with the natural condition where the microbes grow. Enzyme activity was found to be 3–5 fold higher in SSF than in SmF medium which is higher when compared to earlier reports.

Conflict of Interests

The authors declare that there is no conflict of interests regarding the publication of this paper.

Acknowledgments

Authors are thankful to the Director of the Central Food Technological Research Institute, Mysore, India, for permitting them to carry out this research, the Head of the Department of Food Microbiology, CFTRI, for his constant support and encouragement, and University of Mysore for the financial aid.

References

[1] R. Greiner and U. Konietzny, "Phytase for food application," *Food Technology and Biotechnology*, vol. 44, no. 2, pp. 125–140, 2006.

[2] A. Vohra and T. Satyanarayana, "Phytases: microbial sources, production, purification, and potential biotechnological applications," *Critical Reviews in Biotechnology*, vol. 23, no. 1, pp. 29–60, 2003.

[3] M. Ragon, F. Hoh, A. Aumelas, L. Chiche, G. Moulin, and H. Boze, "Structure of *Debaryomyces castellii* CBS 2923 phytase," *Acta Crystallographica Section F*, vol. 65, no. 4, pp. 321–326, 2009.

[4] E. J. Mullaney, C. B. Daly, and A. H. J. Ullah, "Advances in phytase research," *Advances in Applied Microbiology*, vol. 47, pp. 157–199, 2000.

[5] V. Ciofalo, N. Barton, K. Kretz, J. Baird, M. Cook, and D. Shanahan, "Safety evaluation of a phytase, expressed in *Schizosaccharomyces pombe*, intended for use in animal feed," *Regulatory Toxicology and Pharmacology*, vol. 37, no. 2, pp. 286–292, 2003.

[6] A. Ebune, S. Al-Asheh, and Z. Duvnjak, "Production of phytase during solid state fermentation using *Aspergillus ficuum* NRRL 3135 in canola meal," *Bioresource Technology*, vol. 53, no. 1, pp. 7–12, 1995.

[7] Y. W. Han, D. J. Gallagher, and A. G. Wilfred, "Phytase production by *Aspergillus ficuum* on semisolid substrate," *Journal of Industrial Microbiology*, vol. 2, no. 4, pp. 195–200, 1987.

[8] V. C. Nair and Z. Duvnjak, "Reduction of phytic acid content in canola meal by *Aspergillus ficuum* in solid state fermentation process," *Applied Microbiology and Biotechnology*, vol. 34, no. 2, pp. 183–188, 1990.

[9] A. H. Ullah and D. M. Gibson, "Extracellular phytase (E.C. 3.1.3.8) from *Aspergillus ficuum* NRRL 3135: purification and characterization," *Preparative Biochemistry*, vol. 17, no. 1, pp. 63–91, 1987.

[10] M. Becerra and M. I. González Siso, "Yeast β-galactosidase in solid-state fermentations," *Enzyme and Microbial Technology*, vol. 19, no. 1, pp. 39–44, 1996.

[11] A. Pandey, G. Szakacs, C. R. Soccol, J. A. Rodriguez-Leon, and V. T. Soccol, "Production, purification and properties of microbial phytases," *Bioresource Technology*, vol. 77, no. 3, pp. 203–214, 2001.

[12] B. S. Gunashree and G. Venkateswaran, "Effect of different cultural conditions for phytase production by *Aspergillus niger* CFR 335 in submerged and solid-state fermentations," *Journal of Industrial Microbiology and Biotechnology*, vol. 35, no. 12, pp. 1587–1596, 2008.

[13] J. K. Heinonen and R. J. Lahti, "A new and convenient colorimetric determination of inorganic orthophosphate and its application to the assay of inorganic pyrophosphatase," *Analytical Biochemistry*, vol. 113, no. 2, pp. 313–317, 1981.

[14] M. M. Bradford, "A rapid and sensitive method for the quantitation of microgram quantities of protein utilizing the principle of protein dye binding," *Analytical Biochemistry*, vol. 72, no. 1-2, pp. 248–254, 1976.

[15] G. W. Snedecor and W. G. Cochran, *Statistical Methods*, Iowa State University Press, Ames, Iowa, USA, 6th edition, 1976.

[16] C. Krishna and S. E. Nokes, "Predicting vegetative inoculum performance to maximize phytase production in solid-state fermentation using response surface methodology," *Journal of Industrial Microbiology and Biotechnology*, vol. 26, no. 3, pp. 161–170, 2001.

[17] S. Al-Asheh and Z. Duvnjak, "The effect of phosphate concentration on phytase production and the reduction of phytic acid content in canola meal by Aspergillus carbonarius during a solid-state fermentation process," *Applied Microbiology and Biotechnology*, vol. 43, no. 1, pp. 25–30, 1995.

[18] G. E. A. Awad, M. M. M. Elnashar, and E. N. Danial, "Optimization of phytase production by *Penicillium funiculosum* NRC467 under solid state fermentation by using full factorial design," *World Applied Sciences Journal*, vol. 15, no. 11, pp. 1635–1644, 2011.

[19] A. I. El-Batal and H. Abdel Karem, "Phytase production and phytic acid reduction in rapeseed meal by *Aspergillus niger* during solid state fermentation," *Food Research International*, vol. 34, no. 8, pp. 715–720, 2001.

[20] P. Vats and U. C. Banerjee, "Studies on the production of phytase by a newly isolated strain of *Aspergillus niger* var teigham obtained from rotten wood-logs," *Process Biochemistry*, vol. 38, no. 2, pp. 211–217, 2002.

[21] A. Casey and G. Walsh, "Purification and characterization of extracellular phytase from *Aspergillus niger* ATCC 9142," *Bioresource Technology*, vol. 86, no. 2, pp. 183–188, 2003.

[22] B.-C. Oh, W.-C. Choi, S. Park, Y.-O. Kim, and T.-K. Oh, "Biochemical properties and substrate specificities of alkaline and histidine acid phytases," *Applied Microbiology and Biotechnology*, vol. 63, no. 4, pp. 362–372, 2004.

[23] M. Papagianni, S. E. Nokes, and K. Filer, "Production of phytase by *Aspergillus niger* in submerged and solid-state fermentation," *Process Biochemistry*, vol. 35, no. 3-4, pp. 397–402, 1999.

[24] "How Nice, Brown Rice: Study Shows Rice Bran Lowers Blood Pressure in Rats," *Science Daily*, March 2006.

[25] O. A. Davies and N. C. Ezenwa, "Groundnut cake as an alternative protein source in the diet of *Clarias gariepinus* fry," *International Journal of Science and Nature*, vol. 1, pp. 73–76, 2010.

[26] S. Ramachandran, K. Roopesh, K. M. Nampoothiri, G. Szakacs, and A. Pandey, "Mixed substrate fermentation for the production of phytase by *Rhizopus* spp. using oilcakes as substrates," *Process Biochemistry*, vol. 40, no. 5, pp. 1749–1754, 2005.

[27] K. Roopesh, S. Ramachandran, K. M. Nampoothiri, G. Szakacs, and A. Pandey, "Comparison of phytase production on wheat bran and oilcakes in solid-state fermentation by *Mucor racemosus*," *Bioresource Technology*, vol. 97, no. 3, pp. 506–511, 2006.

[28] B. Bogar, G. Szakacs, J. C. Linden, A. Pandey, and R. P. Tengerdy, "Optimization of phytase production by solid substrate fermentation," *Journal of Industrial Microbiology and Biotechnology*, vol. 30, no. 3, pp. 183–189, 2003.

12

Acetic Acid Bacteria and the Production and Quality of Wine Vinegar

Albert Mas,[1] **María Jesús Torija,**[1]
María del Carmen García-Parrilla,[2] **and Ana María Troncoso**[2]

[1] *Facultad de Enología, Universitat Rovira i Virgili, Marcel·lí Domingo s/n, 43003 Tarragona, Spain*
[2] *Facultad de Farmacia, Universidad de Sevilla, Profesor García González 2, 41012 Sevilla, Spain*

Correspondence should be addressed to Albert Mas; albert.mas@urv.cat

Academic Editors: Y. M. Chen and S. J. Suh

The production of vinegar depends on an oxidation process that is mainly performed by acetic acid bacteria. Despite the different methods of vinegar production (more or less designated as either "fast" or "traditional"), the use of pure starter cultures remains far from being a reality. Uncontrolled mixed cultures are normally used, but this review proposes the use of controlled mixed cultures. The acetic acid bacteria species determine the quality of vinegar, although the final quality is a combined result of technological process, wood contact, and aging. This discussion centers on wine vinegar and evaluates the effects of these different processes on its chemical and sensory properties.

1. Introduction

Vinegar production dates back at least to 200 BC, and it is an illustrative example of microbial biotransformation. However, vinegar has always been seen as a "leftover" in the family of fermented products [1]. Vinegar has been part of the human diet as a condiment and food preservative, as well as the basis for simple remedies for people and animals, since remote antiquity. However, its production was always considered a chemical process. As mentioned in the review on vinegar history [1], in 1732, the Dutchman Boerhaave noted that the "mother of vinegar" was a living organism, although he did not specify the role of this organism in the process of acidification. We shall refer to this process as "acetification" instead of the more popular "acetous fermentation" due to its strict requirement for oxygen. Lavoisier in 1789 demonstrated that acetification is the oxidation of ethanol, but he did not suspect a role for living organisms. Persoon in 1822 described the film formed at the surface of wine, beer, or pickled vegetables and the biological nature of such substances, and in "European Mycology," he added new species of *Mycoderma*: *ollare, mesentericum, lagenoe,* and *pergameneum*. Chaptal

also observed that the production of vinegar went well when the "wine flower", whose appearance heralds and precedes the acidification, appeared on the surface of the wine. However, Berzelius warned that all decaying organic matter developed the same type of flora if it was exposed to air. Acetification became part of the controversy between scientists such as Berzelius and Liebig; some argued that the process was purely chemical, and some claimed that this transformation involved an "organized living being." Regarding the "mother of vinegar," Kützing noted in 1837 that the thin film that covered the surface of the liquid was made by "globulles" six times smaller than yeasts; thus, he can be credited with the first microscopic observation of acetic acid bacteria in 1837. Finally, Pasteur in 1864 claimed that the transformation of wine into vinegar was due to the development of the veil of *Mycoderma aceti* on its surface [1].

Despite some small local differences, in general, food regulations consider vinegar to be the result of a double fermentation (alcoholic and acetous or acetification) of any sugar substrate. European countries have specific rules for vinegars sold in different regions. In the European Union, the established limits for acidity and residual ethanol content are strictly set. Thus, the acidity of wine vinegar (acetification obtained exclusively from wine) must be at least 6% (w/v), and the maximum residual ethanol allowed is 1.5% (v/v). However, the variety of raw materials used in the production of vinegar is very great, ranging from byproducts and agricultural surpluses to high-quality substrates for the most unique and prized vinegars, such as Sherry vinegar (Spain) and Aceto Balsamico Tradizionalle (Italy). The quality standard defines up to ten types of vinegars, which include wine vinegar, fruit, cider, alcoholic, cereal, malt, malt distillate, balsamic (with added grape must), and "other balsamic vinegars," which encompass any other substrate of agricultural origin, such as honey or rice. Undoubtedly, wine vinegar is the most common type in Mediterranean countries, although the latest gastronomic trends have led to a considerable expansion of the varieties available in recent years. However, worldwide most of the vinegar produced is "white" vinegar, that is, vinegar produced directly from diluted alcohol. In this review, we will focus mostly on wine vinegar and the role of acetic acid bacteria and the quality of this product.

2. Production Technology of Wine Vinegars

Apart from their different substrates, vinegars can also be differentiated by their production systems. In traditional vinegars, the transformation of ethanol into acetic acid is performed by a static culture of acetic acid bacteria at the interface between the liquid and air. The barrels are filled to 2/3 capacity to leave an air chamber, which is kept in contact with the outside air using one of various types of openings. This production system is called "surface culture," and this process is considered the traditional method. The more standardized version of this method, the "Orleans method," includes side holes for air circulation and adds a funnel with an extension to the base of the barrel to allow wine to be added at the bottom of the barrel, preventing the alteration of the "mother of vinegar," that is, the biofilm formed by acetic acid bacteria on the surface. The vinegars produced by this traditional system are generally considered of high quality because of their organoleptic complexity. In fact, the product quality results from (i) the raw material (wine or other substrate), (ii) the metabolism of the acetic acid bacteria, which produce some additional transformations (mostly oxidation reactions, but also ester formations, e.g.) on top of the basic transformation (ethanol to acetic acid), (iii) the interaction between the vinegar and the wood from the barrels, and (iv) the aging process, which integrates all of the previously mentioned characteristics. However, the characterization of wine vinegar as a byproduct means that its production is often inadequately performed and includes many unnecessary risks. The groups participating in this review, together with the group of the University of Modena and Reggio Emilia and the University of Geneve, in collaboration with 3 vinegar companies (Acetaia Cavalli of Reggio Emilia, Italy, Viticultors Mas den Gil, from Priorat, Spain, and Vianigrerie ala Guinelle, from Banyuls, France) and one barrel making company (Boteria Torner, Penedes, Spain) developed the EU WINEGAR Project (wood solutions

for excessive acetification length in traditional vinegar production 6th Framework Program). The WINEGAR project aimed to find alternative methods to improve and shorten the process without compromising the quality of the end product. The project focused on changes in a number of parameters: (i) the raw material used; (ii) the use of barrels specially designed for the development of the product, including assessment of wood type, barrel shape, volume, and use of new wood; and (iii) the selection of acetic acid bacteria starter cultures. The combination of these changes significantly sped up the process (a process that originally took between six months to a year was reduced to 50 days [2]) and maintained or increased the sensory quality of the product [3, 4].

However, there are also other methods that have been used to reduce the acidification time, such as the Schutzenbach system or systems with submerged cultures. In the first type, the bacteria are immobilized on wood chips, forming a solid bed on which the vinegar spreads. After this vinegar passes through the bed of chips, it is collected in a container at the bottom and pumped back to the same fixed bed. The acidity successively increases, and it is possible to obtain vinegar of reasonable quality within a week.

Submerged culture systems provide a much faster alternative. These systems rely on suitable turbines to generate a flow of air bubbles into the wine or alcoholic solution. The oxidative process occurs in the air-liquid interfaces of the air bubbles. Improvements to this process generally involve engineering (maintenance and persistence of the bubbles in the liquid, uniformity of the bubble size, recovery of lost aromas, etc.). In this type of vinegar, the bacteria become bioreactors for the transformation of alcohol into acetic acid, with only very limited production of other metabolites. The airflow also contributes to considerable loss of the volatile compounds present in the original wine, resulting in more organoleptically limited product that was produced at a significantly lower cost. Although early containers for submerged culture processing were made of wood, the most current containers are stainless steel, which is more hygienic and resistant to wear. Although the wood containers were meant to provide some organoleptic complexity, there was hardly any transfer from the wood to the vinegar because of the imbalance between contact surface and volume and the speed of the process. This limitation can be compensated for by subsequent aging in barrels or incubation with wood fragments or wood chips, which may contribute to the recovery of some missing organoleptic characters. Despite the loss in product quality, this methodology has two important advantages: speed (the vinegar is produced in cycles of 24 hours) and acidity (the product can reach concentrations of acetic acid of up to 23–25%, compared to 6–13% achieved with other systems). Higher acidity helps to reduce transportation costs by reducing water transport.

An important aspect that contributes to the organoleptic quality of vinegars is aging. In fact, this is a fundamental aspect of the integration of the different compounds in vinegars. The increase in organoleptic quality after aging is remarkable; in addition to interactions with the wood, a series of chemical reactions, evaporation, the production of esters, reactions between acids and residual alcohols, and other processes result in better integration of aromas and metabolites and a reduction in the pungency of acetic acid.

3. Acetic Acid Bacteria

Although acetic acid bacteria are feared among oenologists because of their negative effects on grapes and on wine in general, they are the main agents in the production of vinegar. Acetic acid bacteria are Gram negative or Gram variable, ellipsoidal or cylindrical, and can be observed under the microscope alone, in pairs or in aggregates and chains [5]. Acetic acid bacteria have aerobic respiratory metabolism, and oxygen is generally used as the final electron acceptor; however, other compounds may occasionally act as final electron acceptors, allowing the bacteria to survive under nearly anaerobic conditions, such as the ones present during wine fermentation [6]. The bacteria's growth in these media is severely limited, and they may remain viable but not culturable [7]. These bacteria are found on substrates containing sugars and/or alcohol, such as fruit juice, wine, cider, beer, and vinegar. On these substrates, the sugars and alcohols are incompletely oxidized, leading to the accumulation of organic acids, such as the production of acetic acid from ethanol or gluconic acid from glucose. Some of the transformations performed by acetic acid bacteria are of great interest to the biotechnology industry. Despite this interest, the role of these bacteria in vinegar production remains their most familiar and extensively used industrial application.

The metabolism of some acetic acid bacteria may include a tricarboxylic acid cycle function, enabling them to completely transform acetic acid to CO_2 and water [5]. However, because entry into the acetate cycle is inhibited by the presence of ethanol, it is essential to maintain a low concentration of ethanol in the presence of acetic acid bacteria to prevent this full oxidation. In fact, ethanol concentrations between 0.5 and 1% are regularly maintained in vinegars.

Acetic acid bacteria have been considered "fastidious" due to their response to growth in culture media. Their cultivability is often lower and more irregular than that observed under the microscope, and these differences can be of several log units [8]. Many strains lose some features (e.g., the ability to produce appreciable concentrations of acetic acid) after growth in culture media. Species identification has traditionally been performed by physiological and biochemical tests, and only half a dozen species from the genera *Acetobacter* and *Gluconobacter* were identified. These two genera could be differentiated based on their preference for alcohol or glucose as a substrate [5]. However, the use of molecular methods has improved efforts at taxonomy, and there are currently 14 genera and approximately 70 species described [9]. Approximately one dozen species and more than 40 strains have been sequenced. Some of the best-known species in the production of vinegars have been transferred from different genera. For example, three of the oldest species described in the production of vinegar were initially classified as genus *Acetobacter*, reclassified as *Gluconacetobacter* [10], and more recently moved to *Komagataeibacter* [9]. These species, *Komagataeibacter europaeus*, *hansenii*, and *xylinus*,

now appear in the literature or textbooks under three different genera.

These molecular methods and their adaptation to the conditions of routine studies for the analysis of populations and the control of microbiological processes have been studied by a group at the Rovira i Virgili University. We have developed a number of methods for the routine identification of species by restriction analysis of ribosomal genes or their spacers [11, 12], which has allowed us to better understand the process of appearance and resistance during the alcoholic fermentation and vinegar production process. Likewise, we applied methods for strain-level identification, which allowed us to track acetic acid bacterial populations from grape to wine and during the process of vinegar making [13, 14]. However, we routinely applied these methods for the analysis of populations recovered in culture media, which has the disadvantage of low recovery, as mentioned above. In recent years, other molecular applications have allowed us to use independent methodologies such as DGGE culture [15–18] or quantitative PCR [8, 19–21]. These methods provide us with additional opportunities to follow the acetic acid bacterial populations in wine or vinegar.

Focusing on the production of wine vinegar, the use of these techniques has allowed us to observe that the vinegar is produced by a succession of strains and species, depending on the concentration of acetic acid [22]. At low concentrations of acetic acid, species of the genus *Acetobacter* predominate. *A. pasteurianus* seems to be the most common in wine vinegars, although other *Acetobacter*, such as *A. malorum*, *A. cerevisiae*, or *A. aceti*, may also be frequent in other fruit vinegars [14, 23]. However, when acetic acid concentrations exceed 5%, the species from the former *Gluconacetobacter* take over the process, with species such as *Komagataeibacter europaeus* or *Gluconacetobacter intermedius* predominating. This transition has also been observed in processes where we inoculated pure starter cultures of *Acetobacter pasteurianus*, during the WINEGAR project. In these cases, we observed that the starter cultures of *A. pasteurianus* effectively initiated the process but were later replaced by *Komagataeibacter europaeus* [24, 25]. This can be explained by both the differing acetic acid tolerances of the species and the presence of a contaminating population of acetic acid bacteria in the raw material (wine). At present, we believe that the best controlled process should include a starter formed by a mixed culture of a "quick start" acetic acid bacterium (*A. pasteurianus* or similar) and another with a high tolerance to acetic acid (*Komagataeibacter europaeus* or similar) to guarantee the best vinegar production through a rapid start and a good ending for the process.

4. Chemical Composition and Quality of Vinegars

The final quality of vinegars depends on the selection of appropriate starter cultures (generally mixed cultures) to lead the process. However, other factors include the quality of the starting material, the production method, and, if applicable, aging. In general, it is relatively easy to appreciate the sensory differences between products made by traditional methods and those manufactured on an industrial scale. Thorough characterization and quality evaluation requires the determination of the content of a number of compounds and sensory analysis. In recent years, there have been significant advances in the elucidation of the compounds responsible for the sensory quality of the products, and production methods have been changed to obtain vinegars with high acceptance at very competitive prices. The group at the University of Sevilla has been focusing on the characterization of wine vinegars for the last 20 years.

Aromatic compounds have a decisive effect on the quality of vinegars. The aroma is a complex fraction, containing many compounds with a wide range of volatilities, polarities, and concentrations ranging from several mg/L to ng/L. To date, we have identified more than 100 different chemical compounds in the aroma of wine vinegar, including carbonyl compounds, ethers, acetals, lactones, acids, alcohols, phenols, and volatile esters, all of which are involved to different extents in the final flavor [4]. During the aging process, the contact with wood produces a substantial increase in the aromatic complexity [26]. However, not all volatile compounds are responsible for the aroma of the product. They must not only reach odorant receptors but also interact with them in the olfactory epithelium, and not all volatile compounds are active odorants. The use of techniques based on gas chromatography coupled with olfactometry has allowed the contribution of each volatile compound in the final vinegar to be evaluated. For example, it has been determined that the characteristic aroma of Sherry vinegar involves several volatile compounds, such as diacetyl, isoamyl acetate, isovaleric acid, ethyl acetate, and sotolon [27].

Polyphenolic compounds, which are ubiquitous in plant products, are of great interest as quality determinants because, in addition to their antioxidant activity, they are responsible for the color and astringency of vinegar. Acetification is an aerobic process, and oxygen is critical to the growth of the bacteria. The reactivity of phenolic compounds and oxygen is specifically analyzed in winemaking for its relationship to the browning of white wines and the reactions of anthocyanins in red wines. The rate of acetification is also expected to be related to the solubility of oxygen in the medium, a decisive factor in the phenolic composition that can be useful for determining the method by which vinegar is produced. It should be emphasized that submerged systems use excess oxygen to secure and accelerate the process, whereas oxygen availability is limited in superficial cultures because it is continuously taken up by acetic acid bacteria. Additionally, oxygen affects the classes of polyphenolic compounds to different degrees. For example, the flavonol content of vinegars is largely influenced by oxygen availability during submerged fermentation. In contrast, surface acetification vinegars do not affect phenolic aldehydes, which are released from wooden barrels into the product [28].

The evolution of phenolic compounds during acetification in submerged culture systems has been studied in both laboratory and industrial fermenters. In a laboratory fermenter with Sherry wine as the substrate, the phenolic profile was not significantly altered [29]. However, a 50% decrease in phenolic compounds, mainly anthocyanins, has been reported in red wine vinegar [30].

The aging process involves the reaction of compounds over time: both the polymerization and release of compounds from the wood and losses through evaporation. The substances provided by the wood will depend on the type of wood and roasting, the ratio of the contact surface to liquid volume, and the aging time. As a consequence, significant differences have been observed in the phenolic composition of Sherry vinegars aged two or more years in static or traditional solera systems [31]. An observation of the evolution of phenolic compounds in Sherry vinegar aged in oak barrels showed that there were significant differences in the compounds vanillin, syringaldehyde, coniferyl aldehyde, and cinnamic acid after 90 days of aging [32]. Indeed, a 100% correct classification of vinegars aged for different periods of time was achieved by means of linear discriminant analysis using phenolic aldehydes as the variables. Furthermore, certain flavonoids are chemical markers of the wood that the vinegar has been in contact with; (+)-dihydrorobinetin is a characteristic compound released from nontoasted acacia wood, while (−)-taxifolin is typically released from cherry wood [33].

However, the consumer's perception of the product is the most important factor. Vinegar is a difficult product to taste, due to the intense sensations it provokes. The pungency of the high acetic acid content masks other flavors, and some familiarity with the product is required to proceed with a tasting. In fact, there is no consensus on how vinegar should be tasted. A vinegar sensory analysis panel requires well-trained tasters, and the specific attributes that are useful for differentiating among samples must be chosen.

To train a vinegar panel, Tesfaye et al. [34] used solutions of different concentrations of the compounds most typically found in wine vinegar, such as acetic acid, ethyl acetate, and wood extract obtained by maceration. The last two were prepared in 7% acetic acid to provoke a sensation similar to that of vinegar. Acetic acid aggressiveness determines the number of samples that can be examined in each session, and each sample is tasted four times. These four replicate samples should be tasted on different days to avoid sensorial saturation of the tasters. A descriptive analysis of the samples is prepared based on previously selected attributes that can be evaluated by the panel. The attributes used to describe the vinegar samples were color, aromatic intensity, woody scent, herbaceous smell, fruity odor of ethyl acetate, wine smell, and pungent feeling [34, 35]. Higher sensory thresholds for most compounds were obtained in an acetic acid matrix compared with water solutions. Conversely, high quality vinegars contain a large number of these compounds at concentrations higher than their threshold limits, including vanillin, eugenol, and benzaldehyde, and this characteristic could be therefore selected as an attribute of high quality vinegars [36]. Additionally, adequate training and a standardized tasting protocol contribute to the reliability of descriptive sensory analyses of a vinegar and of ascertaining the aging period and wood used in its elaboration [37].

Conflict of Interests

The authors declare that there is no conflict of interests regarding the publication of this paper.

Acknowledgment

This work was supported by the European Project WINEGAR (COOP-CT-2005/017269).

References

[1] L. Solieri and P. Giudici, *Vinegars of the World*, Springer, Berlin, Germany, 2009.

[2] M. J. Torija, E. Mateo, C.-A. Vegas et al., "Effect of wood type and thickness on acetification kinetics in traditional vinegar production," *International Journal of Wine Research*, vol. 1, no. 1, pp. 155–160, 2009.

[3] A. B. Cerezo, W. Tesfaye, M. J. Torija, E. Mateo, M. C. García-Parrilla, and A. M. Troncoso, "The phenolic composition of red wine vinegar produced in barrels made from different woods," *Food Chemistry*, vol. 109, no. 3, pp. 606–615, 2008.

[4] R. M. Callejón, W. Tesfaye, M. J. Torija, A. Mas, A. M. Troncoso, and M. L. Morales, "Volatile compounds in red wine vinegars obtained by submerged and surface acetification in different woods," *Food Chemistry*, vol. 113, no. 4, pp. 1252–1259, 2009.

[5] J. de Ley, F. Gossele, and J. Swings, "Genus I *Acetobacter*," in *Bergey's Manual of Systematic Bacteriology*, vol. 1, pp. 268–274, Williams & Wilkens, Baltimore, Md, USA, 1984.

[6] G. S. Drysdale and G. H. Fleet, "Acetic acid bacteria in winemaking: a review," *American Journal of Enology and Viticulture*, vol. 39, pp. 143–154, 1988.

[7] V. Millet and A. Lonvaud-Funel, "The viable but non-culturable state of wine micro-organisms during storage," *Letters in Applied Microbiology*, vol. 30, no. 2, pp. 136–141, 2000.

[8] M. J. Torija, E. Mateo, J. M. Guillamón, and A. Mas, "Identification and quantification of acetic acid bacteria in wine and vinegar by TaqMan-MGB probes," *Food Microbiology*, vol. 27, no. 2, pp. 257–265, 2010.

[9] Y. Yamada, P. Yukpan, H. T. L. Vu et al., "Description of *Komagataeibacter* gen. nov., with proposals of new combinations (*Acetobacteraceae*)," *The Journal of General and Applied Microbiology*, vol. 58, pp. 397–404, 2012.

[10] Y. Yamada, K.-I. Hoshino, and T. Ishikawa, "The phylogeny of acetic acid bacteria based on the partial sequences of 16S ribosomal RNA: the elevation of the subgenus *Gluconoacetobacter* to the generic level," *Bioscience, Biotechnology and Biochemistry*, vol. 61, no. 8, pp. 1244–1251, 1997.

[11] A. Ruiz, M. Poblet, A. Mas, and J. M. Guillamón, "Identification of acetic acid bacteria by RFLP of PCR-amplified 16S rDNA and 16S-23S rDNA intergenic spacer," *International Journal of Systematic and Evolutionary Microbiology*, vol. 50, no. 6, pp. 1981–1987, 2000.

[12] Á. González and A. Mas, "Differentiation of acetic acid bacteria based on sequence analysis of 16S-23S rRNA gene internal transcribed spacer sequences," *International Journal of Food Microbiology*, vol. 147, no. 3, pp. 217–222, 2011.

[13] Á. González, N. Hierro, M. Poblet, N. Rozès, A. Mas, and J. M. Guillamón, "Application of molecular methods for the differentiation of acetic acid bacteria in a red wine fermentation,"

Journal of Applied Microbiology, vol. 96, no. 4, pp. 853–860, 2004.

[14] C. Hidalgo, E. Mateo, A. Mas, and M. J. Torija, "Identification of yeast and acetic acid bacteria isolated from the fermentation andacetification of persimmon (Diospyros kaki)," *Food Microbiology*, vol. 30, no. 1, pp. 98–104, 2012.

[15] I. Lopez, F. Ruiz-Larrea, L. Cocolin et al., "Design and evaluation of PCR primers for analysis of bacterial populations in wine by denaturing gradient gel electrophoresis," *Applied and Environmental Microbiology*, vol. 69, no. 11, pp. 6801–6807, 2003.

[16] L. de Vero, E. Gala, M. Gullo, L. Solieri, S. Landi, and P. Giudici, "Application of denaturing gradient gel electrophoresis (DGGE) analysis to evaluate acetic acid bacteria in traditional balsamic vinegar," *Food Microbiology*, vol. 23, no. 8, pp. 809–813, 2006.

[17] C. Ilabaca, P. Navarrete, P. Mardones, J. Romero, and A. Mas, "Application of culture culture-independent molecular biology based methods to evaluate acetic acid bacteria diversity during vinegar processing," *International Journal of Food Microbiology*, vol. 126, no. 1-2, pp. 245–249, 2008.

[18] I. Andorrà, S. Landi, A. Mas, J. M. Guillamón, and B. Esteve-Zarzoso, "Effect of oenological practices on microbial populations using culture-independent techniques," *Food Microbiology*, vol. 25, no. 7, pp. 849–856, 2008.

[19] Á. González, N. Hierro, M. Poblet, A. Mas, and J. M. Guillamón, "Enumeration and detection of acetic acid bacteria by real-time PCR and nested PCR," *FEMS Microbiology Letters*, vol. 254, no. 1, pp. 123–128, 2006.

[20] C. Jara, E. Mateo, J.-M. Guillamón, A. Mas, and M. J. Torija, "Analysis of acetic acid bacteria by different culture-independent techniques in a controlled superficial acetification," *Annals of Microbiology*, vol. 63, pp. 393–398, 2013.

[21] C. Vegas, A. Gonzalez, E. Mateo, A. Mas, M. Poblet, and M. J. Torija, "Evaluation of representativity of the acetic acid bacteria species identified by culture-dependent method during a traditional wine vinegar production," *Food Research International*, vol. 51, pp. 404–411, 2013.

[22] C. Vegas, E. Mateo, Á. González et al., "Population dynamics of acetic acid bacteria during traditional wine vinegar production," *International Journal of Food Microbiology*, vol. 138, no. 1-2, pp. 130–136, 2010.

[23] C. Hidalgo, E. Mateo, A. Mas, and M. J. Torija, "Effect of inoculation on strawberry fermentation and acetification processes using native strains of yeast and acetic acid bacteria," *Food Microbiology*, vol. 34, pp. 88–94, 2013.

[24] M. Gullo, L. de Vero, and P. Giudici, "Succession of selected strains of *Acetobacter pasteurianus* and other acetic acid bacteria in traditional balsamic vinegar," *Applied and Environmental Microbiology*, vol. 75, no. 8, pp. 2585–2589, 2009.

[25] C. Hidalgo, C. Vegas, E. Mateo et al., "Effect of barrel design and the inoculation of *Acetobacter pasteurianus* in wine vinegar production," *International Journal of Food Microbiology*, vol. 141, no. 1-2, pp. 56–62, 2010.

[26] R. M. Callejón, M. J. Torija, A. Mas, M. L. Morales, and A. M. Troncoso, "Changes of volatile compounds in wine vinegars during their elaboration in barrels made from different woods," *Food Chemistry*, vol. 120, no. 2, pp. 561–571, 2010.

[27] R. M. Callejón, M. L. Morales, A. C. Silva Ferreira, and A. M. Troncoso, "Defining the typical aroma of Sherry vinegar: sensory and chemical approach," *Journal of Agricultural and Food Chemistry*, vol. 56, no. 17, pp. 8086–8095, 2008.

[28] M. C. García-Parrilla, F. J. Heredia, and A. M. Troncoso, "The influence of the acetification process on the phenolic

composition of wine vinegars," *Sciences des Aliments*, vol. 18, no. 2, pp. 211–221, 1998.

[29] M. L. Morales, W. Tesfaye, M. C. García-Parrilla, J. A. Casas, and A. M. Troncoso, "Sherry wine vinegar: physicochemical changes during the acetification process," *Journal of the Science of Food and Agriculture*, vol. 81, no. 7, pp. 611–619, 2001.

[30] W. Andlauer, C. Stumpf, and P. Fürst, "Influence of the acetification process on phenolic compounds," *Journal of Agricultural and Food Chemistry*, vol. 48, no. 8, pp. 3533–3536, 2000.

[31] M. C. García-Parrilla, F. J. Heredia, and A. M. Troncoso, "Sherry wine vinegars: phenolic composition changes during aging," *Food Research International*, vol. 32, no. 6, pp. 433–440, 1999.

[32] W. Tesfaye, M. L. Morales, M. C. García-Parrilla, and A. M. Troncoso, "Evolution of phenolic compounds during an experimental aging in wood of Sherry vinegar," *Journal of Agricultural and Food Chemistry*, vol. 50, no. 24, pp. 7053–7061, 2002.

[33] A. B. Cerezo, J. L. Espartero, P. Winterhalter, M. C. García-Parrilla, and A. M. Troncoso, "(+)-dihydrorobinetin: a marker of vinegar aging in acacia (*Robinia pseudoacacia*) wood," *Journal of Agricultural and Food Chemistry*, vol. 57, no. 20, pp. 9551–9554, 2009.

[34] W. Tesfaye, M. C. García-Parrilla, and A. M. Troncoso, "Sensory evaluation of Sherry wine vinegar," *Journal of Sensory Studies*, vol. 17, no. 2, pp. 133–144, 2002.

[35] W. Tesfaye, M. L. Morales, B. Benítez, M. C. García-Parrilla, and A. M. Troncoso, "Evolution of wine vinegar composition during accelerated aging with oak chips," *Analytica Chimica Acta*, vol. 513, no. 1, pp. 239–245, 2004.

[36] W. Tesfaye, M. L. Morales, R. M. Callejón et al., "Descriptive sensory analysis of wine vinegar: tasting procedure and reliability of new attributes," *Journal of Sensory Studies*, vol. 25, no. 2, pp. 216–230, 2010.

[37] A. B. Cerezo, W. Tesfaye, M. E. Soria-Díaz et al., "Effect of wood on the phenolic profile and sensory properties of wine vinegars during ageing," *Journal of Food Composition and Analysis*, vol. 23, no. 2, pp. 175–184, 2010.

13

Effect of Selenate on Viability and Selenomethionine Accumulation of *Chlorella sorokiniana* Grown in Batch Culture

Živan Gojkovic,[1,2] **Carlos Vílchez,**[1,3] **Rafael Torronteras,**[4]
Javier Vigara,[1] **Veronica Gómez-Jacinto,**[5] **Nora Janzer,**[6] **José-Luis Gómez-Ariza,**[5]
Ivana Márová,[2] **and Ines Garbayo**[1]

[1] *Algal Biotechnology Group, Department of Chemistry and Material Sciences, Faculty of Experimental Sciences, University of Huelva, Campus el Carmen, Avenida de las Fuerzas Armadas s/n, 21007 Huelva, Spain*
[2] *Department of Food Technology and Biotechnology, Faculty of Chemistry, Brno University of Technology, Purkyňova 118, 61200 Brno, Czech Republic*
[3] *Algal Biotechnology Group, International Centre for Environmental Research (CIECEM), Parque Dunar s/n, Matalascaňas, Almonte, 21760 Huelva, Spain*
[4] *Department of Environmental Biology and Public Health, University of Huelva, Campus el Carmen, 21007 Huelva, Spain*
[5] *Department of Chemistry and Material Sciences, University of Huelva, Campus el Carmen, 21007 Huelva, Spain*
[6] *University of Applied Science, Giessen, Wiesenstrasse 14, 35390 Giessen, Germany*

Correspondence should be addressed to Živan Gojkovic; xcgojkovic@fch.vutbr.cz

Academic Editors: T. Betakova and D. Zhou

The aim of this work was to study the effect of Se(+VI) on viability, cell morphology, and selenomethionine accumulation of the green alga *Chlorella sorokiniana* grown in batch cultures. Culture exposed to sublethal Se concentrations of 40 mg·L^{-1} (212 μM) decreased growth rates for about 25% compared to control. A selenate EC$_{50}$ value of 45 mg·L^{-1} (238.2 μM) was determined. Results showed that chlorophyll and carotenoids contents were not affected by Se exposure, while oxygen evolution decreased by half. Ultrastructural studies revealed granular stroma, fingerprint-like appearance of thylakoids which did not compromise cell activity. Unlike control cultures, SDS PAGE electrophoresis of crude extracts from selenate-exposed cell cultures revealed appearance of a protein band identified as 53 kDa Rubisco large subunit of *Chlorella sorokiniana*, suggesting that selenate affects expression of the corresponding chloroplast gene as this subunit is encoded in the chloroplast DNA. Results revealed that the microalga was able to accumulate up to 140 mg·kg^{-1} of SeMet in 120 h of cultivation. This paper shows that *Chlorella sorokiniana* biomass can be enriched in the high value aminoacid SeMet in batch cultures, while keeping photochemical viability and carbon dioxide fixation activity intact, if exposed to suitable sublethal concentrations of Se.

1. Introduction

Selenium is a trace element that acts either as an essential micronutrient or as a toxic element for fish, birds, animals, and humans depending on its concentration [1, 2]. It is of fundamental importance to human health, it plays a role in mammalian development [3], immune function [4], and in slowing down aging [5]. At low levels, it contributes to normal cell growth and function (a daily intake of 40 μg for adult men and 30 μg for women is recommended by WHO [6]). High concentrations are toxic, causing the generation of reactive oxygen species (ROS), which can induce DNA oxidation, DNA double strand breaks, and cell death [7]. The primary cause of Se deficiency that reduces growth, reproduction, and even causes death is its low amount in soil and consequently in animal feed [8]. Selenium bioeffects are mainly involved in immune function, reproduction, metal toxicity resistance, and other biological functions [9]. Besides, selenium has been proven to be an effective anticancer agent mainly based on statistical and model studies [10], when it is supplied in a suitable bioactive form [11–13].

In nature, inorganic selenium is present in three oxidation states: selenate (+VI), selenite (+IV), and elemental selenium (0) over a range of natural water chemical conditions. Selenate is the dominant dissolved form, representing more than 67% of the total dissolved selenium concentration. Both selenates and selenites are taken up by microalgae and converted to protein-bound selenocysteine (SeCys) and selenomethionine (SeMet) [14]. SeCys is the predominant selenoaminoacid in tissues when inorganic selenium or organic bound as Se-enriched yeast is given to animals [15, 16]. Besides, selenium can substitute sulphur in methionine and forms SeMet. This can be incorporated unspecifically into proteins instead of methionine. SeMet cannot be synthesized by higher animals and humans [17] and is present in plant foods, while SeCys is more common in animal foods. Recent studies have shown that certain Se-compounds, as SeMet, are effective chemoprotective agents, reducing the incidence of breast, liver, prostate, and colorectal cancers in model systems [15, 18, 19]. Because SeMet is the main natural selenium form, synthesized SeMet or SeMet-enriched foods (e.g., selenized yeast) are acceptable as more effective forms of selenium in humans and animals [20, 21]. Some studies have been done to test the effects of Se-enriched animal feed in animal health, which has increased interest for plant enrichment in selenocompounds [19, 22, 23]. Microalgae appears to be the easiest plant-like biomass to be Se-enriched.

Many studies concerning Se toxicity in microalgae can be found in the literature; selenate effect has been studied in *Chlamydomonas reinhardtii* [1, 24–26], *Scenedesmus quadricauda* [27, 28], and cyanobacterium *Spirulina platensis* [29]. Similar studies have been done with *Chlorella zofingiensis* with emphasis on heat-stable selenoproteins [30], on metabolism of Se volatile compounds [14], Se effect on *Chlorella sp.* cultivated on glucose [16], and Se effect on continuous microalgae cultures of *Chlorella pyrenoidosa* [31] and *Chlorella sorokiniana* [32].

Microalgae *Chlorella sorokiniana* was selected for this study as an ideal target microorganism which is ubiquitous, exerts positive effects on human health and biotransforms selenate in selenocompounds such as SeMet [16, 33]. Study of selenate effect was focused on several levels: monitoring various culture parameters and comparing results with those of unexposed cultures (control cultures), ultrastructure examined by transmission electron microscopy, isolation and identification of Se-affected proteins, and Se biotransformation to SeMet and other Se aminoacids. To our knowledge, there are no recent detailed papers describing Se effect on microalgae batch cultures. With this study we offer new insight on SeMet-enriched *Chlorella sorokiniana* biomass production in batch cultures exposed to sublethal Se concentrations, intending to show that SeMet-enriched algal biomass production is feasible in batch systems, keeping both cell photochemical viability and structural stability if suitable selenium conditions are selected.

2. Materials and Methods

2.1. Microalga, Growth Medium, and Experimental Conditions. The microalga *Chlorella sorokiniana* CCAP 211/8 K was obtained from the UTEX culture collection. It was maintained in modified M-8 medium [34] in Erlenmeyer flasks at 25°C and 165 μmol photons m^{-2}s^{-1}. The culture medium was prepared as follows (composition expressed in g·L^{-1}): KH$_2$PO$_4$, 0.74 g·L^{-1}; Na$_2$HPO$_4$ × 2H$_2$O, 0.26 g·L^{-1}; MgSO$_4$ × 7H$_2$O, 0.4 g·L^{-1}; CaCl$_2$ × 2H$_2$O, 0.013 g·L^{-1}; KNO$_3$, 3 g·L^{-1}; EDTA ferric sodium salt, 0.116 g·L^{-1}; Na$_2$EDTA × 2H$_2$O, 0.0372 g·L^{-1}; H$_3$BO$_3$, 6.18 × 10^{-5} g·L^{-1}; MnCl$_2$ × 4H$_2$O, 1.3 × 10^{-2} g·L^{-1}; ZnSO$_4$ × 7H$_2$O, 3.20 × 10^{-3} g·L^{-1}; and CuSO$_4$ × 5H$_2$O, 3.2 × 10^{-3} g·L^{-1}. Chemicals were purchased from Sigma-Aldrich (Germany), unless otherwise indicated. In the prepared fresh medium precalculated amount of selenium was added in the form of aqueous stock solutions of Na-selenate (Na$_2$SeO$_4$). Prior to experiments, cultures were inoculated with cells in the exponential growth phase in order to obtain an initial cell density of approximately 3·10^6 cell·mL^{-1}. The pH was adjusted to 6.7 with concentrated solution of NaOH.

The microalga *Chlorella sorokiniana* was cultivated in 5 L laboratory glass bottles at 25°C and continuously illuminated with white fluorescent lamps (Philips TLD, 30 W, 160 μmol photons m^{-2}s^{-1}), at the surface of the flask. The irradiance was measured with a photoradiometer Delta OHM, model HD 9021, Italy. The culture suspension was mixed both with magnetic stirrer at 150 rpm and by air bubbling containing 5% (v/v) CO$_2$, as unique carbon source. In order to cover a complete algal growth cycle, culture parameters were monitored three times a day. Selenate concentrations in the culture medium used in the experiments were 40 mg·L^{-1} selenate, for Se effect on culture growth and SeMet accumulation studies and 40 mg·L^{-1} and 100 mg·L^{-1} for cell ultrastructure studies.

2.2. Biomass Concentration and Optical Density. Biomass concentration was determined by dry weight measurements. Dry weight was determined by filtration of the culture broth over glass fiber filters with a pore size of 0.7 mm (Whatman GF/F, Kent, UK). The filter weight was determined on a 0.01 mg precision balance. Aliquots of 5 mL of culture broth, diluted 10 times with prefiltered demineralized water in order to remove inorganic salts, were filtered through prewashed, predried, and preweighed filters. Filters were dried at 80°C during at least 16 h and cooled down in a dessicator for 2 hours. Dry weight, expressed as g·L^{-1} of culture broth, was calculated by differential weight.

Optical density was determined spectrophotometrically at 680 nm using UV/Visible spectrophotometer (Ultrospec 3100 pro, Amersham Pharmacia Biotech, Uppsala, Sweden).

2.3. Population Density, Algal Growth, and Statistical Analysis. Population density was determined by counting the number of cells using a Neubauer chamber and light microscopy (Olympus CX41), and calculated based on the equation: $N = 0.25 \cdot 10^4 \cdot (\sum N_i) \cdot D$ and expressed as 10^6 cell·mL^{-1}. Where

N is population density (cell·mL^{-1}), $\sum N_i$ is total sum of the counted cell numbers on Neubauer chamber ($i = 1, 2, 3, 4$), and D is applied dilution of the culture.

To study the effect of Se on the algal growth, a logistic model, defined by Verhlust [35], was used. This model uses three key parameters: initial cell density at time zero (N_0, cell·mL^{-1}), maximal density that can theoretically be reached (N_{max}, cell·mL^{-1}), and maximal culture growth rate (μ_m, h^{-1}). According to the model, cell density, $N(t)$, at any time, (t), is given by the following equation:

$$N(t) = N_{max} \cdot N_0 \cdot \left(N_0 + (N_{max} - N_0) \cdot e^{-\mu t} \right)^{-1}. \quad (1)$$

Maximal cell densities and growth rates were assessed by means of a logistic curve estimation function of SPSS Statistical Package software (v.19). Model curve was fitted to mean values of population density data [36], this model was previously used to describe growth kinetics of microalgae, in general [37] and of species such as *Chlamydomonas reinhardtii* [24–26] and *Chlorella minutissima* [38]. EC$_{50}$ value is defined as half of the maximal effective concentration of a given substance; this is to say, concentration that provokes 50% of maximal effect. In this paper, EC$_{50}$ for selenate is equivalent to 50% of *Chlorella sorokiniana* growth inhibiting selenate concentration. In order to calculate EC$_{50}$ value, maximal growth rates obtained from growth curves previously fitted to logistic model were used to construct the dose-response curve. In order to predict EC$_{50}$ we fitted growth rates data to the parameter log-logistic model (also known as Hill's model), thoroughly explained elsewhere [24–26] using the open source software "R statistical package" [39], according to instructions on fitting a single dose-response curve published by Ritz and Streibig [40]. R package has been previously used by Geoffroy et al. for statistical analysis of data on selenate effect on *Chlamydomonas reinhardtii* [1].

2.4. Chlorophyll and Carotenoids.
The chlorophyll and carotenoids content was determined by methanol extraction and spectrophotometry. After centrifugation (5 min at 4400 rpm), biomass was mixed with methanol and the mixture was placed in an ultrasound bath for 5 min to disrupt the pellet. Subsequently, mixture was incubated at 60°C first and then cooled at 0°C to break the cells. After centrifugation, supernatant was collected and analyzed by UV/Visible spectrophotometry. Modified Arnon's equations [41] were used to calculate the chlorophyll and carotenoid concentrations in the extracts. The cell contents of chlorophyll and carotenoids were expressed per gram of biomass, calculated based on samples dry weight.

2.5. Measurement of Fluorescence.
Another method used for assessing biological activity of the algal population was fluorometry. Maximum fluorescence yield (Y_{op}) was determined by pulse amplitude modulation (PAM) fluorometry with the saturating-pulse technique. A chlorophyll fluorometer (PAM-210, Walz, Germany) was used. The samples were first adapted to dark for 15 min in order to open all photosystems reaction centers. Light of 0.04 μmol photons m^{-2}s^{-1} was used

to measure the zero fluorescence level (F_0). Saturating light pulse (1850 μmol photons m^{-2}s^{-1}) was used to measure the maximum fluorescence (F_m). Then the sample was illuminated with actinic light and series of saturating pulses in order to reach steady (light-adapted) state fluorescence (F') and steady state maximum fluorescence F'_m level. Finally, the actinic light and saturating pulses were switched off to measure F'_0 level. The maximum photochemical yield and effective photochemical yield of photosystem II were calculated using the equations [42]: $Y_{op} = (F_m - F_0)/F_m$ and $\Phi_{PSII} = (F'_m - F')/F'_m$.

2.6. Oxygen Evolution.
The photosynthetic activity was measured to test cell viability; 1 mL of algal cell culture was placed into a Clark-type electrode (Hansatech, UK) to measure O$_2$ evolution. The electrode was equipped with a stirrer bar, a pressure corrector, and a temperature sensor. It was placed in a photosynthetic cylindrical chamber of 15 mm inside diameter and 10 mL capacity, surrounded by an outer water jacket for constant temperature operation. Measurements were made at 25°C under saturating white light (1500 μmol photons m^{-2}s^{-1}) or darkness (endogenous respiration).

2.7. Cell Protein Isolation and Fractionation with Ammonium Sulfate.
Cultures containing 40 mg·L^{-1} of selenite, as well as untreated culture, were grown in batch for 240 h. One liter of each culture was sampled on time zero and 120 h of cultivation as well as at the end of the experiment (240 h). Cells from sampled culture were collected by centrifugation (4400 rpm for 5 min) and resuspended in 20 mM phosphate buffer (pH 7) to a final concentration of 0.67 g·mL^{-1}. Cell disruption was performed on ice with an ultrasonic probe (Lab Sonic) at 40% of power for 10 seconds, followed by a 50 second pause to avoid heat denaturation. This procedure was repeated 10 times. Extracts were centrifuged (13000 rpm for 20 min at 4°C), cell debris was discarded, and supernatant was collected. Prior to ammonium sulfate fractionation, non-protein materials were precipitated with 0.1 M streptomycin sulfate solution in phosphate buffer (pH 7). Ammonium sulfate fractionation procedure was performed using protocol described by Harris [43]. All solid fractions were resuspended in 20 mM P-buffer (pH 7) and kept frozen until electrophoresis was performed.

2.8. SDS-Polyacrylamide Gel Electrophoresis (SDS-PAGE).
SDS-PAGE was performed by the method of Laemmli (1970) [44]. Protein samples were mixed with the sample buffer (0.5 M Tris-HCl, pH 6,8 containing 5% SDS, 20% glycerol) at 1 : 2 ratio at the presence of 10% β-mercaptoethanol. Electrophoresis was performed on 10% resolving gels with 4% stacking gels. Molecular weight marker 14.2–66 kDa (Sigma) was used to estimate the molecular weight of proteins. Volume of 20 μL of sample was loaded in each well containing 15 μg of proteins. Protein concentration was determined by spectrophotometry using BioRad Bradford reagent at 595 nm, with bovine serum albumin as the standard [45]. Electrophoresis run at 180 V for 75 min. Gels were washed three times with distilled water, stained with Coomassie-Blue

stain for 180 min, and destained overnight with 10% acetic acid/30% ethanol aqueous solution.

2.9. Protein Analysis by MALDI-TOF-TOF Mass Spectrometry.
Samples were automatically digested with trypsin according to standard protocols [46]. MALDI-TOF-TOF analysis was performed by Central Services of Research at the University of Cordoba, Spain, using an ABI Applied Biosystems 4700 Proteomics Analyzer (Amersham Biosciences). Mass spectra were obtained using a laser (337 nm, 200 Hz) as desorption ionization source. Data were acquired in the reflection positive mode using delayed extraction. Spectra were calibrated using trypsin autolysis products as internal standards. After MS acquisition, the 10 strongest peptides per spot were selected automatically for MS-MS analysis.

Identification of proteins was carried out by searching against NCBI nonredundant protein sequence database. MASCOT searching engine (Matrixscience, UK) was used for protein identification. Protein identifications with the score value higher than 60 were positively assigned, after considering MW and pI values.

2.10. Extraction and Determination of Selenium Species.
Cultures of *Chlorella sorokiniana* were centrifuged to separate the pellet from the medium. Liquid nitrogen was applied to the pellet to disrupt the cell walls and an amount of 0.020 g was weighted in a centrifuge tube, then 0.02 g of Protease XIV was added. The extraction was performed with the assistance of a ultrasonic probe at 25% power during 2 minutes. After the extraction, the sample was centrifuged for 5 minutes at 6000 rpm and the supernatant collected. Finally the supernatant was filtered through 0.45 μm (PVDF) filters and injected in the HPLC-ICP-MS.

The Se was measured by ICP-MS using the following operational conditions: forward power 1500 W, sampling depth 7-8 mm, auxiliary gas flow rate 0.10–0.15 mL·min^{-1}, extract I: 0–3 V, extract II: −137,5 V, omega Bias-ce −20 V, omega Lens-ce −1.6 V, cell entrance −40 V, QP focus −15 V, cell exit −44 V, octP RF 190 V, octP bias −18 V, H$_2$ flow 3.8 mLmin^{-1}, QP bias −16 V, discriminator 8 m V, and analog HV 1840 V. The ^{77}Se, ^{80}Se, and ^{82}Se were monitored for analysis, but only isotope ^{80}Se was used for quantification. A solution containing Li, Y, Tl, and Ce (1 μg·L^{-1} each) prepared in the mobile phase was used to tune the ICP-MS for sensitivity, resolution, percentage of oxides, and doubly charged ions. The chromatographic separation was performed on the basis of previously described instrumental coupling [47, 48].

2.11. Intracellular Structure Examination by Transmission Electron Microscopy (TEM).
For observations in electron microscopy, cultures containing 40 mg·L^{-1} and 100 mg·L^{-1} of selenate, as well as untreated culture, were cultivated in batch for 240 h. The algal cells were then collected from each culture, washed with culture medium, and collected by centrifugation (2500 rpm, 5 min). The algal cells were fixed with 1% glutaraldehyde in 0.1 M sodium cacodylate buffer (pH 7.4) for 2 h at 4°C. The cells were then washed three

TABLE 1: Values of growth parameters N_{max} (10^6 cell·mL^{-1}) and μ_{max} (day^{-1}) obtained by fitting logistic model equation to experimental data of population density in function of time from selenium exposed culture and control. Model curve was fitted to mean values of population density data (see Figure 1); hence, fitted models give single parameters values, instead of N_{max} and μ_m mean values ± S.D.

Selenate concentration (mg·L^{-1})	0	40
N_{max} (10^6 cell·mL^{-1})	252	145
μ_{max} (day^{-1})	1.72	1.31
Correlation coefficients (R^2)	0.977	0.974

times (5 min each one) using the same buffer. The samples were postfixed with 1% osmium tetroxide in 0.2 M cacodylate buffer at 4°C for 1 h. Samples were washed with the same buffer, dehydrated in a graded ethanol series, and embedded in Epon 812 (EMbed 812 Kit; Electron Microscopy Science, Hatfield, PA, USA). Ultrathin sections of 80–90 nm obtained by an ultramicrotome (UCT, Leica, Wetzlar, Germany) and placed on nickel grids were stained with aqueous 1% (w/v) uranyl acetate and lead citrate. Transmission electron micrographs were observed with a JEM 1011 (JEOL Ltd., Tokyo, Japan) electron microscope using an accelerating voltage of 80 kV. Several photographs of entire cells and of local detailed structures were taken at random, analyzed, and compared to investigate selenium effect in the different subcellular structures of *Chlorella sorokiniana*. All chemicals used for histological preparation were purchased from Electron Microscopy Sciences.

2.12. Statistics.
All experiments were triplicate unless indicated otherwise. Mean values of data are reported with standard deviations (±SD). Statistical analyses were performed using the Statistical Package for Social Sciences, SPSS v. 19 for Windows (SPSS Inc. USA) and open source software "R statistical package" propriety of R Development Core Team [39].

3. Results and Discussion

3.1. Effect of Selenate on Culture Growth.
The above described logistic mathematical model was used to fit data of population density changes as a function of time (Figure 1). Correlation coefficients (R^2) of the fitted models were 0.977 for selenite-exposed cultures and 0.974 for control cultures. Values of the parameters used in model are presented in Table 1. Experimental data of cell numbers and those data calculated from the model are graphically presented in Figures 1(a) and 1(b).

Maximal culture growth rate (μ_m, day^{-1}) is a fundamental growth parameter and if any key metabolic process of the cell is affected by toxins it will result in decreased μ_m values, which makes such parameter a relevant indicator for Se toxicity on microalgal cultures [31]. Maximal growth rate in 40 mg·L^{-1} selenate-exposed *Chlorella sorokiniana* culture (μ_{max}, 1.72 day^{-1}, Table 1) accounted for 76% of the control value, which is comparable to the literature data. Continuously cultivated *Chlorella pyrenoidosa* cells showed μ_{max} values of 1.46 and 0.94 day^{-1} for 0.53 and 1.41 mg·L^{-1} selenate, respectively, in

(a)

(b)

(c)

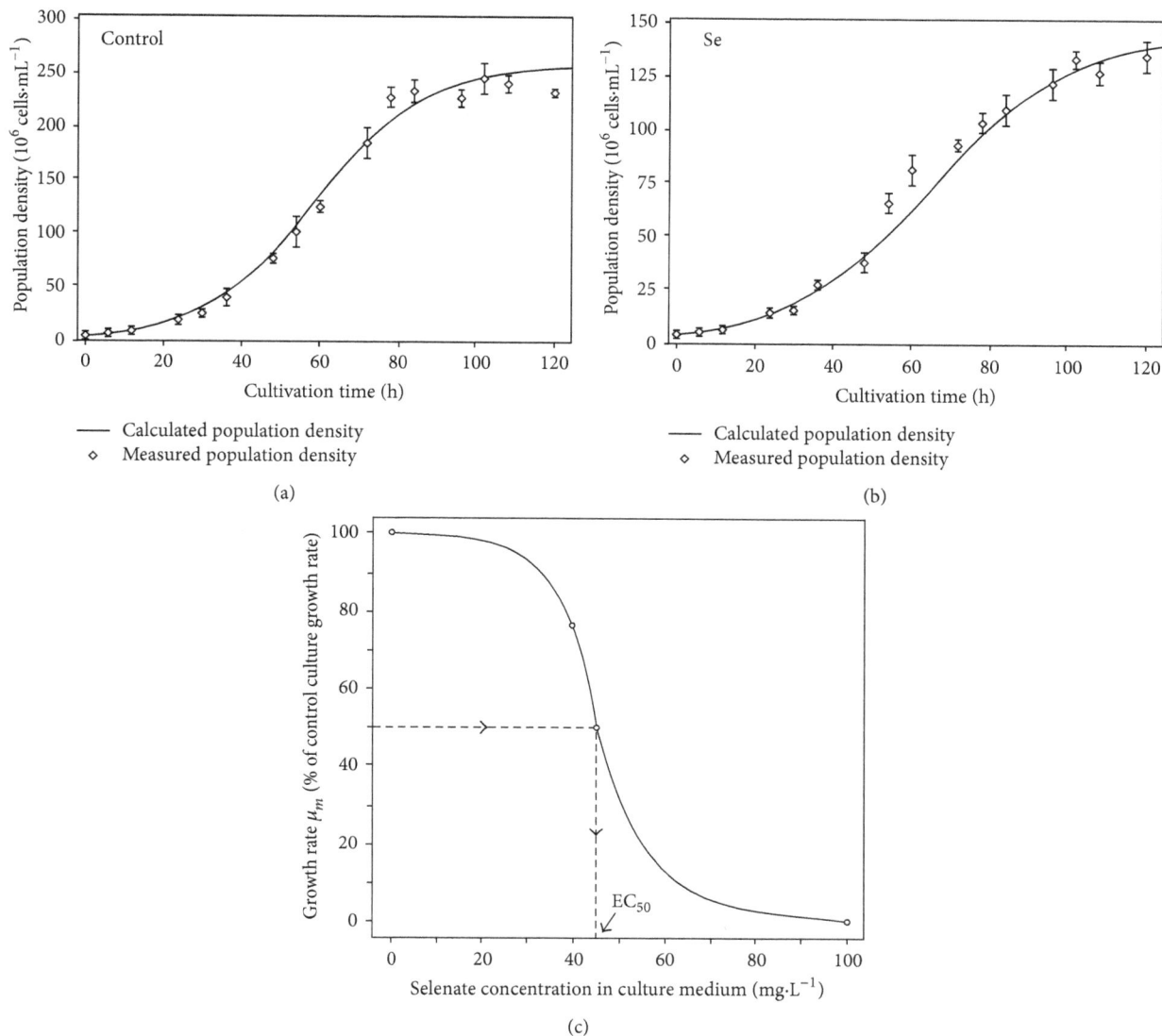

FIGURE 1: Growth curves of *Chlorella sorokiniana* culture with $40\,\mathrm{mg\cdot L^{-1}}$ of selenate (a) and control culture (no selenate) (b). Data are given as mean values $\pm$ S.D. of the means. Both experimental data of population numbers and data calculated from logistic mathematical model fit are graphically presented in Figures 1(a) and 1(b). Model equation used is $N(t) = N_{\max} \cdot N_0 \cdot (N_0 + (N_{\max} - N_0) \cdot e^{-\mu t})^{-1}$, where N_0 is the initial cell density at time zero, $N_{\max}$ is the maximal density that cell population can theoretically reach in the indefinite time, μ_m is and the maximal culture growth rate $(\mathrm{h^{-1}})$, and t is cultivation time (day). Figure 1(c) presents concentration-response relationship between maximal growth rates from logistic growth models and selenate concentrations in medium. Concentration-response curve was fitted to the 2 parameter log-logistic model (Hill's model) using open source software "R statistical package." To predict $\mathrm{EC_{50}}$ presumptions were made that concentration of $100\,\mathrm{mg\cdot L^{-1}}$ selenate corresponds to the maximal (100%) growth inhibition, while zero inhibition corresponds to Se-free culture growth rate. Obtained $\mathrm{EC_{50}}$ value form the log-logistic model curve was $45\,\mathrm{mg\cdot L^{-1}}$ of selenate.

culture medium, which represented 82% and 53% of the control growth rate value [31]. For the microalga *Selenastrum capricornutum* cultivated with $40\,\mathrm{mg\cdot L^{-1}}$, Ibrahim and Spacie found a growth inhibition of 45% and a linear relationship between growth inhibition and selenate concentration [49]. In *Chlamydomonas reinhardtii* cells with $11.5\,\mu\mathrm{M}$ of selenate in the culture medium, growth rate decreased only 5% compared to control cells [26]. Therefore, toxic effect of selenate greatly depends on microalgae genus and species. Consequently, prior to production of Se-enriched biomass,

specific toxicity analysis for the selected algal species will be required.

Based on data published by Morlon et al. for *Chlamydomonas reinhardtii* cultivated in batch systems with 10 to $50\,\mu\mathrm{M}$ selenite, $\mathrm{EC_{50}}$ values vary significantly from one experiment to another, which authors attributed to variations among different batch cultivations (see Table 2) [24, 25]. Toxic effect of Se can be expressed as $\mathrm{EC_{50}}$ value, a measure of the increase in biomass over time, and it is determined from the exponential phase [1, 26, 28, 31, 36, 50]. In our experiments

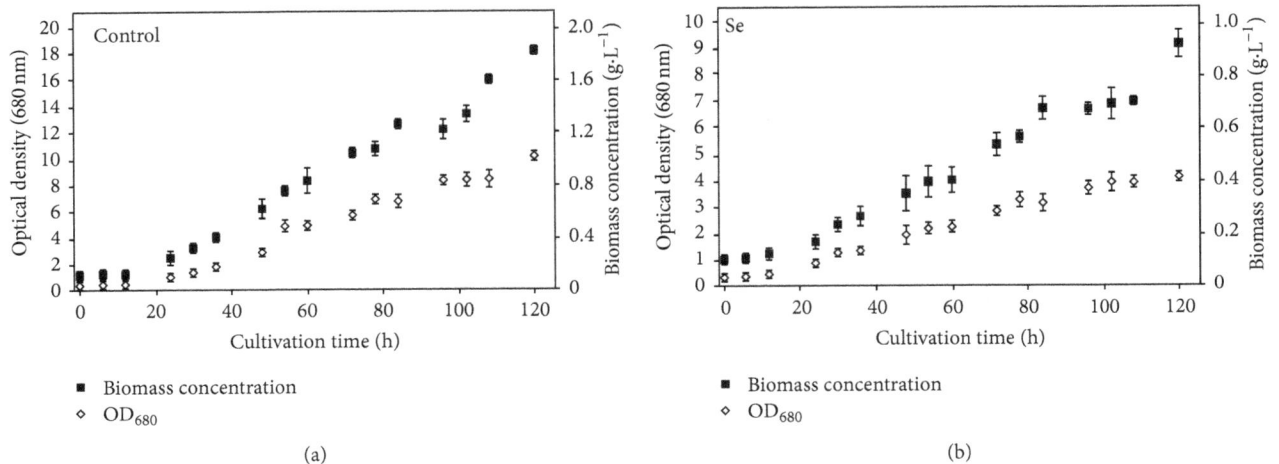

(a)

(b)

FIGURE 2: The optical density and biomass concentration of *Chlorella sorokiniana* as a function of cultivation time for (a) control (no selenate) and (b) culture with 40 mg·L^{-1} of selenate in culture medium. Data are given as mean values ± S.D. of the means.

TABLE 2: Values of EC$_{50}$ parameter for microalga cultures exposed to Se salts, expressed as Se(+IV) and/or Se(+VI) concentrations (mg·L^{-1}/μmol·L^{-1}) in culture medium, as determined and published in the literature. In some studies Se concentrations (in mg·L^{-1}) were expressed based on weight of elemental Se, while in present study it was expressed based on weight of sodium selenate (Na$_2$SeO$_4$). Table adapted from Vítová et al. (2011) [28].

| Reference | Microalga specie | EC$_{50}$ | | | | Cultivation type Cont./Batch |
| | | Selenate | | Selenite | | |
		(mg·L^{-1})	(μmol·L^{-1})	(mg·L^{-1})	(μmol·L^{-1})	
Vítová et al. (2011) [28]	*Scenedesmus quadricauda*	33	418	4	50	B
Geoffroy et al. (2007) [1]	*Chlamydomonas reinhardtii*	0.36	4.5	—	—	B
Morlon et al. (2005a) [24]	*Chlamydomonas reinhardtii*	—	—	1.1	14	B
Morlon et al. (2005b) [25]	*Chlamydomonas reinhardtii*	—	—	6.3	80	B
Fournier et al. (2010) [26]	*Chlamydomonas reinhardtii*	0.032	0.4	—	—	B
Fournier et al. (2010) [26]	*Chlamydomonas reinhardtii*	0.245	3.1	—	—	B
Bennett (1988) [31]	*Chlorella pyrenoidosa*	0.79	10	—	—	C
Present study	*Chlorella sorokiniana*	45	238.2	—	—	B

with *Chlorella sorokiniana*, having in mind that 100 mg·L^{-1} selenate strongly inhibited cell growth and provoked severe cell deformation and death, as proven by ultrastructure microscopy, maximal (100%) growth inhibition was set for 100 mg·L^{-1}, and the obtained EC$_{50}$ value from the log-logistic model curve was 45 mg·L^{-1} (Figure 1(c)). In Table 2, EC$_{50}$ values for different algae species are compared, expressed as Se concentrations (in both chemical forms, selenate Se(+VI), and selenite Se(+IV)) that provoked half of the maximal inhibitory effect, as published in the related literature (Table adopted from Vítová et al. [28]).

In order to obtain SeMet enriched biomass while maintaining cell viability, a sublethal selenate concentration of 40 mg·L^{-1} was used. Time-course evolution of biomass concentration and optical density for control culture and Se-added cultures are presented in Figures 2(a) and 2(b). Data show that both optical density and biomass concentration decreased for about 50% compared to control culture values, and cultures were viable up to 120 h of cultivation.

In the literature, Se concentration range used in experiments varies significantly depending on microalgae species.

Pelah and Cohen reported that *Chlorella zofingiensis* was resistant to selenite concentrations up to 100 mg·L^{-1} [30]. Li et al. found selenium to be an essential trace element at low concentrations and toxic at high levels in cyanobacterium *Spirulina platensis*, for which growth was enhanced when cultivated on 0.5 to 40 mg·L^{-1} selenate [29]. Umysová et al. observed that most of *Scenedesmus quadricauda* wild-type strain cells died within one or two days of cultivation if cultivated with Se (both selenate and selenite forms) at concentrations higher than 50 mg·L^{-1} [27]. Related studies on *Scenedesmus quadricauda* revealed that 50 mg·L^{-1} selenate in medium was not lethal to microalga cultures as cells grew and divided normally [28]. In our case, a selenate concentration of 40 mg·L^{-1} (212 μM) in the culture medium was selected based on previously published data [32].

3.2. *Effects of Selenate on Photosynthesis and Pigment Production.* Chlorophyll fluorescence measurement is used as an economic and sensitive method for rapid detection of photoinhibition on algal cultures [42, 50–52].

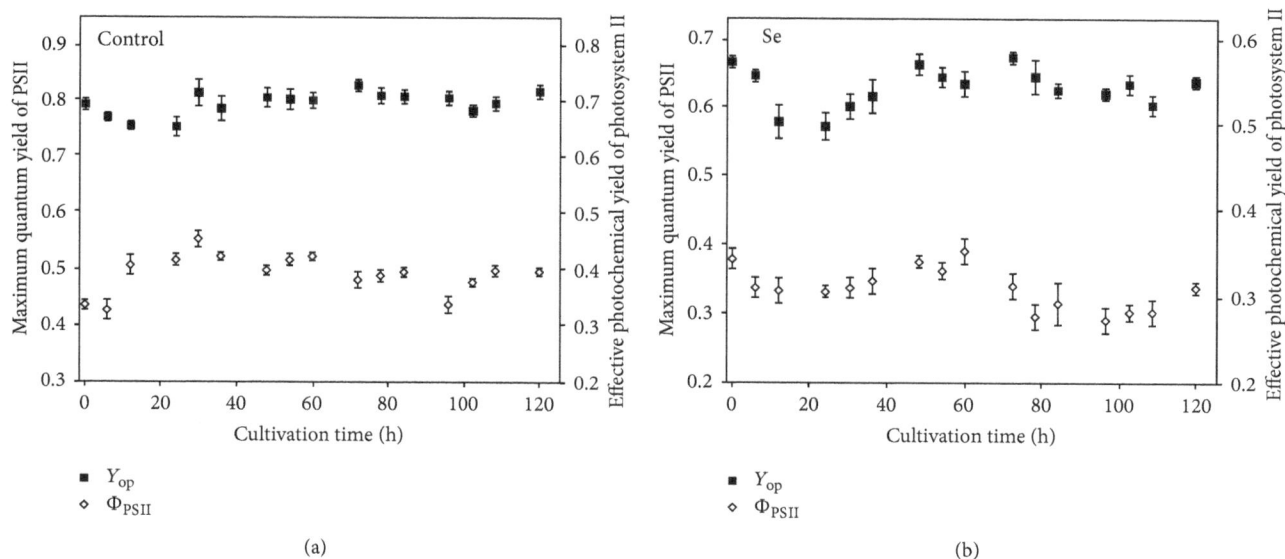

(a)

(b)

FIGURE 3: Maximum quantum yield of PSII (Y_{op}) and effective photochemical yield of PSII (Φ_{PSII}) in function of cultivation time for culture exposed to 40 mg·L^{-1} selenate in culture medium (a) and Se-free culture (b). Data are given as mean values ± S.D. of the means.

In Figures 3(a) and 3(b) it can be shown that for *Chlorella sorokiniana* cultures, 40 mg·L^{-1} (212 μM) of selenate affected maximum quantum yield of PSII (Y_{op}). Even though Y_{op} for Se-exposed cultures was approximately 20% lower than those control culture values, the obtained Y_{op} values were within the typical range for microalgae cells; therefore, cultures were accordingly considered viable throughout the experiment [53]. During the first 24 h of cultivation, values of Y_{op} for Se-exposed and control cultures decreased in approximately 15 and 10%, respectively, compared to initial values, due to culture adaptation phase. Nevertheless, this decrease was only temporary and Y_{op} values stabilized after 24 h, as previously found [32].

During the experiment effective photochemical yield (Φ_{PSII}) values for control remained within the 0.33–0.42 range, while Φ_{PSII} of Se-exposed cultures remained within the 0.27–0.34 range (Figure 3(b)). Throughout the experiment, 25% decrease in Φ_{PSII} was found for Se-exposed cultures compared to control cultures.

The photosynthetic light reactions are located in the thylakoid membrane of the chloroplast, forming a closed system of stacked membranes surrounding the intrathylakoidal space, the lumen, thus separating it from outer chloroplast's area, and the stroma [53]. The thylakoid membrane contains both PSII and PSI systems with their respective reaction centers, which are connected by a series of electron carriers. PSII complexes are usually located in the stacked thylakoid region and the grana, while PSI complexes are located at the grana margins, facing the surrounding stroma [1].

It has been suggested that sublethal selenate concentrations in the culture medium can damage thylakoid membrane structures, thus affecting photosynthesis by both impairing PS II function resulting in decreased Y_{op} and by limiting electron transport between PSII and PSI, with a decrease in

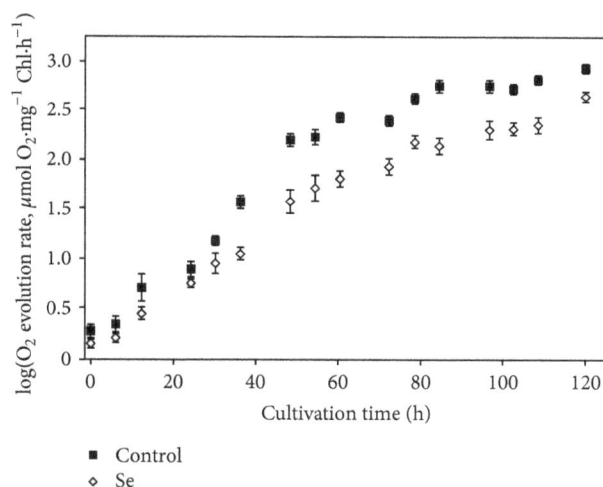

FIGURE 4: Photosynthetic oxygen evolution (μmol O$_2$ mg^{-1} Chl. h^{-1}) by *Chlorella sorokiniana* in function of time for culture with 40 mg·L^{-1} of selenate and control culture (no selenate). Data are given as mean values ± S.D. of the means.

Φ_{PSII} which inhibits photosynthesis and decreases growth rate [1]. In that respect, Geoffroy et al. reported a decrease of 22% in Y_{op} after 24 h of cultivation of *Chlamydomonas reinhardtii* cells growing in 9.3 μM of selenite. After 96 h the decrease arose 66% compared to control culture values. Effective photochemical yield (Φ_{PSII}) decreased 52% after 24 h and 18% after 48 h exposure. These results evidence strong inhibition of the photosynthetic electron transport [1]. In our results, photosynthetic activity was less affected.

Oxygen evolution rates decreased 50% in cells exposed to 40 mg·L^{-1} Se compared to control cells after 48 h cultivation (Figure 4). That difference is similar to values reported by

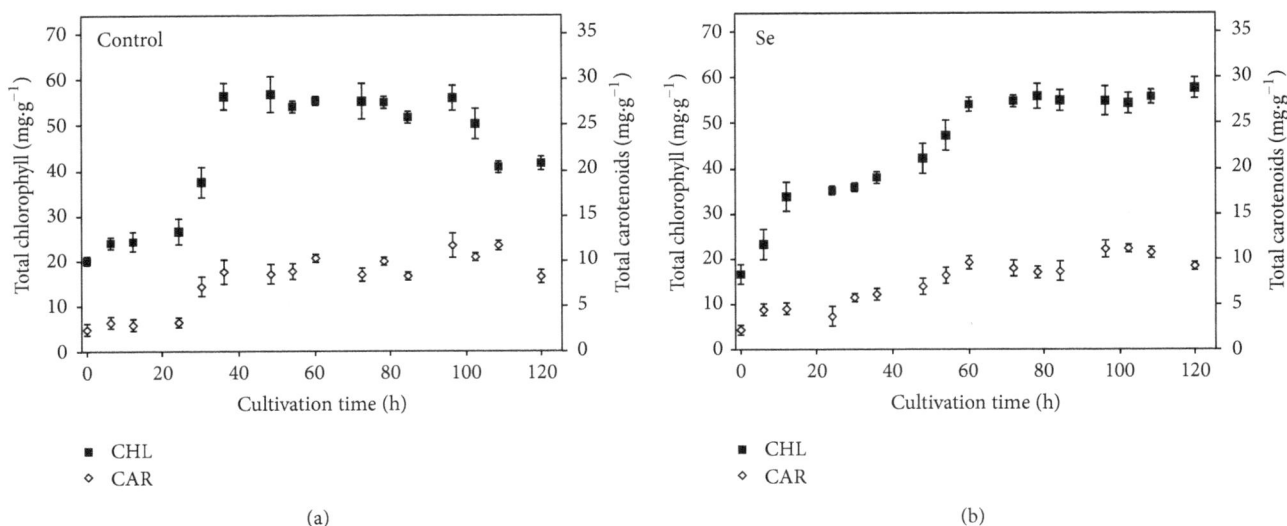

FIGURE 5: Total chlorophyll ($mg \cdot g^{-1}$) and total carotenoids ($mg \cdot g^{-1}$) content per biomass weight as a function of cultivation time for (a) control (no selenate) and (b) culture with $40\,mg \cdot L^{-1}$ of selenate in culture medium. Data are given as mean values ± S.D. of the means.

Schiavon et al. [54], in which a decrease of 44% compared to control cultures was determined for macroalga *Ulva sp.* after 10 days exposure to 100 mM selenite; once *Ulva sp.* thalli were transferred to fresh Se-free seawater, cultures restored to their original oxygen evolution rates [54].

Total chlorophyll and carotenoid content of control culture (Figure 5) increased during the first 48 h of cultivation up to values of $60\,mg \cdot g^{-1}$ and $20\,mg \cdot g^{-1}$, respectively, and remained almost constant until 96 h of the experiment. From then on, total pigments content of control culture decreased due to self-shading effect (Figure 5(a)) [51, 55]. No significant differences in total pigment content were found for Se-exposed cultures. Based on PSII fluorescence, oxygen evolution, and pigment production data, it can be concluded that $40\,mg \cdot L^{-1}$ of selenate was a sublethal concentration for *Chlorella sorokiniana* culture and can therefore be used for SeMet accumulation studies. Similar findings for *Chlorella vulgaris* were reported in which chlorophyll production was not significantly affected by exposure to selenate during 9 days of batch cultivation [9].

3.3. Impact of Selenate on Ultrastructure of Chlorella sorokiniana.
Electron microscopy studies on the ultrastructure of *Chlorella sorokiniana* were carried out at the end of the experiment (240 h). Figure 6(a) shows longitudinal and cross-sections through control cells (Se-free culture). This alga is about 3 μm length and 2 μm wide. The nucleus is about 1 μm length and 1 μm wide located in the central portion of the microalga. The cell has a prominent cup-shaped chloroplast that partially surrounds the nucleus, and the thylakoids inside are compressed and very dense which makes them indistinguishable. The pyrenoid (PY) is surrounded by four layers of starch. At 40 and $100\,mg \cdot L^{-1}$ of selenate in the culture medium, the stroma of the chloroplast became granule and less dense, and the thylakoids had a fingerprint-like appearance (Figures 6(b) and 6(c)). Fingerprint-like

appearance of the chloroplast in Se-exposed microalga cultures was previously observed and reported in the related literature [1, 28]. Analysis of chloroplast ultrastructure by electron microscopy in cultures incubated with selenate 40 and $100\,mg \cdot L^{-1}$ revealed the presence of lipoprotein particles called "plastoglobules" in the stroma of chloroplasts that appeared as small black globules in close proximity to thylakoids (Figures 6(b), and 6(d)). Plastoglobules are involved in stress responses. Several studies have reported their presence in chloroplasts from plants grown under diverse stress conditions [56, 57]. Plastoglobules observed in cells treated with Se (Figure 6(d)) are plastid-localized lipoprotein particles that contain tocopherols and other lipid isoprenoid derived metabolites of commercial value, as well as structural proteins [58, 59]. In addition to vascular plants, plastoglobules are found in nonvascular species such as moss [60] and algae [61]. Some publications show that plastoglobules contain enzymes involved in the metabolism of secondary metabolites, as well as enzymes of unknown function [62]. At the highest Se concentration added to the culture medium ($100\,mg \cdot L^{-1}$), some of the Se-exposed cells had large vacuoles (V) (Figure 6(e)) indicating a process of autophagy, a housekeeping mechanism, in which damaged or unwanted cellular components get degraded in vacuoles.

Autophagic vacuole (VA) and its compounds get recycled [63]. Structure of the cells exposed to $100\,mg \cdot L^{-1}$ became severely disrupted and normal cell organelles were often hardly distinguishable at the end of the experiment (Figure 6(f)). Therefore, from the results in can be inferred that $100\,mg \cdot L^{-1}$ selenate or higher concentrations are not compatible with cell viability, in good agreement with the biochemical results showed above in this paper.

3.4. Impact of Selenate on Total Protein of Chlorella sorokiniana.
Electrophoresis of total protein isolate fractions revealed a band of approximately 50 kDa present in 50%

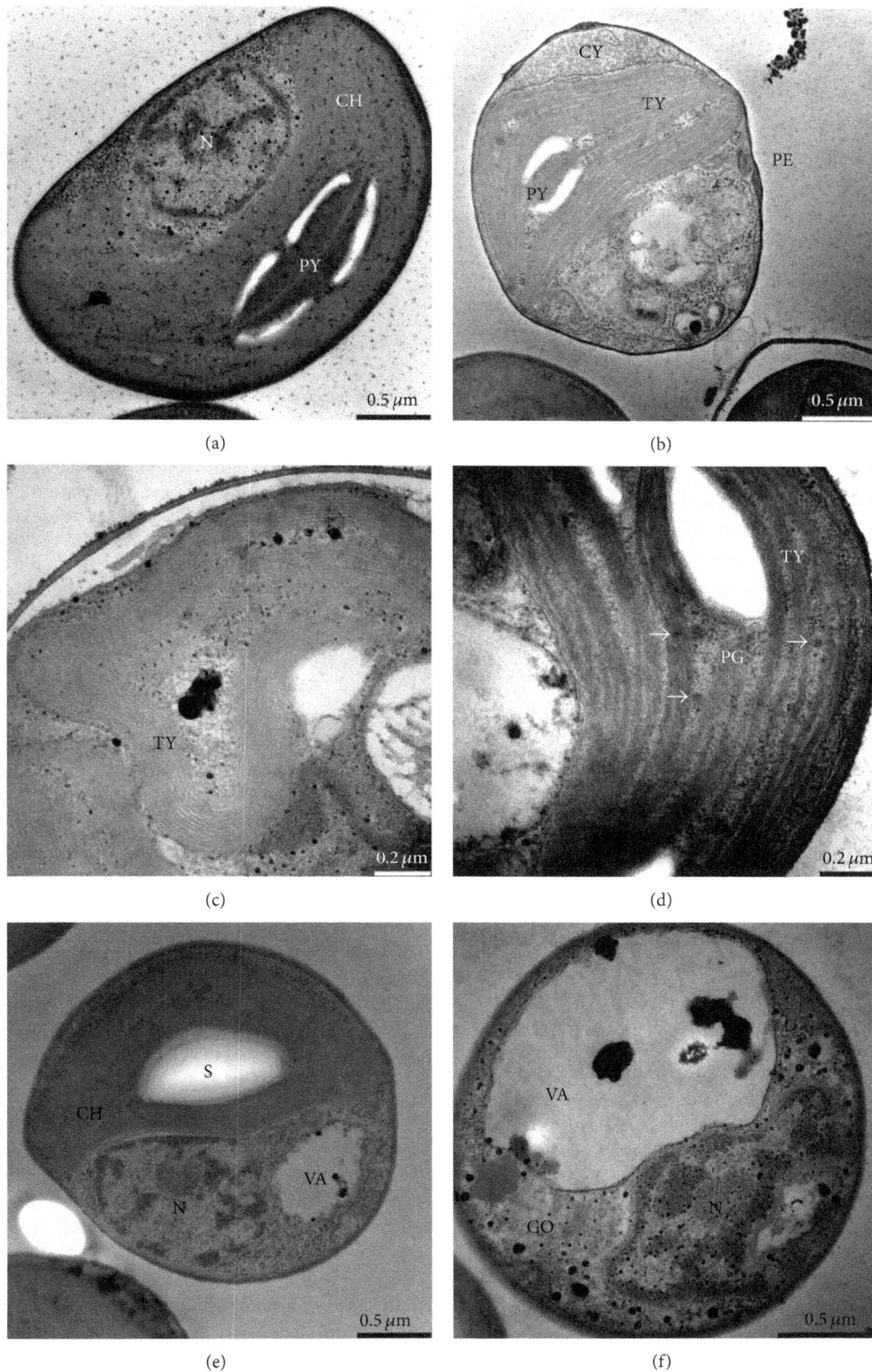

FIGURE 6: Ultrastructure images made by transmission electron microscopy of: Se-free single cell (a) and cell exposed to 40 mgL^{-1} selenate (b); fingerprint-like thylakoids (c) and plastoglobules (indicated by arrows) (d) in cell exposed to 100 mgL^{-1} selenate; autophagy in cell exposed to 40 mgL^{-1} selenate (e, f); CH: chloroplast; CY: cytoplasm; CW: cell wall; N: nucleus; PE: periplasm; PG: plastoglobules; PY: pyrenoid; TY: thylakoids; VA: autophagic vacuole; and V: vacuole.

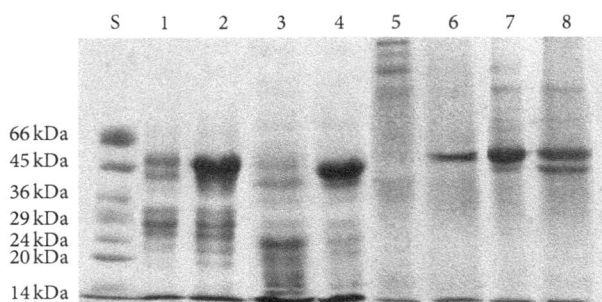

FIGURE 7: SDS-PAGE analysis of proteins extracted from biomass of *Chlorella sorokiniana*: 50% ammonium sulfate fractions of control (lanes 1 and 3) and $40 \, mg \cdot L^{-1}$ selenate-exposed (lanes 2 and 4) cultures, 60% ammonium sulfate fractions of control (lane 5), $40 \, mg \cdot L^{-1}$ selenate-exposed (lane 6) cultures, 40% ammonium sulfate fractions of control (lane 7), and $40 \, mg \cdot L^{-1}$ selenate-exposed (lane 8) cultures. All lanes were extracted from cells of *Chlorella sorokiniana* after 120 h of cultivation. Lanes 1 to 4 origin from initial cultivation, while lanes 5 to 8 belong to repeated cultivation. Lanes 1 and 2 were loaded with 15 μg of proteins, lane 3 with 20 μg, lane 4 with 10 μg, while lanes 5 to 8 contained 7.5 μg of proteins each. S = protein ladder (molecular weight marker).

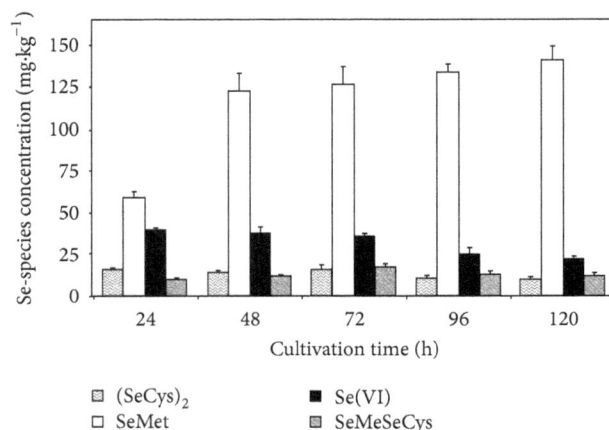

FIGURE 8: Selenium species concentration in *Chlorella sorokiniana* dry biomass ($mg \cdot kg^{-1}$) in function of cultivation time for culture exposed to $40 \, mg \cdot L^{-1}$ selenate in culture medium. During cultivation 750 mL of culture volume was collected daily from each replicate bottle and centrifuged to obtain sufficient biomass to perform the necessary analyses. Data are given as mean values of at least three measurements ± S.D. of the means. Average difference between repeated measurements did not exceed 5% as can be seen from standard deviation indicators. The difference SeMet = selenomethionine; $(SeCys)_2$ = selenocystine; SeMeSeCys = Se-methylselenocysteine; and Se(+VI) = selenate (SeO_4^{2-}).

ammonium sulfate fraction of Se-exposed cells, while the same band was absent in control culture fractions (Figure 7—lanes 1, 2). In order to locate in a more precise way that fraction where such a protein band appears in Se-exposed culture, the experiment was repeated and the results confirmed (Figure 7—lanes 3, 4). Total protein extract was fractionated with ammonium sulfate in the range of 30%–70%. After SDS PAGE electrophoresis, protein band of Se-exposed culture appeared in 60% ammonium sulfate fraction, while the same band was absent again in Se-free culture (Figure 7—lanes 5, 6). To identify proteins in Se-treated culture, bands were cut from SDS PAGE gels and analyzed by MS-TOF-TOF mass spectrometry (see Materials and Methods). Proteins from these bands were identified as 53 kDa large subunit of *Chlorella sorokiniana* Rubisco enzyme, suggesting that Se may interfere with proteins located in the chloroplast. Rubisco (ribulose 1,5-bisphosphate carboxylase/oxygenase), the most abundant enzyme in nature and responsible for CO_2 fixation by photosynthetic organisms, is a complex protein composed from eight identical large subunits (M_r 53000) that are encoded in the chloroplast genome, each one with catalytic site and eight identical small subunits (M_r 14000) that are encoded in the nuclear genome [64]. Having in mind that selenate strongly affects chloroplast morphology and function [1, 26, 28], these results open the possibility that Se exposure could modify large subunit of *Chlorella sorokiniana* Rubisco by incorporation or association to it, as it is encoded in the chloroplast DNA. Se enters the microalga cell by competing with S metabolism and finally getting incorporated into aminoacids (SeMet and SeCys) [14]. Therefore, Se-aminoacids biosynthesis and its further incorporation into proteins might be among the biochemical reasons that explain the appearance of that probable Se-protein band.

Binding of Se on microalga proteins is reported by various authors [16, 30, 65, 66]. Boisson et al. described an increase of cytosolic selenium and total cell protein of the marine microalga *Cricosphaera elongate* when selenium concentrations in the culture increased, suggesting that these proteins take part in detoxifying process [65]. *Spirulina platensis* could accumulate 85% of selenium in organic form of which 25% was integrated with proteins [29]. Novoselov et al. identified selenoproteins present in a 20 to 80% ammonium sulfate fraction of *Chlamydomonas reinhardtii* with molecular weight of 7 to 52 kDa [66].

In spite of these results mentioned above, the mechanism of interaction between Se and large subunits of Rubisco is still unknown and merits further investigation.

3.5. Accumulation of SeMet during Chlorella sorokiniana Batch Cultivation in Presence of Selenate. As can be seen in Figure 8 *Chlorella sorokiniana* accumulated about $60 \, mg \cdot kg^{-1}$ of SeMet in first 24 h of cultivation and doubled this value after 48 h, and increased moderately (approximately 5% daily), while intracellular Se(+VI) concentration gradually decreased, due to its transformation to SeMet and intermediates $(SeCys)_2$ and SeMeSeCys [48]. At the end of the experiment total accumulation of SeMet was $140 \, mg \cdot kg^{-1}$, while intracellular Se(+VI) concentration decreased by 50% from the initial value. Concentrations of intermediates SeMeSeCys and $(SeCys)_2$ did not vary significantly throughout the experiments and maintained values between 10 and $20 \, mg \cdot kg^{-1}$.

Neumann et al. found *Chlorella sp.* metabolized up to 87% of selenate to SeMet after 24 h of cultivation, and suggested that this microalga can develop an important capacity for rapid cellular conversion of selenate to SeMet in order to avoid toxic effect on long-term cell development [14]. More than 70% of the protein-bound Se in *Chlorella* biomass is found to be present in the form of SeMet, likely Se in Se-enriched yeast [16, 19]. Umysová et al. reported that wild type of *Scenedesmus quadricauda* accumulated 300 mg·kg^{-1} SeMet in the presence of 50 mg·L^{-1} of selenate [27]. Authors suggested that *Scenedesmus quadricauda* tolerance mechanism is an internal way to detoxify Se inside the cell [27]. Bottino et al. identified SeMeSeCys and SeCys amino acids present in *Chlorella* and *Dunaliella sp.* microalga exposed to 10 mg·L^{-1} selenite [67]. In our study *Chlorella sorokiniana* exposed to 40 mg·L^{-1} (212 μM calculated as sodium selenate) accumulated 140 mg·kg^{-1} SeMet, which is approximately 20 mg·kg^{-1} more per 100 μM of selenate than those values reported by Umysová et al. [27] for *Scenedesmus quadricauda* cells exposed to 50 mg·L^{-1} selenate (633 μM calculated as elemental Se). Results obtained in this study proved SeMet to be prevailing selenoaminoacid accumulated by Se-exposed *Chlorella sorokiniana* culture.

4. Conclusions

Microalga *Chlorella sorokiniana* was cultivated for 120 h in batch culture with sublethal selenate concentration of 40 mg·L^{-1} in order to evaluate the effect of selenate on culture growth, photosynthetic efficiency, cell ultrastructure, protein expression, and SeMet production. The goal was to prove that with this sublethal selenate concentration cultures were viable and able to accumulate significant amounts of SeMet in batch systems. Exposure of *Chlorella sorokiniana* to 40 mg·L^{-1} selenate decreased culture growth and oxygen evolution rates but had no effect on pigment content. Ultrastructural examination showed typical changes on chloroplast structure provoked by selenate exposure. Selenoproteins which appeared in the protein pool of Se-treated cells, but not in Se-free cells, were identified as 53 kDa large subunit of *Chlorella sorokiniana* Rubisco enzyme, suggesting that Se interferes with proteins located in chloroplast and might be incorporated into proteins as Se aminoacids. Microalga *Chlorella sorokiniana* exposed to 40 mg·L^{-1} selenate accumulated up to 140 mg·kg^{-1} of SeMet after 120 h of cultivation. Data obtained from this study open possibilities for larger culture volume trials in order to obtain biomass enriched in high value aminoacid SeMet.

Nomenclature

μ: Specific growth rate (h^{-1})
μ_m: Maximal specific growth rate (h^{-1})
Φ_{PSII}:   Effective photochemical yield of PSII
Car$_{tot}$:   Total carotenoids per dry weight of biomass (mg·g^{-1})
Chl$_{tot}$:   Total chlorophyll per dry weight of biomass (mg·g^{-1})

EC$_{50}$:   Selenate concentration at which growth rate has 50% of the unexposed culture value (mg·L^{-1})
OD$_{680}$:   Optical density measured at wavelength of 680 nm
PS II: Photosystem II electron carrier complex of thylakoid membrane
Rubisco: Ribulose 1,5-bisphosphate carboxylase/oxygenase enzyme
SeMet: Selenomethionine
(SeCys)$_2$:   Selenocystine
SeMeSeCys: Se-methylselenocysteine
Se(+VI):   Selenate (SeO$_4$$^{2-}$);
Y_{op}:   Maximum quantum yield of PSII.

Conflict of Interests

The authors have no conflict of interest to declare, financial or otherwise.

References

[1] L. Geoffroy, R. Gilbin, O. Simon et al., "Effect of selenate on growth and photosynthesis of *Chlamydomonas reinhardtii*," *Aquatic Toxicology*, vol. 83, no. 2, pp. 149–158, 2007.

[2] K. M. Brown and J. R. Arthur, "Selenium, selenoproteins and human health: a review," *Public Health Nutrition*, vol. 4, no. 2, pp. 593–599, 2001.

[3] L. Schomburg, U. Schweizer, and J. Köhrle, "Selenium and selenoproteins in mammals: extraordinary, essential, enigmatic," *Cellular and Molecular Life Sciences*, vol. 61, no. 16, pp. 1988–1995, 2004.

[4] J. R. Arthur, R. C. McKenzie, and G. J. Beckett, "Selenium in the immune system," *Journal of Nutrition*, vol. 133, no. 5, pp. 1457–1459, 2003.

[5] M. P. Rayman, "The use of high-selenium yeast to raise selenium status: how does it measure up?" *British Journal of Nutrition*, vol. 92, no. 4, pp. 557–573, 2004.

[6] M. P. Rayman, "Dietary selenium: time to act," *British Medical Journal*, vol. 314, no. 7078, pp. 387–388, 1997.

[7] L. Letavayova, V. Vlckova, and J. Brozmanova, "Selenium: from cancer prevention to DNA damage," *Toxicology*, vol. 227, no. 1-2, pp. 1–14, 2006.

[8] T. G. Sors, D. R. Ellis, and D. E. Salt, "Selenium uptake, translocation, assimilation and metabolic fate in plants," *Photosynthesis Research*, vol. 86, no. 3, pp. 373–389, 2005.

[9] D. B. D. Simmons and R. J. N. Emery, "Phytochelatin induction by selenate in *Chlorella vulgaris*, and regulation of effect by sulfate levels," *Environmental Toxicology and Chemistry*, vol. 30, no. 2, pp. 469–476, 2011.

[10] J. Nève, "New approaches to assess selenium status and requirement," *Nutrition Reviews*, vol. 58, no. 12, pp. 363–369, 2000.

[11] R. Ebert, M. Ulmer, S. Zeck et al., "Selenium supplementation restores the antioxidative capacity and prevents cell damage in bone marrow stromal cells in vitro," *Stem Cells*, vol. 24, no. 5, pp. 1226–1235, 2006.

[12] W.-P. Chang, G. F. Combs Jr., C. G. Scanes, and J. A. Marsh, "The effects of dietary vitamin E and selenium deficiencies on plasma thyroid and thymic hormone concentrations in the chicken,"

Developmental and Comparative Immunology, vol. 29, no. 3, pp. 265–273, 2005.

[13] L. Patrick, "Selenium biochemistry and cancer: a review of the literature," *Alternative Medicine Review*, vol. 9, no. 3, pp. 239–258, 2004.

[14] P. M. Neumann, M. P. De Souza, I. J. Pickering, and N. Terry, "Rapid microalgal metabolism of selenate to volatile dimethylselenide," *Plant, Cell and Environment*, vol. 26, no. 6, pp. 897–905, 2003.

[15] P. D. Whanger, "Selenocompounds in plants and animals and their biological significance," *Journal of the American College of Nutrition*, vol. 21, no. 3, pp. 223–232, 2002.

[16] J. Doucha, K. Lívanský, V. Kotrbáček, and V. Zachleder, "Production of Chlorella biomass enriched by selenium and its use in animal nutrition: a review," *Applied Microbiology and Biotechnology*, vol. 83, no. 6, pp. 1001–1008, 2009.

[17] G. N. Schrauzer, "The nutritional significance, metabolism and toxicology of selenomethionine," *Advances in Food and Nutrition Research*, vol. 47, pp. 73–112, 2003.

[18] J. V. Vadgama, Y. Wu, D. Shen, S. Hsia, and J. Block, "Effect of selenium in combination with adriamycin or taxol on several different cancer cells," *Anticancer Research*, vol. 20, no. 3, pp. 1391–1414, 2000.

[19] M. P. Rayman, H. G. Infante, and M. Sargent, "Food-chain selenium and human health: spotlight on speciation," *British Journal of Nutrition*, vol. 100, no. 2, pp. 238–253, 2008.

[20] G. N. Schrauzer, "Selenomethionine: a review of its nutritional significance, metabolism and toxicity," *Journal of Nutrition*, vol. 130, no. 7, pp. 1653–1656, 2000.

[21] P. F. Surai and J. E. Dvorska, "Effect of selenium and vitamin E on lipid peroxidation in thigh muscle tissue of broiler breeder hens during storage," *Archiv Fur Geflugelkunde*, vol. 66, p. 120, 2002.

[22] C. S. Orser, D. E. Salt, I. J. Pickering et al., "Brassica plants to provide enhanced human mineral nutrition: selenium phytoenrichment and metabolic transformation," *Journal of Medicinal Food*, vol. 1, no. 1, pp. 253–261, 1999.

[23] D. R. Ellis, T. G. Sors, D. G. Brunk et al., "Production of Se-methylselenocysteine in transgenic plants expressing selenocysteine methyltransferase," *BMC Plant Biology*, vol. 4, article 1, 2004.

[24] H. Morlon, C. Fortin, C. Adam, and J. Garnier-Laplace, "Cellular quotas and induced toxicity of selenite in the unicellular green alga *Chlamydomonas reinhardtii*," *Radioprotection*, vol. 40, pp. 101–106, 2005.

[25] H. Morlon, C. Fortin, M. Floriani, C. Adam, J. Garnier-Laplace, and A. Boudou, "Toxicity of selenite in the unicellular green alga *Chlamydomonas reinhardtii*: comparison between effects at the population and sub-cellular level," *Aquatic Toxicology*, vol. 73, no. 1, pp. 65–78, 2005.

[26] E. Fournier, C. Adam-Guillermin, M. Potin-Gautier, and F. Pannier, "Selenate bioaccumulation and toxicity in *Chlamydomonas reinhardtii*: influence of ambient sulphate ion concentration," *Aquatic Toxicology*, vol. 97, no. 1, pp. 51–57, 2010.

[27] D. Umysová, M. Vítová, I. Doušková et al., "Bioaccumulation and toxicity of selenium compounds in the green alga *Scenedesmus quadricauda*," *BMC Plant Biology*, vol. 9, pp. 58–74, 2009.

[28] M. Vítová, K. Bišová, M. Hlavová et al., "Glutathione peroxidase activity in the selenium-treated alga *Scenedesmus quadricauda*," *Aquatic Toxicology*, vol. 102, pp. 87–94, 2011.

[29] Z.-Y. Li, S.-Y. Guo, and L. Li, "Bioeffects of selenite on the growth of Spirulina platensis and its biotransformation," *Bioresource Technology*, vol. 89, no. 2, pp. 171–176, 2003.

[30] D. Pelah and E. Cohen, "Cellular response of *Chlorella zofingiensis* to exogenous selenium," *Plant Growth Regulation*, vol. 45, no. 3, pp. 225–232, 2005.

[31] W. N. Bennett, "Assessment of selenium toxicity in algae using turbidostat culture," *Water Research*, vol. 22, no. 7, pp. 939–942, 1988.

[32] Z. Gojkovic, I. Garbayo-Nores, V. Gomez-Jacinto et al., "Continuous production of selenomethionine-enriched Chlorella sorokiniana biomass in a photobioreactor," *Process Biochemistry*, vol. 48, pp. 1235–1241, 2013.

[33] M. Svoboda, V. Kotrbáček, R. Ficek, and J. Drábek, "Effect of organic selenium from se-enriched alga (*Chlorella spp.*) on selenium transfer from sows to their progeny," *Acta Veterinaria Brno*, vol. 78, no. 3, pp. 373–377, 2009.

[34] R. Mandalam and B. Palsson, "Elemental balancing of biomass and medium composition enhances growth capacity in high-density *Chlorella vulgaris* cultures," *Biotechnology & Bioengineering*, vol. 59, no. 5, pp. 605–611, 1998.

[35] A. Tsoularis, "Analysis of logistic growth model," *Research Letters in the Information and Mathematical Sciences*, vol. 2, pp. 23–46, 2001.

[36] C. Rioboo, O. González, C. Herrero, and A. Cid, "Physiological response of freshwater microalga (*Chlorella vulgaris*) to triazine and phenylurea herbicides," *Aquatic Toxicology*, vol. 59, no. 3-4, pp. 225–235, 2002.

[37] M. Bagus, *Identification for algae growth kinetics [Ph.D. thesis]*, Agrotechnology and Food Sciences Group, Wageningen University, Amsterdam, The Netherlands, 2009.

[38] J. Yang, E. Rasa, P. Tantayotai, K. M. Scow, H. Yuan, and K. R. Hristova, "Mathematical model of Chlorella minutissima UTEX2341 growth and lipid production under photoheterotrophic fermentation conditions," *Bioresource Technology*, vol. 102, no. 3, pp. 3077–3082, 2011.

[39] R Development Core Team, *R: A Language and Environment for Statistical Computing*, R Foundation for Statistical Computing, Vienna, Austria, 2013.

[40] C. Ritz and J. C. Streibig, "Bioassay analysis using R," *Journal of Statistical Software*, vol. 12, pp. 1–22, 2005.

[41] H. K. Lichtenthaler, "[34] Chlorophylls and carotenoids: pigments of photosynthetic biomembranes," *Methods in Enzymology*, vol. 148, pp. 350–382, 1987.

[42] K. Maxwell and G. N. Johnson, "Chlorophyll fluorescence—a practical guide," *Journal of Experimental Botany*, vol. 51, no. 345, pp. 659–668, 2000.

[43] E. L. V. Harris, "Concentration of the extract," in *Protein Purification Techniques*, S. Roe, Ed., Chapter 6, pp. 138–139, Oxford University Press, 2001.

[44] U. K. Laemmli, "Cleavage of structural proteins during the assembly of the head of bacteriophage T4," *Nature*, vol. 227, no. 5259, pp. 680–685, 1970.

[45] D. Baines, "Analysis of protein purity," in *Protein Purification Techniques*, S. Roe, Ed., Chapter 3, pp. 31–32, Oxford University Press, 2001.

[46] M. Ramirez-Boo, J. J. Garrido, S. Ogueta, J. J. Calvete, C. Gómez-Díaz, and A. Moreno, "Analysis of porcine peripheral blood mononuclear cells proteome by 2-DE and MS: analytical and biological variability in the protein expression level and protein identification," *Proteomics*, vol. 6, pp. S215–S225, 2006.

[47] F. Moreno, T. García-Barrera, and J. L. Gómez-Ariza, "Simultaneous analysis of mercury and selenium species including chiral forms of selenomethionine in human urine and serum by HPLC column-switching coupled to ICP-MS," *Analyst*, vol. 135, no. 10, pp. 2700–2705, 2010.

[48] V. Gómez-Jacinto, T. García-Barrera, I. Garbayo-Nores, C. Vilchez-Lobato, and J. L. Gómez-Ariza, "Metal-metabolomics of microalga Chlorella sorokiniana growing in selenium- and iodine-enriched media," *Chemical Papers*, vol. 66, no. 9, pp. 821–828, 2012.

[49] A. M. Ibrahim and A. Spacie, "Toxicity of inorganic selenium to the green alga Selenastrum capricornutum printz," *Environmental and Experimental Botany*, vol. 30, no. 3, pp. 265–269, 1990.

[50] J. Masojídek, P. Souček, J. Máchová et al., "Detection of photosynthetic herbicides. Algal growth inhibition test vs. electrochemical photosystem II biosensor," *Ecotoxicology and Environmental Safety*, vol. 74, pp. 117–122, 2011.

[51] M. Cuaresma Franco, M. F. Buffing, M. Janssen, C. Vílchez Lobato, and R. H. Wijffels, "Performance of Chlorella sorokiniana under simulated extreme winter conditions," *Journal of Applied Phycology*, vol. 24, no. 4, pp. 693–699, 2012.

[52] J. Masojídek, J. Kopecký, L. Giannelli, and G. Torzillo, "Productivity correlated to photobiochemical performance of Chlorella mass cultures grown outdoors in thin-layer cascades," *Journal of Industrial Microbiology and Biotechnology*, vol. 38, no. 2, pp. 307–317, 2011.

[53] J. Masojídek, M. Koblížek, and G. Torzillo, "Photosynthesis in Microalga," in *Handbook of Microalgal Culture: Biotechnology and Applied Phycology*, A. Richmond, Ed., pp. 57–82, Blackwell Science, London, UK, 2005.

[54] M. Schiavon, I. Moro, E. A. -H. Pilon-Smits et al., "Accumulation of selenium in Ulva sp. and effects on morphology, ultrastructure and antioxidant enzymes and metabolites," *Aquatic Toxicology*, no. 122-123, pp. 222–231, 2012.

[55] H. Tang, M. Chen, K. Y. Simon Ng, and S. O. Salley, "Continuous microalgae cultivation in a photobioreactor," *Biotechnology and Bioengineering*, 2012.

[56] L. Sallas, E.-M. Luomala, J. Utriainen, P. Kainulainen, and J. K. Holopainen, "Contrasting effects of elevated carbon dioxide concentration and temperature on Rubisco activity, chlorophyll fluorescence, needle ultrastructure and secondary metabolites in conifer seedlings," *Tree Physiology*, vol. 23, no. 2, pp. 97–108, 2003.

[57] L. Giacomelli, A. Rudella, and K. J. Van Wijk, "High light response of the thylakoid proteome in Arabidopsis wild type and the ascorbate-deficient mutant vtc2-2. A comparative proteomics study," *Plant Physiology*, vol. 141, no. 2, pp. 685–701, 2006.

[58] Y. Laizet, D. Pointer, R. Mache, and M. Kuntz, "Subfamily organization and phylogenetic origin of genes encoding plastid lipid-associated proteins of the fibrillin type," *Journal of Genome Science and Technology*, vol. 3, pp. 19–28, 2004.

[59] D. Steinmüller and M. Tevini, "Composition and function of plastoglobuli—I. Isolation and purification from chloroplasts and chromoplasts," *Planta*, vol. 163, no. 2, pp. 201–207, 1985.

[60] R. Rinnan and T. Holopainen, "Ozone effects on the ultrastructure of peatland plants: sphagnum mosses, Vaccinium oxycoccus, Andromeda polifolia and Eriophorum vaginatum," *Annals of Botany*, vol. 94, no. 4, pp. 623–634, 2004.

[61] A. Katz, C. Jimenez, and U. Pick, "Isolation and characterization of a protein associated with carotene globules in the alga Dunaliella bardawil," *Plant Physiology*, vol. 108, no. 4, pp. 1657–1664, 1995.

[62] C. Bréhélin, F. Kessler, and K. J. van Wijk, "Plastoglobules: versatile lipoprotein particles in plastids," *Trends in Plant Science*, vol. 12, no. 6, pp. 260–266, 2007.

[63] F. Li, "Autophagy: a multifaceted intracellular system for bulk and selective recycling," *Trends in Plant Science*, vol. 17, no. 9, pp. 526–537, 2012.

[64] D. L. Nelson and M. M. Cox, *Lehninger's Principles of Biochemistry*, Freeman, 5th edition, 2008.

[65] F. Boisson, M. Gnassia-Barelli, and M. Romeo, "Toxicity and accumulation of selenite and selenate in the unicellular marine alga Cricosphaera elongata," *Archives of Environmental Contamination and Toxicology*, vol. 28, no. 4, pp. 487–493, 1995.

[66] S. V. Novoselov, M. Rao, N. V. Onoshko et al., "Selenoproteins and selenocysteine insertion system in the model plant cell system, Chlamydomonas reinhardtii," *EMBO Journal*, vol. 21, no. 14, pp. 3681–3693, 2002.

[67] N. R. Bottino, C. H. Banks, K. J. Irgolic, P. Micks, A. E. Wheeler, and R. A. Zingaro, "Selenium containing amino acids and proteins in marine algae," *Phytochemistry*, vol. 23, no. 11, pp. 2445–2452, 1984.

Isolation and Characterization of *Paracoccus* sp. GSM2 Capable of Degrading Textile Azo Dye Reactive Violet 5

Mallikarjun C. Bheemaraddi, Santosh Patil, Channappa T. Shivannavar, and Subhashchandra M. Gaddad

Department of Post Graduate Studies and Research in Microbiology, Gulbarga University, Gulbarga, Karnataka 585106, India

Correspondence should be addressed to Channappa T. Shivannavar; ctshiv@gmail.com

Academic Editors: L. Beneduce, J. B. Gurtler, and J. Yoon

A potential bacterial strain GSM2, capable of degrading an azo dye Reactive Violet 5 as a sole source of carbon, was isolated from textile mill effluent from Solapur, India. The 16S rDNA sequence and phenotypic characteristics indicated an isolated organism as *Paracoccus* sp. GSM2. This strain exhibited complete decolorization of Reactive Violet 5 (100 mg/L) within 16 h, while maximally it could decolorize 800 mg/L of dye within 38 h with 73% decolorization under static condition. For color removal, the most suitable pH and temperature were pH 6.0–9.0 and 25–40°C, respectively. The isolate was able to decolorize more than 70% of five structurally different azo dyes within 38 h. The isolate is salt tolerant as it can bring out more than 90% decolorization up to a salt concentration of 2% (w/v). UV-Visible absorption spectra before and after decolorization suggested that decolorization was due to biodegradation and was further confirmed by FT-IR spectroscopy. Overall results indicate the effectiveness of the strain GSM2 explored for the treatment of textile industry effluents containing various azo dyes. To our knowledge, this could be the first report on biodegradation of Reactive Violet 5 by *Paracoccus* sp. GSM2.

1. Introduction

In 1856, the world's first commercially successful synthetic dye, mauveine, was discovered for practical uses. Over 10,000 different dyes with an annual production of over 7×10^5 metric tons worldwide are commercially available [1, 2]. Azo dyes are the diverse group of synthetic organic compounds accounting for the majority of all textile dyestuffs produced and are the most extensively used in a number of industries such as textile dyeing, paper, food, leather, cosmetics, and pharmaceutical industries [3]. The amount of dye lost depends upon the class of dye application, varying from 2% loss while using basic dyes to 50% loss in certain reactive sulfonated dyes, leading to severe contamination of surface and ground waters in the vicinity of dyeing industries [4]. In India, an average mill discharges about 1.5 million liters of contaminated effluent per day, which leads to chronic and acute toxicity [5]. Improper textile dye effluent disposal in aqueous ecosystems leads to the reduction in sunlight penetration which in turn decreases photosynthetic activity, dissolved oxygen concentration, and water quality and depicts acute toxic effects on aquatic flora and fauna, causing severe environmental problems worldwide [6]. They can also cause human health disorders such as nausea, hemorrhage, ulceration of the skin and mucous membranes, and severe damage to kidneys, the reproductive system, liver, brain, and central nervous system [7]. In addition, azo dyes also have an adverse impact in terms of total organic carbon (TOC), biological oxygen demand (BOD), and chemical oxygen demand (COD) [8]. Many synthetic azo dyes and their metabolites are toxic, carcinogenic, and mutagenic [9]. Therefore, the treatment of industrial effluents containing azo dyes and their metabolites is necessary prior to their final discharge to the environment.

Various physicochemical methods like adsorption, chemical precipitation and flocculation, photolysis, chemical oxidation and reduction, electrochemical treatment, and ion pair extraction have been used for the removal of dyes from wastewater [10]. The major drawbacks of these methods have been largely due to the high cost, low efficiency, limited versatility, interference by other wastewater constituents, and the handling of the waste generated [11]. Conversely, biological processes provide an alternative to existing technologies because they are more cost-effective, environmental friendly and do not produce large quantities of sludge. Many microorganisms belonging to the different taxonomic groups of bacteria, fungi, actinomycetes, and algae have been reported for their ability to decolorize azo dyes [12]. Pure fungal cultures have been used to develop bioprocesses for the mineralization of azo dyes, but the long growth cycle and moderate decolorization rate limit the performance of fungal decolorization system [13]. In contrast, bacterial decolorization is normally faster. Bacteria capable of dye decolorization/biodegradation either in pure cultures or in consortia have been reported [11, 14–17]. However, comprehensive solutions for sulfonated azo dyes removal are far from reality, which calls for continued search for new organisms and technologies.

This study aimed to isolate and characterize an efficient bacterial strain, which exhibited the remarkable ability to degrade Reactive Violet 5 as a sole source of carbon. Various physicochemical parameters have been optimized for efficient dye decolorization. The dye degraded products were characterized by ultraviolet-visible (UV-Vis) and Fourier transformed infrared spectroscopy (FT-IR) techniques. Very few reports are available on Reactive Violet 5 degradation. After survey of the literature, this could be the first report on biodegradation of Reactive Violet 5 by *Paracoccus* sp. GSM2.

2. Materials and Methods

2.1. Dyes and Chemicals.
Six textile azo dyes Reactive Violet 5, Reactive Red 2, Reactive Orange 16, Reactive Blue 4, Reactive Black 5, and Reactive Green 19 A were generous gifts from Colors India Inc. Pvt. Ltd., Ahmedabad, India. All these dyes were of industrial grade and are widely used in textile industries. Reactive Violet 5 was used as a model azo dye in this study (Figure 1). All required chemicals were obtained from S.D. Fine chemicals (India) and Sigma-Aldrich, (USA). Biochemical and physiological test kits were obtained from Hi-Media, India. All chemicals used during the study were of analytical grade.

2.2. Culture Medium.
The mineral salts medium (MSM) was prepared as per Brilon et al. [18] with some modifications. The MSM consisted of the following constituents (g/L): $Na_2HPO_4 \cdot 2H_2O$ (12.0), KH_2PO_4 (2.0), NH_4NO_3 (0.50), $MgCl_2 \cdot 6H_2O$ (0.10), $Ca(NO_3)_2 \cdot 4H_2O$ (0.050), and $FeCl_2 \cdot 4H_2O$ (0.0075) with 10 mL of trace element solution per liter. The trace element solution was prepared as follows (mg/L): $ZnSO_4 \cdot 7H_2O$ (10.0), $MnCl_2 \cdot 4H_2O$ (3.0), $CoCl_2 \cdot 6H_2O$ (1.0), $NiCl_2 \cdot 6H_2O$ (2.0), $Na_2MoO_4 \cdot 2H_2O$ (3.0), H_3BO_3 (30.0), and $CuCl_2 \cdot 2H_2O$ (1.0). Further, MSM was blended with different concentrations of Reactive Violet 5 and used throughout the study as a test medium and uninoculated flasks were also incubated as control. The final pH of the medium was adjusted to 7.0 ± 0.2. The MSM with agar (1.9% w/v) was used for isolation and maintenance of pure culture. The media were sterilized at 121°C for 20 min before use.

2.3. Screening, Isolation, and Identification of Dye Decolorizing Bacteria.
Textile mill effluent collected from Solapur, India, was brought to the laboratory for isolation of dye degrading bacteria. 10 mL of sample was added to 100 mL of MSM broth containing Reactive Violet 5 (100 mg/L) as a sole source of carbon and incubated at 30°C for 15 days under static as well as shaking conditions. The flasks were checked for change in initial color and turbidity. Then 10 mL of culture broth from the decolorized culture flask was transferred to 100 mL of fresh MSM broth containing Reactive Violet 5 and incubated for one week under static condition. 0.5 mL of decolorized culture was taken out and spread over the agar plates of MSM containing dye and incubated at 30°C until prominent dye degrading bacterial colonies appeared. Further the prominent colonies were streaked onto the MSM agar plates amended with dye and 0.1% (w/v) yeast extract. The obtained colonies formed were screened out and further were checked for the purity by streaking twice on agar medium. Finally, purified cultures were individually tested for their dye degrading capabilities in MSM under static condition. The potential isolate was preserved at −20°C in 15% (w/v) glycerol and used for further investigation. The potential isolate was selected and preliminarily characterized based on its morphological and biochemical properties [19]. Furthermore, various sugar utilization tests were performed using HiCarbo kit (Hi-Media, India).

2.3.1. 16S rDNA Sequencing and Analysis.
The 16S rDNA fragment was amplified and from the pure genomic DNA of isolated bacterial strain was sequenced at Royal Life Sciences Pvt. Ltd., Hyderabad, India. The genomic DNA was extracted using QIAGEN bacteria DNA purification kit according to manufacturer's instructions. The universal primers, namely, a forward primer, Eub27F ($5'-3'$: AGA GTT TGA TCC TGG CTC AG), and a reverse primer, Eub1492R ($5'-3'$: ACG GCT ACC TTG TTA CGA CTT), were used to amplify bacterial 16S rDNAs by PCR which yielded a product of approximately 1500 bp. After an initial denaturation at 95°C for 10 min, the DNA was amplified during 25 cycles of 95°C for 1 min, 55°C for 1 min, and 72°C for 1.5 min and the final extension (72°C) time was 10 min. Then the purified PCR products were run on ABI-PRISM automated sequencer (ABI-3730 DNA Analyzer). A resultant of 1311 bases was compared with nine closely related taxa of the isolate, retrieved from the GenBank database using BLAST (blastn) program on the NCBI website (http://www.ncbi.nlm.nih.gov). The alignment of the sequences was done using CLUSTALW program V 1.6 at European bioinformatics site (http://www.ebi.ac.uk/Tools/msa/). The sequence was refined manually after crosschecking with the raw data to remove ambiguities and submitted to GenBank. To see the phylogenetic position of bacterial isolate evolutionary history was inferred using the neighbor-joining method [20]. The optimal tree with the sum of branch length =

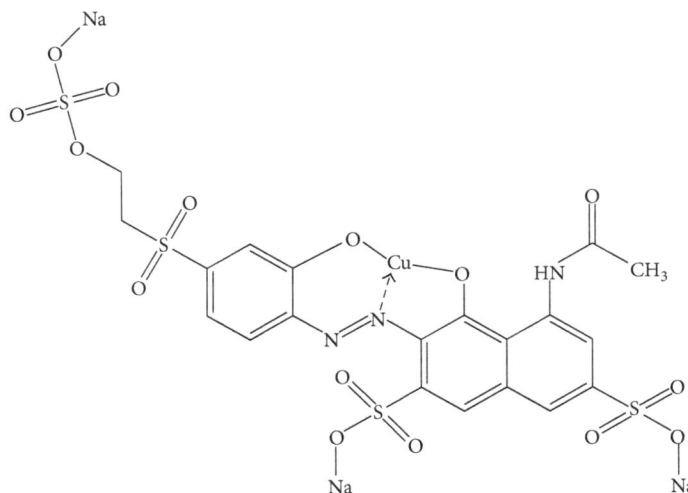

FIGURE 1: Structure of Reactive Violet 5.

0.28801765 is shown. The percentage of replicate trees in which the associated taxa clustered together in the bootstrap test (1000 replicates) is shown next to the branches [21]. The phylogenetic tree was linearized assuming equal evolutionary rates in all lineages [22]. The clock calibration to convert distance to time was 0.02 (time/node height). The tree is drawn to scale, with branch lengths in the same units as those of the evolutionary distances used to infer the phylogenetic tree. The evolutionary distances were computed using the maximum composite likelihood method [23] and are in the units of the number of base substitutions per site. Codon positions included were 1st + 2nd + 3rd + Noncoding. All positions containing gaps and missing data were eliminated. Evolutionary analyses were conducted in MEGA5 software [24].

2.4. Decolorization Experiment.

The dye decolorization experiments were performed in 250 mL Erlenmeyer flasks containing 100 mL of sterilized MSM broth supplemented with yeast extract (0.1% w/v) and Reactive Violet 5 (100 mg/L). We recorded complete decolorization of Reactive Violet 5 in MSM with yeast extract within 16 h as compared to 56 h without yeast extract under the static condition (data not shown). Reports suggest that the inclusion of yeast extract was found to be most effective supplement for growth of azo dye degrading bacteria as well as increasing the dye decolorization efficiency [13, 25]. Therefore all further decolorization experiments were performed using MSM broth supplemented with 0.1% (w/v) yeast extract as a cosubstrate. The flasks were inoculated with 5 mL of cultures broth in test and uninoculated controls were used to compare abiotic color loss during the decolorization studies. The flasks were incubated at 30°C under static as well as shaking (120 rpm) conditions till the decolorization was completed. The 5 mL of cultures was withdrawn at different intervals for color measurement. The supernatant was collected by centrifuging at 10,000 rpm for 15 min. Decolorization was monitored spectrophotometrically by measuring absorbance

of culture supernatant at 558 nm. Growth of bacteria in dye containing medium was determined spectrophotometrically. The cell pellet obtained upon centrifugation of 5 mL culture was resuspended in 5 mL distilled water and its absorbance was studied at 660 nm. The percentage of decolorization was calculated as mentioned by R. Dave and H. Dave [26]:

$$\text{Decolorization (\%)} = \frac{I - F}{I} \times 100, \tag{1}$$

where I = initial absorbance and F = absorbance of decolorized sample.

2.5. Optimization of Physicochemical Parameters.

The decolorization efficiency of *Paracoccus* sp. GSM2 on Reactive Violet 5 was studied at different pH (4–10) and temperatures values (20–50°C). The obtained optimum pH 7.0 and temperature at 30°C were selected to study the decolorization activity under various physicochemical factors such as initial dye concentration (100–800 mg/L), salt concentration (1–6%), and yeast extract concentration (0.1–2.0 g/L). Further, the decolorization of various azo dyes was studied by incubating MSM containing respective dye with bacterial strain GSM2 under static condition.

2.6. Decolorization and Biodegradation Studies.

The Reactive Violet 5 degraded products formed during biodegradation after 16 h of incubation under static condition were studied by following the change in the UV-Vis spectra (200 to 800 nm) using a UV-Vis spectrophotometer (Systronics, AU-2700). To know that the decolorization was due to biodegradation of Reactive Violet 5 was confirmed by FT-IR by analyzing dye degraded products in the decolorized medium. After complete decolorization, culture medium was centrifuged at 10,000 rpm for 15 min to remove the suspended particles. The supernatant was once again centrifuged to ensure the supernatant was free of bacterial cells and was used for extraction of metabolites using an equal volume of ethyl acetate. The extracts were dried over anhydrous Na_2SO_4 and

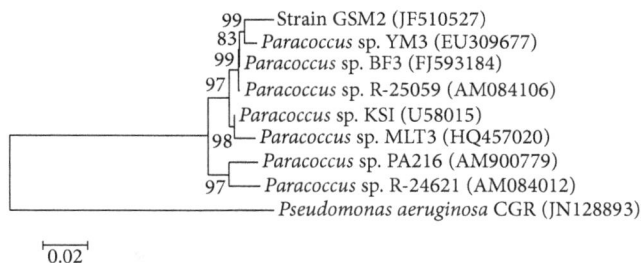

FIGURE 2: Phylogenetic tree of *Paracoccus* sp. GSM2 based on 16S rDNA analysis.

concentrated in a rotary evaporator. The crystals obtained were dissolved in a small volume of high performance liquid chromatography (HPLC) grade methanol and the same sample was used for FT-IR analysis. The FT-IR analysis of extracted metabolites was done using Fischer Scientific (Nicolet, iH5) Spectrophotometer and compared with control dye in the IR region of 550–4000 cm^{-1} with 32 scan speed.

3. Results and Discussion

3.1. Isolation, Screening, and Identification. A total of seven morphologically distinct colonies were observed on the MSM agar plates (data not shown). Amongst positive strains subjected to screening, the potential bacterial strain GSM2 showed a rapid and complete decolorization of Reactive Violet 5 within 16 h under static condition and was selected for further study. The selected strain was gram negative, nonspore former, and nonmotile coccoid. Its colony was white to light yellow, round hunch, and slick. The potential strain was identified as *Paracoccus* sp. GSM2 on the basis of 16S rDNA gene sequence and biochemical characteristics (Table 1). The 16S rDNA sequence (1311 base pairs) was deposited in GenBank with accession number JF510527. Its 16S rDNA sequence did not show any similarity to known azo dye degrader and had the greatest similarity to members of *Paracoccus* sp. group. The phylogenetic relationship between the *Paracoccus* sp. GSM2 and other related microorganisms using MEGA5 software can be depicted from Figure 2. *Pseudomonas aeruginosa* CGR has been taken as out-group and the numbers shown in parentheses are accession numbers of different species.

The homology assay result indicated that the *Paracoccus* sp. GSM2 in phylogenetic branch showed maximum similarity (99%) to *Paracoccus* sp. YM3 which is already known for degradation of carbofuran [27]. Few reports are published on degradation of polycyclic aromatic hydrocarbons by *Paracoccus* sp. [28, 29]. To our knowledge, this could be the first report on biodegradation of textile azo dye Reactive Violet 5 by *Paracoccus* sp. GSM2.

3.2. Effect of Static and Shaking Conditions. *Paracoccus* sp. GSM2 showed that 100% decolorization of added Reactive Violet 5 (100 mg/L) within 16 h under static condition when compared to only 16% decolorization was observed under shaking condition, while the growth of bacterium was greater

TABLE 1: Biochemical and sugar utilization tests of bacterial strain GSM2.

Characteristics	Bacterial strain GSM2
Catalase	−
Oxidase	+
Urease	+
Citrate utilization	−
Nitrate reduction	+
Glucose	+
Adonitol	+
Lactose	−
Sorbitol	+
Esculin hydrolysis	−
Xylose	−
Maltose	+
Fructose	+
Galactose	−
Raffinose	−
Trehalose	+
Melibiose	−
Sucrose	+
L-Arabinose	+
Mannose	+
Inulin	+
Sodium gluconate	+
Glycerol	+
Salicin	+
Dulcitol	+
Inositol	+
Mannitol	[+]
Arabitol	+
Erythritol	[+]
α-Methyl-D-glucoside	+
Rhamnose	+
Cellobiose	+
Melezitose	+
α-Methyl-D-mannoside	−
Xylitol	−
ONPG	−
D-Arabinose	+
Malonate utilization	+
Sorbose	−

+: positive, [+]: weakly positive, and −: negative.

under shaking condition as compared to static condition (Figure 3). To confirm whether this decolorization was due to microbial action or due to change in pH, the change in pH was recorded, which was in the range of 6.0–7.5 at static condition, thus confirming that the biodegradation of dye was due to microbial action. Under aerobic conditions azo dyes are generally resistant to attack by bacteria [30]. Similar findings were reported by other researchers [14, 15]. During dye decolorization in shaking environment electrons released

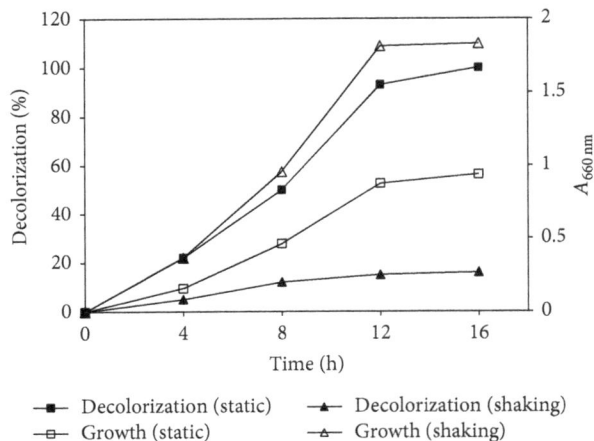

FIGURE 3: Decolorization of Reactive Violet 5 by *Paracoccus* sp. GSM2 in MSM under static and shaking condition (120 rpm) at 30°C.

by oxidation of electron donors are preferentially utilized to reduce free oxygen rather than azo dyes [31]. Hence, in this study static conditions were maintained in the following experiments.

3.3. Effect of pH. The effect of pH on decolorization of Reactive Violet 5 by *Paracoccus* sp. GSM2 was determined over a wide range of pH 4.0 to 10.0 with an interval of pH 1. The isolate showed the maximum of 100% decolorization at pH 7.0 at 30°C. Following the increases from either side of neutral pH, the percentage of decolorization decreased steadily from 97% to 40% on the alkaline side while steep decline in percent decolorization from 89% to less than 15% on acidic side was found. More than 88% of decolorization was observed in a wide range of pH 6.0 to 8.0 (Figure 4). Similar optimum pH was observed in the decolorization of the same dye Reactive Violet 5 by bacterial consortium SB4 [14]. Chan and Kuo [32] reported that the neutral pH would be more favorable for decolorization of the azo dyes and is suitable for industrial applications.

3.4. Effect of Temperature. Similarly in the temperature optimization study, the dye decolorization activity of *Paracoccus* sp. GSM2 was found to increase with increase in incubation temperature from 20 to 30°C. Further increase in temperature, decolorization was decreased by 23% and 44% at 40°C and 45°C, respectively, and almost no activity was found at 50°C (Figure 5). This might be attributed to the adverse effect of high temperature on enzyme activities [33]. Tamboli et al. [16] also found the decrease in dye decolorization efficiency of *Sphingobacterium* sp. ATM for color removal beyond 30°C, which may be due to the thermal inactivation of the decolorizing enzymes.

3.5. Effect of Initial Dye Concentration. The decolorization performance of Reactive Violet 5 by the *Paracoccus* sp. GSM2 was studied by increasing initial dye concentration (100–800 mg/L). We observed that the percentage of decolorization

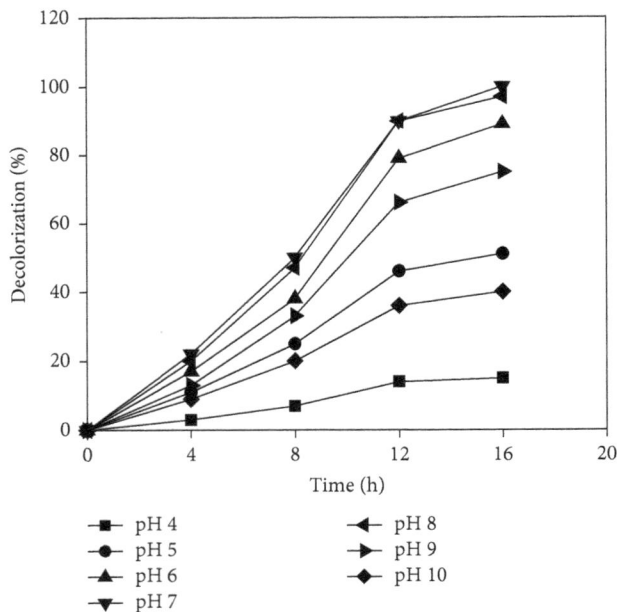

FIGURE 4: Effect of pH on decolorization of Reactive Violet 5.

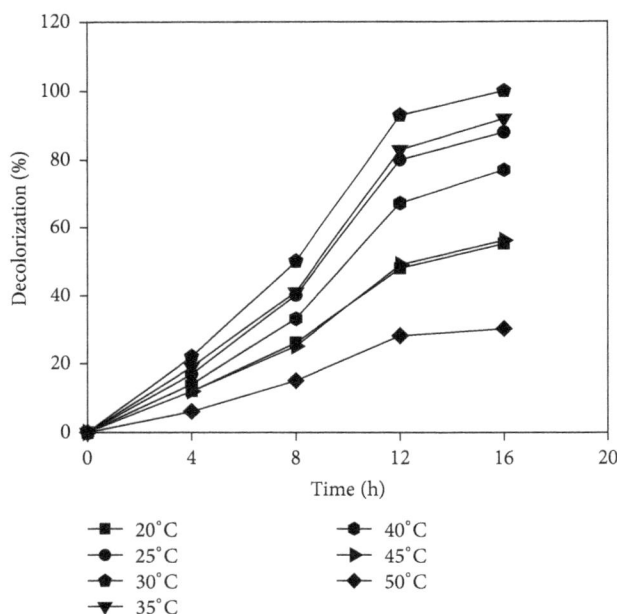

FIGURE 5: Effect of temperature on decolorization of Reactive Violet 5.

was decreased slowly with increasing dye concentration (Figure 6). It could effectively decolorize up to 100 mg/L of Reactive Violet 5 (100%) within 16 h and is decreased to 63%, when dye concentration increased to 800 mg/L and decolorization time increases from 16 h to 38 h, respectively. Lower percentage of decolorization and enhanced time period at high dye concentration may be attributed to the presence of four sulfonic acid groups on Reactive Violet 5 which acts as

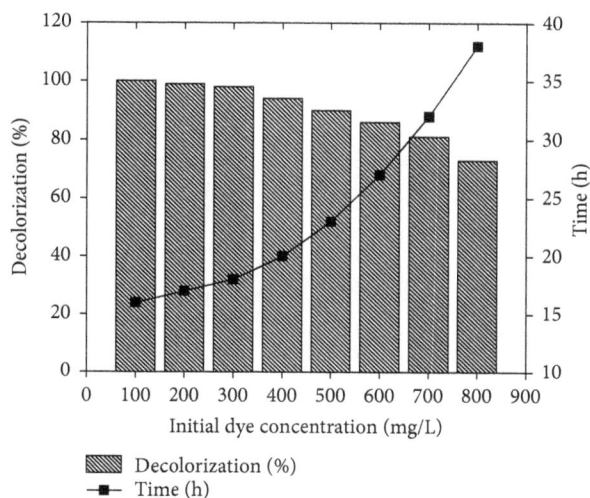

FIGURE 6: Effect of initial dye concentration on decolorization of Reactive Violet 5.

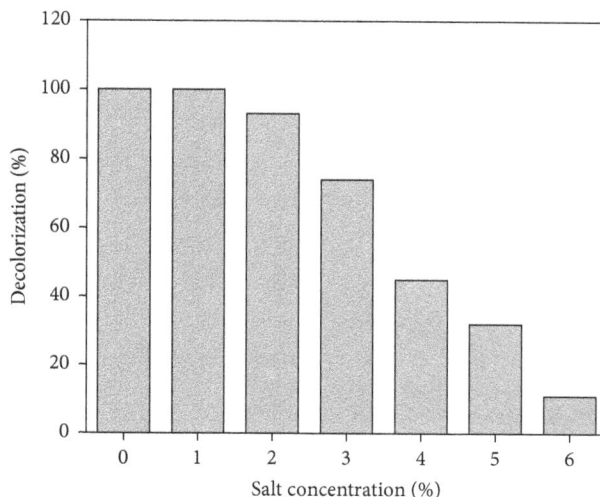

FIGURE 7: Effect of salt concentration on decolorization of Reactive Violet 5.

detergent exerting inhibitory effect on growth of *Paracoccus* sp. GSM2 [14].

3.6. Effect of Salt Concentration.

Textile industry effluents generally contain chloride salts of sodium and potassium which are most widely used for salting out of dyes and discharged along with unused dyes. Hence, the present investigation was undertaken to study the effect of salt concentration (1–6% w/v) on decolorization of Reactive Violet 5 by the strain GSM2. The organism showed satisfactory decolorization up to maximum of 2% salt concentration in MSM under optimum conditions after 16 h of incubation (Figure 7). Previously De Baere et al. [34] have stated that sodium concentration higher than 3 g/L can cause inhibition of most of the bacterial metabolism. But, contrary to the above statement, we could notice 74% and 45% of decolorization at 3% and 4% of salt concentration, respectively. Which agree with the earlier report [14]. Negligible activity was observed when 6% of salt concentration was employed into the medium. This may be attributed to the inhibition of bacteria at high salt concentration due to plasmolysis or loss of activity of cells [35].

3.7. Effect of Different Concentrations of Yeast Extract.

Paracoccus sp. GSM2 was able to degrade Reactive Violet 5 (100 mg/L) efficiently in the presence of yeast extract as a cosubstrate. Among all other nitrogen sources, only yeast extract that served as better nitrogen source for decolorization of Reactive Violet 5 within less time was selected for further experiments (data not shown). Effect of different concentrations of yeast extract (0.1–2.0 g/L) in MSM on the decolorization efficacy of GSM2 was evaluated (Figure 8). Addition of 1 g/L of yeast extract enhanced the decolorization activity and complete decolorization of Reactive Violet 5 was recorded within 16 h. Further increase in yeast extract concentration has no effect on decolorization activity. Thus,

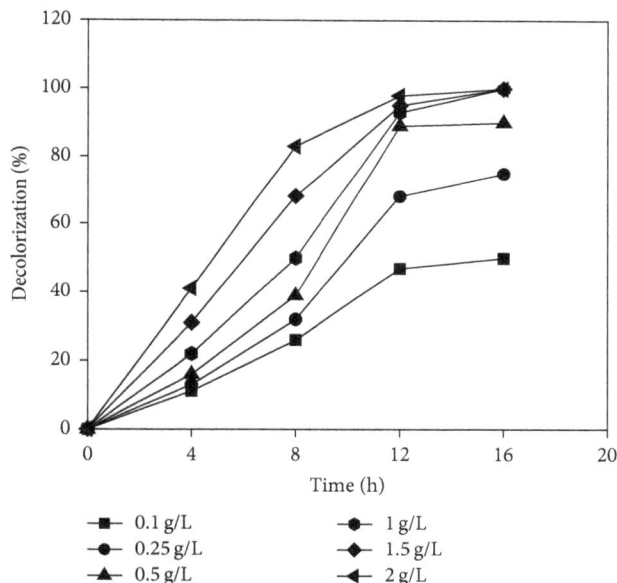

FIGURE 8: Effect of yeast extract concentration on decolorization of Reactive Violet 5.

to make the process economical 1 g/L of yeast extract concentration was found to be optimum. Similar results were reported by Jain et al. [14] where their findings proposed that yeast extract was essential for regeneration of NADH which acts as an electron donor in azo bond reduction.

3.8. Decolorization of Structurally Different Azo Dyes.

Structurally different azo dyes were most widely used in textile processing industries, and, therefore, the effluents from the industry are markedly diverse in composition [36]. Thus, *Paracoccus* sp. GSM2 was tested for its ability to decolorize five other structurally different azo dyes such as Reactive Red 2, Reactive Orange 16, Reactive Black 5, Reactive Blue 4, and

TABLE 2: Decolorization of structurally different azo dyes by *Paracoccus* sp. GSM2.

Reactive dyes	λ_{max} (nm)	% Decolorization	Time (hours)
Reactive Red 2	538	100	12
Reactive Orange 16	416	99	18
Reactive Blue 4	595	92	22
Reactive Black 5	597	82	30
Reactive Green 19A	540	73	38

Reactive Green 19 A. The organism effectively decolorized all structurally different azo dyes within 38 h (Table 2). The efficiency was 100% for Reactive Red 2, followed by 99% for Reactive Orange 16, 92% for Reactive Blue 4, 82% for Reactive Black 5, and 73% for Reactive Green 19 A. We presume that decolorization of structurally different azo dyes by *Paracoccus* sp. GSM2 within 38 h might be the first. This variation in the decolorization of different dyes might be attributed to the structural differences, high molecular weight, and presence of inhibitory groups like -NO$_2$ and -SO$_3$Na in the dyes [36]. The present study confirms the ability of strain GSM2 to decolorize different azo dyes with decolorization efficiency of more than 70%. Thus, the strain GSM2 could be successfully employed for the treatment of textile industry effluents containing various azo dyes.

3.9. Decolorization and Biodegradation Studies. To disclose the possible mechanism of dye decolorization, we also analyzed the degraded products of Reactive Violet 5 by UV-Vis and FT-IR techniques. UV-Vis absorbance of 200–800 nm of Reactive Violet 5 in MSM showed single peak in visible region at 558 nm (λ_{max}) and two intense peaks in UV region near 250 and 325 nm, respectively, correspond to phenyl and naphthyl rings of Reactive Violet 5 (Figure 9) [37]. During decolorization azo bond in Reactive Violet 5 was broken down and peak at 558 nm continuously decreased and completely disappeared within 16 h, without any shift in λ_{max}. Similar observations have been recorded by Jain et al. [14]. According to Asad et al. [12] decolorization of dyes by bacteria could be due to adsorption by microbial cells or to biodegradation. In the case of adsorption, the UV-Vis absorption peaks decrease approximately in proportion to each other, whereas, in biodegradation, either the major visible light absorbance peak disappears completely or a new peak appears. The observation of cell mass showed that *Paracoccus* sp. GSM2 retained their natural color after decolorization of Reactive Violet 5. FT-IR spectrum of control dye with metabolites extracted after complete decolorization (16 h) clearly indicated the biodegradation of the parent dye compound by *Paracoccus* sp. GSM2 (Figure 10). FT-IR analysis of control and decolorized samples showed significant differences in specific peaks of Reactive Violet 5 fingerprint region (550–4000 cm^{-1}). FT-IR spectra of control Reactive Violet 5 show specific peaks for multisubstituted benzene ring, where peaks at 1140.83, 1338.36, 1186.18, and 1549.25 cm^{-1} corresponded to two SO$_3$H groups, symmetric SO$_2$, and –N=N– (azo bond), whereas parasubstituted azo

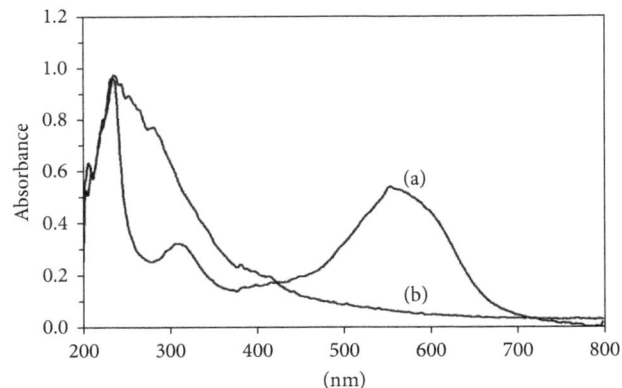

FIGURE 9: UV-Vis spectra of Reactive Violet 5 before and after decolorization by *Paracoccus* sp. GSM2 (a, 0 hour; b, 16 hours).

benzene showed bands near 1433.48 cm^{-1}. Reactive Violet 5 is a metal containing textile azo dye where carbonic ion is bonded with a central copper metal ion giving rise to asymmetrical stretching which was observed near 1614.85 cm^{-1} [38]. Peak at 1650.89 cm^{-1} represents the primary amide of the parent structure of Reactive Violet 5. The FT-IR spectra of 16 h extracted metabolites of degraded Reactive Violet 5 showed peaks at 1651.91 and 3252.48 cm^{-1} which indicates the production of primary amine and secondary amine, respectively, during biodegradation of Reactive Violet 5. Absence of peaks at 671, 721, 763, and 817 cm^{-1} indicates the breakdown of benzene ring or the loss of aromatic nature of the compound. Jain et al. [14] reported similar kind of benzene ring fission in the same dye Reactive Violet 5 by bacterial consortium SB4. Correspondingly, breakdown of azo bond was confirmed by the absence of spectral peak at 1549.25 cm^{-1}, while the absence of the peaks around 1300 and 1165–1150 cm^{-1} clearly indicates the degradation of S=O bonds. Peaks at 2927.40 and 2960.03 cm^{-1} in control Reactive Violet 5 and degraded metabolites, respectively, show the asymmetrical stretching of C–H in CH$_3$. Similar kind of asymmetrical stretching of peak at 2856 cm^{-1} in control Reactive Violet 5 shows that the asymmetrical stretching for C–H stretching was observed in degradation of the same dye Reactive Violet 5 [39]. On the basis of above results, it can be concluded that *Paracoccus* sp. GSM2 has ability to mineralize Reactive Violet 5 completely.

4. Conclusion

The present study showed that a bacterial strain *Paracoccus* sp. GSM2 is capable of degrading Reactive Violet 5 as a sole source of carbon with minimal nutritional requirements under static condition. The potential of this strain has ability to decolorize Reactive Violet 5 in a wide range of pH, temperature, salt, and initial dye concentrations, which is significant for its commercial application. The FT-IR results showed complete loss of the aromatic nature of the dye Reactive Violet 5 by *Paracoccus* sp. GSM2. Furthermore, strain GSM2 had the ability to decolorize five other structurally different

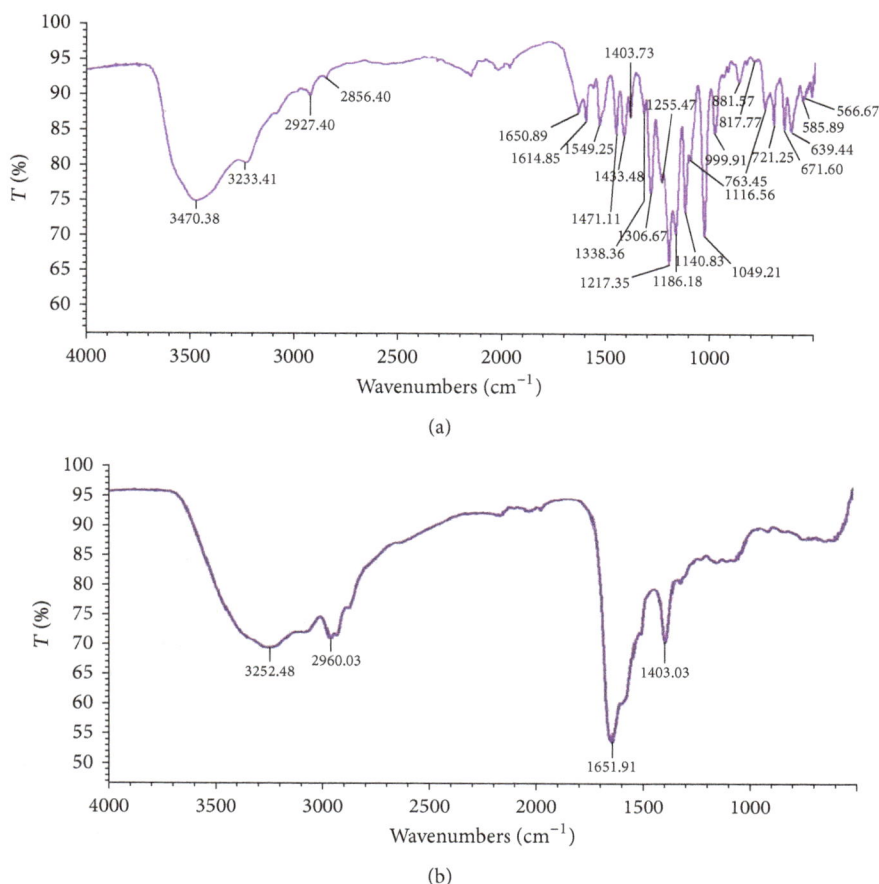

FIGURE 10: FT-IR spectra of (a) control Reactive Violet 5 and (b) its degradation metabolites.

azo dyes indicating its field applicability in the treatment of textile effluents. Therefore, *Paracoccus* sp. GSM2 is the highly promising bacterium that can be used for the treatment of textile industry effluents containing various reactive azo dyes.

Conflict of Interests

The authors declare that there is no conflict of interests regarding the publication of this paper.

Acknowledgment

The authors are thankful to Professor A. Venkataraman, Chairman, Department of Materials Science, Gulbarga University, Gulbarga, for providing FT-IR facility. Authors are also thankful to Gulbarga University, Gulbarga for providing research student fellowship during the research work.

References

[1] H. Zollinger, *Color Chemistry—Synthesis, Properties and Application of Organic Dyes and Pigments*, VCH Publishers, New York, NY, USA, 1987.

[2] V. K. Gupta, R. Jain, A. Nayak, S. Agarwal, and M. Shrivastava, "Removal of the hazardous dye-Tartrazine by photodegradation on titanium dioxide surface," *Materials Science and Engineering C*, vol. 31, no. 5, pp. 1062–1067, 2011.

[3] J.-S. Chang, C. Chou, Y.-C. Lin, P.-J. Lin, J.-Y. Ho, and T. Lee Hu, "Kinetic characteristics of bacterial azo-dye decolorization by *Pseudomonas luteola*," *Water Research*, vol. 35, no. 12, pp. 2841–2850, 2001.

[4] C. O'Neill, F. R. Hawkes, D. L. Hawkes, N. D. Lourenco, H. M. Pinheiro, and W. Delee, "Color in textile effluents sources, measurement, discharge consents and simulation: a review," *Journal of Chemical Technology and Biotechnology*, vol. 74, no. 11, pp. 1009–1018, 1999.

[5] S. Sandhya, S. Padmavathy, K. Swaminathan, Y. V. Subrahmanyam, and S. N. Kaul, "Microaerophilic-aerobic sequential batch reactor for treatment of azo dyes containing simulated wastewater," *Process Biochemistry*, vol. 40, no. 2, pp. 885–890, 2005.

[6] P. C. Vandevivere, R. Bianchi, and W. Verstraete, "Treatment and reuse of wastewater from the textile wet-processing Industry: review of emerging technologies," *Journal of Chemical Technology and Biotechnology*, vol. 72, pp. 289–302, 1998.

[7] P. Verma and D. Madamwar, "Decolourization of synthetic dyes by a newly isolated strain of *Serratia marcescens*," *World Journal of Microbiology and Biotechnology*, vol. 19, no. 6, pp. 615–618, 2003.

[8] R. G. Saratale, G. D. Saratale, D. C. Kalyani, J. S. Chang, and S. P. Govindwar, "Enhanced decolorization and biodegradation

of textile azo dye Scarlet R by using developed microbial consortium-GR," *Bioresource Technology*, vol. 100, no. 9, pp. 2493–2500, 2009.

[9] Z. W. Myslak and H. M. Bolt, "Occupational exposure to azo dyes and risk of bladder cancer," *ZblArbeitsmed*, vol. 38, pp. 310–321, 1988.

[10] H. Wang, X.-W. Zheng, J.-Q. Su, Y. Tian, X.-J. Xiong, and T.-L. Zheng, "Biological decolorization of the reactive dyes Reactive Black 5 by a novel isolated bacterial strain *Enterobacter* sp. EC3," *Journal of Hazardous Materials*, vol. 171, no. 1–3, pp. 654–659, 2009.

[11] P. Kaushik and A. Malik, "Fungal dye decolourization: recent advances and future potential," *Environment International*, vol. 35, no. 1, pp. 127–141, 2009.

[12] S. Asad, M. A. Amoozegar, A. A. Pourbabaee, M. N. Sarbolouki, and S. M. M. Dastgheib, "Decolorization of textile azo dyes by newly isolated halophilic and halotolerant bacteria," *Bioresource Technology*, vol. 98, no. 11, pp. 2082–2088, 2007.

[13] S. Moosvi, X. Kher, and D. Madamwar, "Isolation, characterization and decolorization of textile dyes by a mixed bacterial consortium JW-2," *Dyes and Pigments*, vol. 74, no. 3, pp. 723–729, 2007.

[14] K. Jain, V. Shah, D. Chapla, and D. Madamwar, "Decolorization and degradation of azo dye—reactive Violet 5R by an acclimatized indigenous bacterial mixed cultures-SB4 isolated from anthropogenic dye contaminated soil," *Journal of Hazardous Materials*, vol. 213-214, pp. 378–386, 2012.

[15] D. C. Kalyani, A. A. Telke, R. S. Dhanve, and J. P. Jadhav, "Ecofriendly biodegradation and detoxification of Reactive Red 2 textile dye by newly isolated Pseudomonas sp. SUK1," *Journal of Hazardous Materials*, vol. 163, no. 2-3, pp. 735–742, 2009.

[16] D. P. Tamboli, M. B. Kurade, T. R. Waghmode, S. M. Joshi, and S. P. Govindwar, "Exploring the ability of *Sphingobacterium* sp. ATM to degrade textile dye Direct Blue GLL, mixture of dyes and textile effluent and production of polyhydroxyhexadecanoic acid using waste biomass generated after dye degradation," *Journal of Hazardous Materials*, vol. 182, no. 1–3, pp. 169–176, 2010.

[17] J. Cheriaa, M. Khaireddine, M. Rouabhia, and A. Bakhrouf, "Removal of triphenylmethane dyes by bacterial consortium," *The Scientific World Journal*, vol. 2012, Article ID 512454, 9 pages, 2012.

[18] C. Brilon, W. Beckmann, M. Hellwig, and H.-J. Knackmuss, "Enrichment and isolation of naphthalenesulfonic acid-utilizing pseudomonads," *Applied and Environmental Microbiology*, vol. 42, no. 1, pp. 39–43, 1981.

[19] J. G. Holt, N. R. Krieg, P. H. A. Sneath, J. T. Staley, and S. T. Williams, *Bergey's Manual of Determinative Bacteriology*, Williams & Wilkins, Baltimore, Md, USA, 9th edition, 1994.

[20] N. Saitou and M. Nei, "The neighbor-joining method: a new method for reconstructing phylogenetic trees," *Molecular Biology and Evolution*, vol. 4, no. 4, pp. 406–425, 1987.

[21] J. Felsenstein, "Confidence limits on phylogenies: an approach using the bootstrap," *Evolution*, vol. 39, pp. 783–791, 1985.

[22] N. Takezaki, A. Rzhetsky, and M. Nei, "Phylogenetic test of the molecular clock and linearized trees," *Molecular Biology and Evolution*, vol. 12, no. 5, pp. 823–833, 1995.

[23] K. Tamura, M. Nei, and S. Kumar, "Prospects for inferring very large phylogenies by using the neighbor-joining method," *Proceedings of the National Academy of Sciences of the United States of America*, vol. 101, no. 30, pp. 11030–11035, 2004.

[24] K. Tamura, D. Peterson, N. Peterson, G. Stecher, M. Nei, and S. Kumar, "MEGA5: molecular evolutionary genetics analysis using maximum likelihood, evolutionary distance, and maximum parsimony methods," *Molecular Biology and Evolution*, vol. 28, no. 10, pp. 2731–2739, 2011.

[25] A. Khalid, M. Arshad, and D. E. Crowley, "Accelerated decolorization of structurally different azo dyes by newly isolated bacterial strains," *Applied Microbiology and Biotechnology*, vol. 78, no. 2, pp. 361–369, 2008.

[26] S. R. Dave and R. H. Dave, "Isolation and characterization of *Bacillus thuringiensis* for Acid red 119 dye decolourisation," *Bioresource Technology*, vol. 100, no. 1, pp. 249–253, 2009.

[27] X. Peng, J. S. Zhang, Y. Y. Li, W. Li, G. M. Xu, and Y. C. Yan, "Biodegradation of insecticide carbofuran by Paracoccus sp. YM3," *Journal of Environmental Science and Health— B: Pesticides, Food Contaminants, and Agricultural Wastes*, vol. 43, no. 7, pp. 588–594, 2008.

[28] Y. Bai, Q. Sun, C. Zhao, D. Wen, and X. Tang, "Microbial degradation and metabolic pathway of pyridine by a Paracoccus sp. strain BW001," *Biodegradation*, vol. 19, no. 6, pp. 915–926, 2008.

[29] H. Zhang, A. Kallimanis, A. I. Koukkou, and C. Drainas, "Isolation and characterization of novel bacteria degrading polycyclic aromatic hydrocarbons from polluted Greek soils," *Applied Microbiology and Biotechnology*, vol. 65, no. 1, pp. 124–131, 2004.

[30] T. L. Hu, "Decolourization of reactive azo dyes by transformation with *Pseudomonas luteola*," *Bioresource Technology*, vol. 49, no. 1, pp. 47–51, 1994.

[31] C. I. Pearce, J. R. Lloyd, and J. T. Guthrie, "The removal of colour from textile wastewater using whole bacterial cells: a review," *Dyes and Pigments*, vol. 58, no. 3, pp. 179–196, 2003.

[32] J. Chan and T. Kuo, "Kinetics of bacterial decolorization of azo dye with *Escherichia coli* NO3," *Bioresource Technology*, vol. 75, no. 2, pp. 107–111, 2000.

[33] D. Cetin and G. Donmez, "Decolorization of reactive dyes by mixed cultures isolated from textile effluent under anaerobic condition," *Enzyme and Microbial Technology*, vol. 38, pp. 926–930, 2006.

[34] L. A. De Baere, M. Devocht, P. Van Assche, and W. Verstraete, "Influence of high NaCl and NH4Cl salt levels on methanogenic associations," *Water Research*, vol. 18, no. 5, pp. 543–548, 1984.

[35] T. Panswad and C. Anan, "Specific oxygen, ammonia, and nitrate uptake rates of a biological nutrient removal process treating elevated salinity wastewater," *Bioresource Technology*, vol. 70, no. 3, pp. 237–243, 1999.

[36] P. S. Patil, U. U. Shedbalkar, D. C. Kalyani, and J. P. Jadhav, "Biodegradation of Reactive Blue 59 by isolated bacterial consortium PMB11," *Journal of Industrial Microbiology and Biotechnology*, vol. 35, no. 10, pp. 1181–1190, 2008.

[37] F. He, W. Hu, and Y. Li, "Biodegradation mechanisms and kinetics of azo dye 4BS by a microbial consortium," *Chemosphere*, vol. 57, no. 4, pp. 293–301, 2004.

[38] R. Silverstein and F. Webster, *Spectrometric Identification of Organic Compounds*, Wiley, New York, NY, USA, 1998.

[39] C. Desai, K. Jain, B. Patel, and D. Madamwar, "Efficacy of bacterial consortium-AIE2 for contemporaneous Cr(VI) and azo dye bioremediation in batch and continuous bioreactor systems, monitoring steady-state bacterial dynamics using qPCR assays," *Biodegradation*, vol. 20, no. 6, pp. 813–826, 2009.

15

Antimicrobial Effect of *Lippia sidoides* and Thymol on *Enterococcus faecalis* Biofilm of the Bacterium Isolated from Root Canals

H. N. H. Veras,[1] F. F. G. Rodrigues,[1] M. A. Botelho,[2] I. R. A. Menezes,[3] H. D. M. Coutinho,[4] and J. G. M. da Costa[1]

[1] Laboratory of Natural Products Research (LPPN), Regional University of Cariri, Crato, CE, Brazil

[2] Northeast Biotechnology Network, Brazil

[3] Laboratory of Pharmacology and Molecular Chemistry (LFQM), Regional University of Cariri, Crato, CE, Brazil

[4] Laboratory of Microbiology and Molecular Biology (LMBM), Department of Biological Chemistry, Regional University of Cariri, Cel. Antonio Luis Street, 1161 Pimenta, 63105-000 Crato, CE, Brazil

Correspondence should be addressed to H. D. M. Coutinho; hdmcoutinho@gmail.com

Academic Editors: A. I. Vela and J. Yoon

The species *Lippia sidoides* Cham. (Verbenaceae) is utilized in popular medicine as a local antiseptic on the skin and mucosal tissues. *Enterococcus faecalis* is the bacterium isolated from root canals of teeth with persistent periapical lesions and has the ability to form biofilm, where it is responsible for the failure of endodontic treatments. Essential oil of *L. sidoides* (EOLS) and its major component, thymol, were evaluated for reducing the CFU in biofilms of *E. faecalis in vitro*. The essential oil was obtained by hydrodistillation and examined with respect to the chemical composition, by gas chromatography-mass spectrometry (GC-MS). The GC-MS analysis has led to the identification of thymol (84.9%) and p-cymene (5.33%). EOLS and thymol reduced CFU in biofilms of *E. faecalis in vitro* (time of maturation, 72 h), with an exposure time of 30 and 60 min at concentrations of 2.5 and 10%. There was no statistical difference in effect between EOLS and thymol, demonstrating that this phenolic monoterpene was the possible compound responsible for the antimicrobial activity of EOLS. This study provides a basis for the possible utilization of EOLS as an adjuvant in the treatment of root canals that show colonization by *E. faecalis*.

1. Introduction

Advances in technology dealing with the problematics of the emergence of microorganisms increasingly more resistant to conventional antimicrobials have prompted the search for new substances of natural origin, with greater or equal efficacy as those drugs usually used [1]. One of the greatest problems faced is the formation of biofilm by bacteria. Organisms living in communities as in biofilms can tolerate changes in pH and the action of oxygen radicals, disinfectants, and antibiotics better than cells living in a planktonic manner [2].

Enterococcus faecalis is a Gram-positive coccus which generally occurs as pairs and short chains, and it is catalase negative [3]. It is the most common and, occasionally, single bacterium most often isolated from root canals of teeth with persistent periapical lesions [4]. The ability to form biofilm by this genus allows the colonization of inert and biological surfaces, protection against antimicrobial agents and the action of phagocytes, mediating adhesion, and invasion of host cells [5]. Therefore, the formation of biofilm is responsible for the failure of endodontic treatments [6].

Therefore, studies are conducted in order to determine the efficacy of natural products in the control of biofilm. The majority of these studies investigate the control of dental biofilm (bacterial plaque), such as the gel and extract of *Punica granatum* (pomegranate) [7, 8], *Lippia sidoides* (alecrim pimenta) [9], *Coffea arabica* (coffee) [10], and *Rheedia brasiliensis* (bacupari) [11].

Lippia is the second largest genus of the family Verbenaceae and includes many medicinal and aromatic species, which are found in South America (approximately 70–75% of the known species are in Brazil), Central America, and tropical Africa [12]. *Lippia sidoides* Cham., popularly known as "alecrim pimenta," is a shrub native to the semiarid region of Northeast Brazil. It is widely employed in popular medicine as a local antiseptic on skin and mucosal tissues. The therapeutic effect of *L. sidoides* is attributed mainly to the presence of thymol, a substance with high antimicrobial activity, which is the major component in the plant's essential oil and is also found in hydroalcoholic extracts of the plant [13]. This study investigated the *in vitro* antimicrobial activity of the essential oil of *Lippia sidoides* and of its major component, thymol, on biofilm of *Enterococcus faecalis*.

2. Material and Methods

2.1. Plant Material. Leaves of *Lippia sidoides* Cham. were collected in August 2010, from the Small Aromatic and Medicinal Plants Garden of the Natural Products Research Laboratory (LPPN) at Regional University of Cariri (URCA), Crato County, Ceara State, Brazil. A voucher specimen was deposited in the Herbarium Caririense Dardano of Andrade Lima of the Department of Biological Sciences (URCA) under registration number 3038.

2.2. Drug. Thymol was obtained from Sigma Chemical Corporation, St. Louis, MO, USA.

2.3. Essential Oil Isolation. Samples of *L. sidoides* fresh leaves (140 g) were triturated and submitted to hydrodistillation process, in a Clevenger-type apparatus for 2 h. The collected essential oil was subsequently dried with anhydrous sodium sulfate (Na_2SO_4) and stored refrigerated at <10°C until analyzed and tested.

2.4. Analysis of the Essential Oil. Analysis by CG/MS of the essential oil was carried out on a Hewlett-Packard Model 5971 GC/MS using a nonpolar DB-1 fused silica capillary column (30 m × 0.25 mm i.d., 0.25 m film thickness). Helium was the carrier gas, and flow rate was 0.8 mL/min, using split mode. The injector temperature and detector temperature were 250°C and 200°C, respectively. The column temperature was programmed from 35°C to 180°C at 4°C/min and then from 180°C to 250°C at 10°C/min. Mass spectra were recorded from 30 to 450 m/z. Individual components were identified by matching their 70 eV mass spectra with those of the spectrometer database using the Wiley L-built library and two other MS library searches using retention indices as a preselection routine [14], as well as by visual comparison of the fragmentation pattern with those reported in the literature [15].

2.5. Evaluation of the Inhibition of Biofilm Formation. A pure culture of *Enterococcus faecalis* ATCC 4083 was subcultured on a BHI agar plate for 24 h at 35 ± 2°C under aerobic conditions. After growth, isolated colonies were suspended in tubes containing 5 mL of BHI broth. After mixing, the suspension was adjusted to a concentration equivalent to

TABLE 1: Chemical components of *Lippia sidoides* fresh leaves essential oil.

Compounds	Tr (min)	IK*	(%)
p-cymene	4.2	1020	5.33
1,8-Cineol	4.4	1031	1.68
γ-Terpinene	5.0	1060	1.32
Ether ethyl carvacrol	9.7	1164	3.01
Thymol	11.8	1288	84.9
Carvacrol	12.9	1292	0.41
β-Caryophyllene	15.1	1418	1.17
Total			97.82

*Relative retention indices experimental: n-alkanes were used as reference points in the calculation of relative retention indices.

6.0 on the McFarland scale. Nitrocellulose membrane filters (0.22 μm porosity, 13 mm in diameter) were placed on the BHI agar plates, and then 50 μL of the bacterial suspension was placed on each membrane. The plates were incubated for 72 h in air at 35 ± 2°C. The essential oil of *Lippia sidoides* (EOLS) and thymol were dissolved separately in DMSO and were then diluted with sterile distilled water at concentrations of 2.5% and 10%. Sodium hypochlorite was used as the positive control and DMSO as the negative control, both at the same concentrations as the samples analyzed. After the incubation period, the biofilms were immersed in 3 mL of each solution, at different concentrations, for 30 and 60 min. After the exposure time, the membranes were carefully transferred to 3 mL of neutralization broth D/E (for EOLS, thymol, and DMSO) or to 3 mL of 1% sodium thiosulfate (for sodium hypochlorite) to stop the possible antimicrobial action of the test agent. Next, the membranes were vortexed for 30 s to resuspend the microorganisms [16, 17]. Finally, the suspensions were diluted 10 times for counting of colony forming units (CFU/mL) utilizing D/E agar, in triplicate [18].

2.6. Statistical Analysis. The results were expressed as means ± standard error of mean (S.E.M.) and statistical significance was determined by two-way ANOVA followed by Bonferroni's test, with the level of significance set at $P < 0.05$ using the program *GraphPadPrism 5.0*.

3. Results

The essential oil obtained by hydrodistillation of fresh leaves of *L. sidoides* gave a yield of 1.06% (w/w). The major constituents of the essential oil of *L. sidoides* were thymol (84.9%), *p*-cymene (5.33%), and ethyl methyl carvacrol (3.01%) (see Table 1). The means of colony forming units (CFU) per disk of *E. faecalis* biofilm after the exposure time (30 or 60 min) with 2.5% and 10% solutions of EOLS, thymol, DMSO, and sodium hypochlorite (NaOCl) are shown in Figures 1 and 2, respectively. NaOCl was the most effective antimicrobial agent, eliminating 99.99% of the bacteria with the concentrations and exposure times utilized in this study.

Figure 1 shows that 2.5% DMSO, the negative control, had no significant effect on cell viability for both times tested,

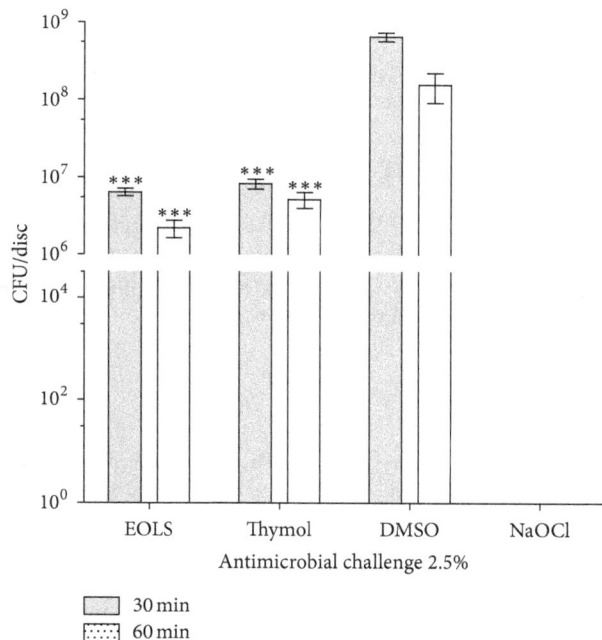

FIGURE 1: Susceptibility of biofilms of *Enterococcus faecalis* to antimicrobial challenge at 2.5% (v/v) for 30 or 60 min exposure times. EOLS, essential oil of *Lippia sidoides*; DMSO, dimethyl sulfoxide (negative control); and NaOCl, sodium hypochlorite (positive control). The vertical bars indicate the standard deviation ($n = 3$). $^{***}P < 0.001$ compared with DMSO (two-way ANOVA followed by the Bonferroni test).

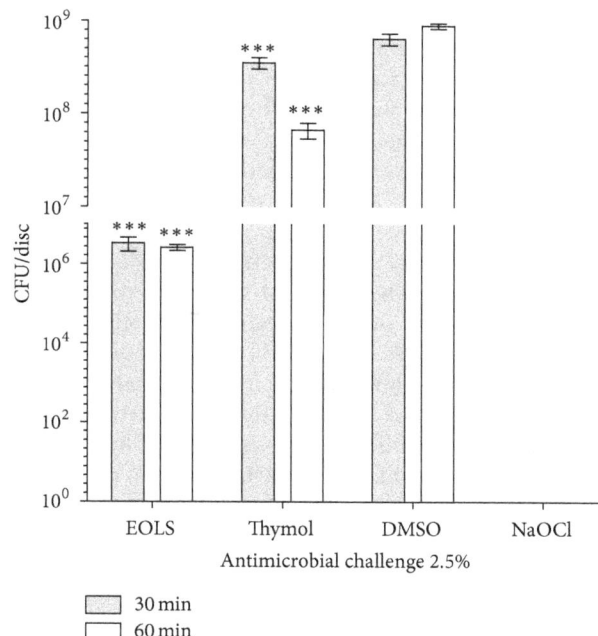

FIGURE 2: Susceptibility of biofilms of *Enterococcus faecalis* to antimicrobial challenge at 10.0% (v/v) for 30 or 60 min exposure times. EOLS, essential oil of *Lippia sidoides*; DMSO, dimethylsulfoxide (negative control); and NaOCl, sodium hypochlorite (positive control). The vertical bars indicate the standard deviation ($n = 3$). $^{***}P < 0.001$ compared with DMSO (two-way ANOVA followed by the Bonferroni test).

resulting in 6.5×10^8 and 1.5×10^8 CFU in 30 and 60 min of exposure. After exposure to EOLS for 30 and 60 min, CFU count in relation to DMSO control was significantly reduced ($P < 0.001$) to 6.4×10^6 and 2.2×10^6 CFU. Thymol decreased significantly ($P < 0.001$) the CFU count to 8.3×10^6 and 5.2×10^6, respectively. There were no statistical differences ($P > 0.05$) between EOLS and thymol effects for the designated exposure times.

After 30 and 60 min of exposure, 10% DMSO had no significant effect on cell viability, resulting in 6.4×10^8 and 9.0×10^8 CFU, respectively. CFU counts for biofilms exposed to EOLS and thymol at 10% in relation to the negative control were significantly reduced ($P < 0.001$) to 3.3×10^6 and 2.6×10^6 and 3.5×10^8 and 6.7×10^7 CFU, respectively. There was a statistical difference ($P < 0.001$) in mean CFU counts between EOLS and thymol for 30 min exposure. On the other hand, exposure of biofilms to EOLS and thymol for 60 min showed no difference ($P > 0.05$) (Figure 2).

4. Discussion

In some studies, the level of thymol present in the essential oil of the leaves can vary from 34.2 to 95.1% [19, 20]. This variation in level of constituents in essential oil can be influenced by the cultivation and development conditions (type of soil and climate), harvest and postharvest processing (time of day and season) [21] (Gil et al. 2002). The majority of

microorganisms do not exist as a culture of free-living cells, but rather associated with a living or inert surface, forming a structured community of cells surrounded by a polysaccharide matrix [22] (Costerton et al. 1999). There are various *in vitro* methods that are used to evaluate the effectiveness of antimicrobial agents against biofilms, but the results are conflicting in works utilizing the same test substances and the same microorganisms but different methods [16, 17, 23, 24]. The protocols utilized in this study were adapted from Abdullah et al. and Enright et al. studies This method is feasible and rapid, besides allowing the comparison of various antimicrobial challenges against microorganisms present in biofilm [16, 17].

The virulence of *E. faecalis* in root canals can be related to its capacity to resist intracanal drug treatment and to its ability to survive in the root canal as the only microorganism without the support of other bacteria, forming biofilms [25]. The irrigation of root canals is an important step in disinfection and is an integral part of procedures of endodontic treatment. Currently, the irrigant most often used is sodium hypochlorite (NaOCl) due its strong antimicrobial activity, but the main disadvantage of its use in dental treatment is its toxicity to patient's tissues [26].

Structured bacteria in biofilm behave differently when exposed to chemical substances, because polymeric substances that make up the biofilm matrix hamper the diffusion of chemical substances and antibiotics [27, 28]. The susceptibility of biofilm is directly related to time of exposure and to the concentration of the substance, besides the phase of biofilm development [17]. The speed of penetration of the substance varies according to the microorganism and composition of exopolysaccharide matrix [22]. Therefore, our results demonstrate that EOLS and thymol are capable of reducing *E. faecalis* CFU in biofilms *in vitro* (time of maturation, 72 h) with an exposure time of 30 and 60 min, at concentrations of 2.5 and 10%. At 2.5%, there were no statistical differences ($P > 0.05$) between exposure time and the samples tested, where thymol was responsible for the antimicrobial activity of EOLS against the biofilm. On the other hand, the higher concentration of thymol (10%) was not as effective as the lower concentration (2.5%), which was not the case for EOLS, showing the same activity at both concentrations and with both exposure times. This is the first report on the action of EOLS against biofilms of *E. faecalis*.

The mechanisms by which EOLS and thymol kill microorganisms present in biofilms are still not well elucidated. However, studies of the mechanism of action of carvacrol and thymol on biofilms remain unclear; their amphipathic nature could account for the observed effects. The relative hydrophilicity of carvacrol and thymol may allow their diffusion through the polar polysaccharide matrix, whilst the prevalent hydrophobic properties of these compounds could lead to specific interactions with the bacterial membrane causing the dispersion of the polypeptide chains of the cell membrane and destabilizing the cell [29–31]. This hypothesis is supported by the electron micrographs of damaged cells and the significant increase of the cell constituents' release demonstrated that thymol and other essential oil combinations affected the cell membrane integrity [32].

A preparation based on essential oils of *Eucalyptus globulus*, *Melaleuca alternifolia*, *Thymus* sp., and *Syzygium aromaticum*, containing mainly monoterpenes, demonstrated, *in vitro*, reduced adherence of *Staphylococcus epidermidis* and formation of biofilm [33]. The combination of thymol and chlorhexidine gluconate demonstrated synergistic activity against *S. epidermidis* biofilm [34]. Braga et al. found that thymol also interferes with the adherence of *C. albicans* on mucosal cells, and they suggested that this compound can significantly interfere not only with the initial phases of biofilm formation but also with its maturation, since it effectively inhibits the metabolic activity of biofilm.

According to Nostro et al., thymol is as much hydrophilic as hydrophobic, which can favor the diffusion of this compound through the polysaccharide layer of biofilm and reach the bacterial cells to exert its antimicrobial effect by altering membrane permeability [31]. This hypothesis is supported by the results obtained in various clinical studies with mouthwashes or toothpastes containing EOLS, which have demonstrated a decrease in bacterial plaque [35, 36].

Therefore, our results provide a basis for the possible utilization of EOLS or its major component, thymol, as adjuvants in the treatment of root canals that show colonization by *E. faecalis*. However, preclinical studies are necessary to evaluate the true efficacy of these products and the concentration needed to kill biofilm bacteria *in vivo*.

Conflict of Interests

The authors declare that there is no conflict of interests regarding the publication of this paper.

Acknowledgments

The authors thank CAPES, FUNCAP, and CNPq for financial support, Dr. Sidney Gonçalves Lima (UFPI) for the analysis of the essential oil, and Dr. A. Leyva who helped with the translation and editing of the paper.

References

[1] H. Brotz-Oesterhelt, D. Beyer, and H. P. Kroll, "Dysregulation of bacterial proteolytic machinery by a new class of antibiotics," *Natural Medicines*, vol. 11, no. 10, pp. 1082–1087, 2005.

[2] K. K. Jefferson, "What drives bacteria to produce a biofilm?" *FEMS Microbiology Letters*, vol. 236, no. 2, pp. 163–173, 2004.

[3] L. M. Teixeira and R. R. Facklam, "*Enterococcus*," in *Topley and Wilson's Microbiology and Microbial Infections*, John Wiley & Sons, Chichester, UK, 2005.

[4] I. N. Rôças, J. F. Siqueira Jr., and K. R. N. Santos, "Association of Enterococcus faecalis with different forms of periradicular diseases," *Journal of Endodontics*, vol. 30, no. 5, pp. 315–320, 2004.

[5] L. Baldassarri, R. Creti, L. Montanaro, G. Orefici, and C. R. Arciola, "Pathogenesis of implant infections by enterococci," *International Journal of Artificial Organs*, vol. 28, no. 11, pp. 1101–1109, 2005.

[6] L. M. Lin, J. E. Skribner, and P. Gaengler, "Factors associated with endodontic treatment failures," *Journal of Endodontics*, vol. 18, no. 12, pp. 625–627, 1992.

[7] I. E. Alsaimary, "Efficacy of some antibacterial agents against Streptococcus mutans associated with tooth decay," *The Internet Journal of Microbiology*, vol. 7, no. 2, 2010.

[8] A. D. Y. Salgado, J. L. Maia, S. L. D. S. Pereira, T. L. G. de Lemos, and O. M. D. L. Mota, "Antiplaque and antigingivitis effects of a gel containing Punica granatum Linn extract. A double-blind clinical study in humans," *Journal of Applied Oral Science*, vol. 14, no. 3, pp. 162–166, 2006.

[9] M. A. Botelho, R. A. dos Santos, J. G. Martins et al., "Comparative effect of an essential oil mouthrinse on plaque, gingivitis and salivary Streptococcus mutans levels: a double blind randomized study," *Phytotherapy Research*, vol. 23, no. 9, pp. 1214–1219, 2009.

[10] L. F. Landucci, L. D. Oliveira, E. H. S. Brandão, C. Y. Koga-Ito, E. G. Jardim Júnior, and A. O. C. Jorge, "Efeitos de Coffea arabica sobre a aderência de Streptococcus mutans à superfície de vidro," *Brazilian Dental Science*, vol. 6, no. 3, 2010.

[11] L. S. B. Almeida, R. M. Murata, R. Yatsuda et al., "Antimicrobial activity of Rheedia brasiliensis and 7-epiclusianone against Streptococcus mutans," *Phytomedicine*, vol. 15, no. 10, pp. 886–891, 2008.

[12] L. F. Viccini, P. M. O. Pierre, M. M. Praça et al., "Chromosome numbers in the genus Lippia (Verbenaceae)," *Plant Systematics and Evolution*, vol. 256, no. 1–4, pp. 171–178, 2005.

[13] F. Matos and F. Oliveira, "Lippia sidoides Cham.: farmacognosia, química e farmacologia," *Revista Brasileira de Farmácia*, vol. 79, pp. 84–87, 1998.

Antimicrobial Effect of Lippia sidoides and Thymol on Enterococcus faecalis Biofilm of the Bacterium Isolated
from Root Canals

133

[14] J. Alencar, A. Craveiro, F. Matos, and M. Machado, "Kovats indices simulation in essential oils analysis," *Química Nova*, vol. 13, no. 4, pp. 282–284, 1990.

[15] R. Adams, *Identification of Essential Oil Components by Gas*, Allured Publishing Corporation, 3rd edition, 2001.

[16] M. Abdullah, Y.-L. Ng, K. Gulabivala, D. R. Moles, and D. A. Spratt, "Susceptibilties of two Enterococcus faecalis phenotypes to root canal medications," *Journal of Endodontics*, vol. 31, no. 1, pp. 30–36, 2005.

[17] M. C. Enright, D. A. Robinson, G. Randle, E. J. Feil, H. Grundmann, and B. G. Spratt, "The evolutionary history of methicillin-resistant Staphylococcus aureus (MRSA)," *Proceedings of the National Academy of Sciences of the United States of America*, vol. 99, no. 11, pp. 7687–7692, 2002.

[18] B. P. Dey and F. B. Engley Jr., "Methodology for recovery of chemically treated Staphylococcus aureus with neutralizing medium," *Applied and Environmental Microbiology*, vol. 45, no. 5, pp. 1533–1537, 1983.

[19] V. C. C. Girão, D. C. S. Nunes-Pinheiro, S. M. Morais, J. L. Sequeira, and M. A. Gioso, "A clinical trial of the effect of a mouth-rinse prepared with Lippia sidoides Cham essential oil in dogs with mild gingival disease," *Preventive Veterinary Medicine*, vol. 59, no. 1-2, pp. 95–102, 2003.

[20] R. O. S. Fontenelle, S. M. Morais, E. H. S. Brito et al., "Chemical composition, toxicological aspects and antifungal activity of essential oil from Lippia sidoides Cham," *Journal of Antimicrobial Chemotherapy*, vol. 59, no. 5, pp. 934–940, 2007.

[21] A. Gil, E. B. de la Fuente, A. E. Lenardis et al., "Coriander essential oil composition from two genotypes grown in different environmental conditions," *Journal of Agricultural and Food Chemistry*, vol. 50, no. 10, pp. 2870–2877, 2002.

[22] J. W. Costerton, P. S. Stewart, and E. P. Greenberg, "Bacterial biofilms: a common cause of persistent infections," *Science*, vol. 284, no. 5418, pp. 1318–1322, 1999.

[23] E. J. G. Seabra, I. P. C. Lima, S. V. Barbosa, and K. C. Lima, "Atividade antimicrobiana "in vitro" de compostos a base de hidróxido de cálcio e tergentol em diferentes concentrações sobre bactérias orais," *Acta Cirurgica Brasileira*, vol. 20, no. 1, pp. 12–18, 2005.

[24] M. T. Arias-Moliz, C. M. Ferrer-Luque, M. Espigares-García, and P. Baca, "Enterococcus faecalis biofilms eradication by root canal irrigants," *Journal of Endodontics*, vol. 35, no. 5, pp. 711–714, 2009.

[25] T. C. Paradella, C. Y. Koga-Ito, and A. O. C. Jorge, "Enterococcus faecalis: considerações clínicas e microbiológicas," *Revista de Odontologia da UNESP*, vol. 36, no. 2, pp. 163–168, 2007.

[26] J. R. Bowden, M. Ethunandan, and P. A. Brennan, "Life-threatening airway obstruction secondary to hypochlorite extrusion during root canal treatment," *Oral Surgery, Oral Medicine, Oral Pathology, Oral Radiology and Endodontology*, vol. 101, no. 3, pp. 402–404, 2006.

[27] P. Stoodley, K. Sauer, D. G. Davies, and J. W. Costerton, "Biofilms as complex differentiated communities," *Annual Review of Microbiology*, vol. 56, pp. 187–209, 2002.

[28] J. Xavier, C. Picioreanu, J. Almeida, and M. Loosdrecht, "Monitorização e modelação da estrutura de biofilmes," *Boletim de Biotecnologia*, vol. 76, pp. 2–13, 2003.

[29] M. M. Cowan, "Plant products as antimicrobial agents," *Clinical Microbiology Reviews*, vol. 12, no. 4, pp. 564–582, 1999.

[30] E. A. Soumya, I. K. Saad, L. Hassan, Z. Ghizlane, M. Hind, and R. Adnane, "Carvacrol and thymol components inhibiting Pseudomonas aeruginosa adherence and biofilm formation," *African Journal of Microbiology Research*, vol. 5, no. 20, pp. 3229–3232, 2011.

[31] A. Nostro, A. R. Blanco, M. A. Cannatelli et al., "Susceptibility of methicillin-resistant staphylococci to oregano essential oil, carvacrol and thymol," *FEMS Microbiology Letters*, vol. 230, no. 2, pp. 191–195, 2004.

[32] F. Lv, H. Liang, Q. Yuan, and C. Li, "In vitro antimicrobial effects and mechanism of action of selected plant essential oil combinations against four food-related microorganisms," *Food Research International*, vol. 44, no. 9, pp. 3057–3064, 2011.

[33] J. Al-Shuneigat, S. Cox, and J. Markham, "Effects of a topical essential oil-containing formulation on biofilm-forming coagulase-negative staphylococci," *Letters in Applied Microbiology*, vol. 41, no. 1, pp. 52–55, 2005.

[34] T. J. Karpanen, T. Worthington, E. R. Hendry, B. R. Conway, and P. A. Lambert, "Antimicrobial efficacy of chlorhexidine digluconate alone and in combination with eucalyptus oil, tea tree oil and thymol against planktonic and biofilm cultures of Staphylococcus epidermidis," *Journal of Antimicrobial Chemotherapy*, vol. 62, no. 5, pp. 1031–1036, 2008.

[35] R. S. Nunes, A. A. M. Lira, C. M. Lacerda, D. O. B. da Silva, J. A. da Silva, and D. P. de Santana, "Obtenção e avaliação clínica de dentifrícios à base do extrato hidroalcoólico da Lippia sidoides Cham (Verbenaceae) sobre o biofilme dentário," *Revista de Odontologia da UNESP*, vol. 35, no. 4, pp. 275–283, 2006.

[36] P. L. D. Lobo, C. S. R. Fonteles, C. B. M. de Carvalho et al., "Dose-response evaluation of a novel essential oil against Mutans streptococci in vivo," *Phytomedicine*, vol. 18, no. 7, pp. 551–556, 2011.

16

Yarrowia lipolytica and Its Multiple Applications in the Biotechnological Industry

F. A. G. Gonçalves,[1] **G. Colen,**[1] **and J. A. Takahashi**[2]

[1] *Faculdade de Farmácia, Universidade Federal de Minas Gerais, 31270-901 Belo Horizonte, MG, Brazil*
[2] *Departamento de Química, Instituto de Ciências Exatas, Universidade Federal de Minas Gerais, 31270-901 Belo Horizonte, MG, Brazil*

Correspondence should be addressed to J. A. Takahashi; jat@qui.ufmg.br

Academic Editors: N. Dumais, N. Heng, and Y. Yan

Yarrowia lipolytica is a nonpathogenic dimorphic aerobic yeast that stands out due to its ability to grow in hydrophobic environments. This property allowed this yeast to develop an ability to metabolize triglycerides and fatty acids as carbon sources. This feature enables using this species in the bioremediation of environments contaminated with oil spill. In addition, *Y. lipolytica* has been calling the interest of researchers due to its huge biotechnological potential, associated with the production of several types of metabolites, such as bio-surfactants, γ-decalactone, citric acid, and intracellular lipids and lipase. The production of a metabolite rather than another is influenced by the growing conditions to which *Y. lipolytica* is subjected. The choice of carbon and nitrogen sources to be used, as well as their concentrations in the growth medium, and the careful determination of fermentation parameters, pH, temperature, and agitation (oxygenation), are essential for efficient metabolites production. This review discusses the biotechnological potential of *Y. lipolytica* and the best growing conditions for production of some metabolites of biotechnological interest.

1. Introduction

The use of microorganisms to obtain different types of food, such as beer, wine, bread, cheeses, and fermented milk is very old. There have been reports of application of fermentation processes for the production of foods from times before Christ. In the 20th century, the industrial microbiology expanded even more, because they perceived new possibilities for obtaining large variety and quantity of products by fermentative processes. At this time, there was also a boom in the industrial scale production of solvents, antibiotics, enzymes, vitamins, amino acids, and polymers, among many other compounds formed by microbial action [1, 2]. The development of molecular biology techniques in the 1970's gave new impulse in this area, with significant innovations, which have resulted in the emergence of new useful industrial biotechnological processes.

Yarrowia lipolytica is an excellent example of a microorganism with multiple biotechnological applications. This nonconventional, aerobic, dimorphic yeast can usually be found in environments containing hydrophobic substrates, rich in alkanes and fats. It can be isolated from cheeses, yoghurts, kefir, soy sauce, meat and shrimp salads. Its maximum temperature of growth is lower than 34°C [3]. It is used in biotechnological and industrial processes for obtaining various products such as citric and isocitric acids and enzymes (acid or alkaline proteases, lipases, and RNase), for bioremediation and production of biosurfactants [4–6]. *Y. lipolytica* is also capable of producing γ-decalactone, a compound that features a fruity aroma of great industrial interest, obtained by conversion of methyl ricinoleate [7]. Among the metabolites produced by *Y. lipolytica*, one of the most important is lipase, an enzyme that has gained interest of scientists due to its broad technological applications in food, pharmaceutical and detergent production areas. However, the largest industrial application of this yeast species seems to be in the production of biomass to be used as single cell protein (SCP) [3, 8, 9]. Currently there is a great interest in the ability of *Y. lipolytica* to produce and store lipids, which can be used in the production of biofuels, or in the production of oils enriched with essential fatty acids, which has wide application in pharmaceutical and food industries [10].

This multiplicity of applications has resulted in many studies related to lipase production by both wild and naturally occurring strains of *Y. lipolytica* in the presence of lipids [11]. These characteristics enable *Y. lipolytica* to stand among the microorganisms promising for the production of biodiesel. Use of microbial oil in replacement of vegetable oil presents several advantages such as easiness of reaching increased scale, requiring a much smaller area for production, as well as independence of seasonality [12].

Y. lipolytica has been considered as a suitable model for studies on the yeasts dimorphism since it produces pseudohyphae filaments in nitrogen-limited conditions [13]. Its growth in fatty acids, as carbon sources, with casein, yeast, or meat extracts induces the formation of hyphae, which is inhibited by the deficiency of magnesium sulfate and ferric chloride, by the presence of cysteine or reduced glutathione levels [3].

The ability of this yeast to grow in alkanes and to hydrolyze triglycerides and fatty acids used as extracellular carbon sources makes it an interesting oleaginous yeast model for studying the metabolism of fatty acids [14]. *Y. lipolytica* has developed mechanisms for efficient use of these substrates as carbon sources [15, 16]. *Y. lipolytica* produces no ethanol but uses it as a source of carbon, in concentrations up to 3%. In higher concentrations, ethanol becomes toxic to the yeast [8]. It has great potential for production of various intracellular metabolites of industrial importance that are exported by the cells [16].

The growth of *Y. lipolytica* and the secretion of metabolites are affected by different organic ingredients and minerals (and their relative amounts) employed as sources of carbon, nitrogen, and micronutrients and by the pH of the growth medium, incubation temperature, inoculum and intensity of oxygenation [13].

The amount of oxygen available for microbial cells can be provided in the growth medium with the use of compressed air in bioreactors [17], by aeration and agitation in jar flasks [18] and by the addition of oxygen vectors, or by adding a nonaqueous, organic phase to induce a significant increase in oxygen transfer rate [19].

2. *Yarrowia lipolytica* and Biotechnological Applications

2.1. Bioremediation and Production of Biosurfactants. Bioremediation is a technique that uses microorganisms to speed up the degradation of environmental contaminants into less toxic forms or to promote contaminants reduction to acceptable concentration levels [20–22]. This technique became the principal method used for the restoration of environments contaminated with oil and waste water treatments from the oil industry [15, 23].

Pollution by oil/oil spill is one of the leading causes of environmental damage and can occur in both terrestrial and marine environments or in freshwater [24]. Because of its ability to use alkanes, fatty acids, and oils, *Y. lipolytica* is regarded as a potential agent in bioremediation of environments for the degradation of vegetal and mineral oil waste [15, 24].

There are reports in the literature on the evaluation of indigenous yeasts associated with natural detoxification processes of a wide variety of pollutants. Strains of *Y. lipolytica* have been isolated from several polluted environments. The indigenous microbial populations present in such locations are constantly threatened by the presence of pollutants and, therefore, they have evolved so that their enzyme configuration conquered effectiveness for detoxification [24].

Under the evolutionary point of view, it is believed that microorganisms that multiply in aqueous environments rich in materials of hydrophobic nature, dispersed in the environment in the form of drops, developed a mechanism to facilitate the use of such substrates as carbon sources. The ability of *Y. lipolytica* in degrading a variety of organic compounds, including aliphatic and aromatic hydrocarbons, is always accompanied by the production of bio-surfactants, molecules made up predominantly of glycolipids, which increases the contact surface. The growth of microorganisms in hydrophobic substrates requires a contact between the hydrophobic substrate present in the organic phase and the cell surface. This contact may occur by direct adsorption of hydrophobic droplets on the surface of the cell or by the action of bio-surfactants and both mechanisms are reported for *Y. lipolytica*. The interaction between the hydrophobic molecules and cells is mediated by proteins or glycoproteins present in the cell wall and the surfactant that is secreted can facilitate this interaction [13, 15]. According to Beopoulos et al. [25], the extracellular lipase produced by *Y. lipolytica* also acts on the hydrophobic substrate, permitting the hydrolysis of triglycerides. The action of bio-surfactant and lipase occurs progressively, after formation of several droplets that facilitate the transport of the substrate (Figure 1).

The bio-surfactants are molecules that possess both hydrophilic and hydrophobic groups, which determine properties such as adsorption, formation of micelles, macro- and micro emulsions, and detergency solubilization capacity, indispensable for the process of bioremediation [26].

The composition and characteristics of bio-surfactants produced by microorganisms are influenced by the nature of the carbon and nitrogen sources used, by the presence of phosphorus, iron, manganese, and magnesium in the means of production, temperature, pH, and agitation. The production can be spontaneous or induced by the presence of lipophilic compounds, by changes in pH, temperature, aeration and agitation speed and subjection to stress conditions (e.g., low concentration of nitrogen) [27].

Cirigliano and Carman [28] studied different carbon sources (hexadecane, paraffin, soy oil, olive oil, corn oil, and cottonseed oil) for the production of bio-surfactants by *Y. lipolytica* and realized that there was greater productivity when hexadecane was used. Sarubbo et al. [29] studied the production of bio-surfactants by *Y. lipolytica* from glucose

Triacylglycerol (1)

↓ Lipases
  Biosurfactants

(2)

↓

Substrate droplets

↓

Cell membrane

Fatty acid

        Triacylglycerol synthesis

(3)   (5)

β-Oxidation ⟵ Lipid bodies

Mobilization of triacylglycerol
(4)

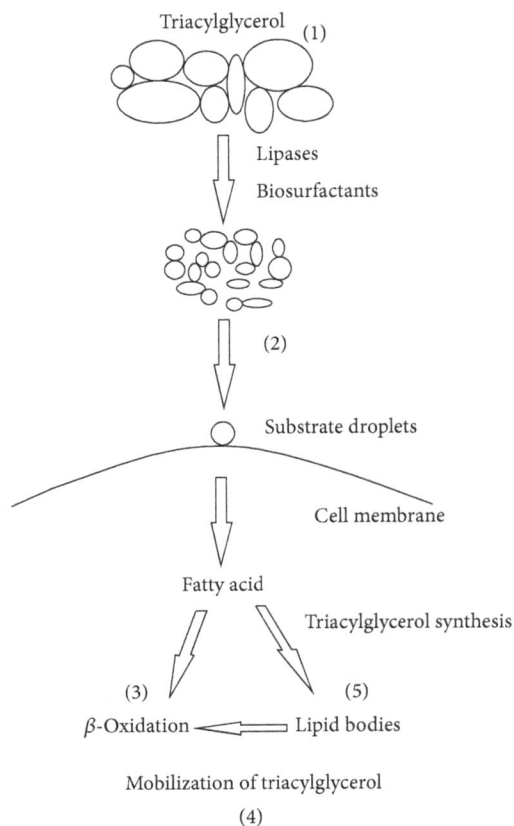

FIGURE 1: Representation of triglyceride and fatty acid (examples of hydrophobic substrates) assimilation by *Y. lipolytica*. (1) Reduction of droplet size of the hydrophobic substrates by the action of biosurfactants and extracellular lipase that hydrolyses the triglycerides. (2) Droplets of hydrophobic substrate bound to cell surface protrusions. (3) Fatty acid degradation by β-oxidation or (5) triacylglycerol storage lipid bodies. (4) Mobilization of triacylglycerol by lipases (adapted from Beopoulos et al. [25]).

and reported that the presence of hydrocarbons was not a prerequisite for biosynthetic induction of surfactants under the conditions used.

2.2. γ-Decalactone. The γ-decalactone is a peach-scented compound widely used in food and beverages, which is the reason of the great interest in its biotechnological production [13]. This can be accomplished by biotransformation of ricinoleic acid by *Y. lipolytica* [30]. Ricinoleic acid is a hydroxylated fatty acid (C18) and, in its esterified form, is the main constituent of castor oil. This fatty acid is the precursor of γ-decalactone [13]. To increase the availability of ricinoleic acid to cells, castor oil can be hydrolyzed by lipases [30], generating esters such as methyl ricinoleate [13]. The process involves the degradation of ricinoleic acid into 4-hydroxy-decanoic acid, a precursor of γ-decalactone, which is obtained by the action of peroxisomal β-oxidation enzymes (Figure 2) [31]. Peroxisomal β-oxidation occurs in four cyclic oxidation reactions catalyzed by the enzymes acyl-CoA oxidase and 3-ketoacyl-CoA thyilase [31, 32].

Y. lipolytica produces high yield of γ-decalactone and also has a large number of genes encoding enzymes that degrade hydrophobic substrates and, therefore, became a model for this metabolic route. *Y. lipolytica* has a family of six medium-chain acyl-CoA oxidases, the enzyme that catalyzes the first reaction of β-oxidation in this series [31].

2.3. Production of Citric Acid. Citric acid is much employed in industry as acidulant to stiff aromas in food, beverages, and pharmaceuticals as well as preservative in cosmetics. To feed the industrial demand, citric acid has been produced by fermentation processes since the beginning of the 20th century [34]. Its annual production reached approximately 1.6 million tons in 2008 [35], being one of the metabolites most produced in industrial scale [36].

The most commonly used microorganism for citric acid production is *Aspergillus niger* [34]. Species of yeasts as *Y. lipolytica*, *C. guilliermondii* and *C. oleophila* have also been used [37], *Y. lipolytica* being the most used yeast [38].

The microorganism, the type of substrate, and growing conditions (temperature, aeration, concentration and source of carbon and nitrogen, phosphate, trace elements, and pH) are factors that influence directly the production of citric acid [39–41]. In general, this production is conducted by batch, fed batch or continuous process [37]. The water content of the culture medium or, in other words, the type of fermentation, solid substrate or submerged fermentation, also influences the production of citric acid. Therefore, to obtain optimum performance, it is mandatory to evaluate the performance of the strain in different conditions [42].

According to Papanikolaou and Aggelis [36], the production of citric acid mainly occurs when the concentration of nitrogen in the culture medium becomes limited and using sugars or glycerol as carbon sources. Nitrogen limitation causes rapid decrease in the concentration of AMP (adenosine monophosphate), responsible for the formation of intracellular enzyme NAD^+ (nicotinamide adenine dinucleotide) isocitrate dehydrogenase (that turns isocitric acid in alpha-ketoglutaric acid), causing loss of enzyme activity. As a consequence, in the build-up of citric acid in the mitochondria of the cell, upon reaching a critical level, it is secreted in the cytosol.

2.3.1. Influence of Carbon Source on Production of Citric Acid. Carbon is the chemical element present in higher amounts in the cell. The accumulation of citric acid is strongly influenced by both the carbon source and its concentration. Yeasts are able to produce citric acid using a variety of carbon sources of various types [43].

Many species of yeasts that grow in carbohydrate substrates are capable of producing high concentrations of citric acid [43]. *Y. lipolytica* is the unique yeast able to maximize the production of citric acid using both carbohydrate and fat carbon sources [37]. Many carbon sources have been screened for production of citric acid by *Y. lipolytica* such as glucose [40], n-paraffin [44], ethanol [38, 45], glycerol [37, 46, 47], sunflower oil [48], and wastewater from the production of olive oil [49].

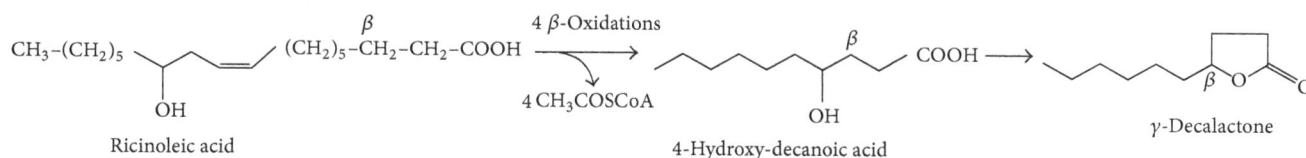

FIGURE 2: Ricinoleic acid bioconversion to γ-decalactone (adapted from Schrader et al. [33]).

Glucose and yeast extract are considered preferred sources of various microorganisms, which is not surprising since yeast extract contains all the necessary micronutrients, including metal ions, for microbial growth [50]. The presence of carbohydrates easily absorbed by microorganisms is considered to be a key point for citric acid production in significant quantities. Among these carbohydrates, sucrose is considered the best source of carbon, followed by glucose, fructose and galactose [42].

Some authors support the use of cheaper substrates, such as agroindustrial wastes, to produce citric acid [36, 42, 44]. Glycerol is also an interesting substrate to be used in the production of citric acid since in the process of esterification, for each 10 kg of biodiesel produced, 1 kg of glycerol is formed as a by-product [36, 47]. Molasses, a by-product of the sugar and alcohol industry, containing high content of sugars (around 40–55%) can also be used for the production of citric acid [42]. The use of wastewater from the processing of olives, enriched with glucose, was proposed by Sarris et al. [49], as a substrate for producing citric acid. Crolla and Kennedy [44] studied the effect of n-paraffin in the production of citric acid by C. lipolytica NRRL-Y-1095 and obtained significant amounts of citric acid and biomass. The use of n-alkanes usually provides great citric acid performance. In addition, paraffinic substrates are very suitable sources for obtaining biomass (single-cell proteins). To obtain high concentration of citric acid, the concentration of the carbon source is also important [34, 40]. This is corroborated by Sarris et al. [49]; evaluating different strains of Y. lipolytica for citric acid production in carbon limiting conditions, this group noted that although there is formation of biomass, only low yield of citric acid was obtained due to the insufficient amount of carbon present in the growth medium. Finogenova et al. [45] tested the effect of ethanol on the production of citric acid by a mutant strain of Y. lipolytica N1, getting maximum output when using ethanol concentrations in the range between 0.01–1.0 g/L.

On a growth medium containing 200 g/L of crude glycerol, the mutant strain of Y. lipolytica Wratislavia AWG7 produced an exceeding of 0.69 g of citric acid/g glycerol consumed. A lower yield was obtained by Y. lipolytica Wratislavia K1 (about 0.45 g/g) that also produced erythritol in high yield, thus reducing the production of citric acid [51].

2.3.2. Influence of Nitrogen Source on Production of Citric Acid. The nitrogen source (type and concentration) also influences the production of citric acid by microorganisms. Physiologically, there is a preference for salts, such as ammonium nitrate, urea and ammonium sulfate, peptones, and malt extract, among others. Acidic compounds of ammonia are also well accepted, since its consumption causes decrease in pH of the medium, which is essential for citric fermentation. However, in the early hours of fermentation, the ideal is that the pH value does not change much to ensure biomass formation [42].

It is consensus in the literature that the production of citric acid by yeasts is favored under nitrogen-limiting conditions, a situation that leads the organism to citric acid production and lipids [34, 36, 45, 49, 52]. Production of citric acid requires nitrogen concentration in the range of 0.1 to 0.4 g/L. Higher concentrations promote cell growth, but decrease the production of citric acid [42].

Anastassiadis et al. [34] identified the ammoniac nitrogen as limiting substrate for citrate production. According to the results obtained, the production of citric acid begins a few hours after nitrogen exhaustion. The authors even contradict many reports in the literature, reporting that is not the extracellular nitrogen exhaustion that takes the production of citric acid but the limitation of intracellular nitrogen, accompanied by increasing intracellular levels of ammonium ion and energy (ATP). These conditions would induce a specific active transport system for the secretion of citrate. According to the authors, this event may be explained by analyzing Candida oleophila biomass. During the phase of secondary metabolites production (idiophase) which corresponds, in general, to the stationary phase of microbial growth, the yeast cells increase in size. This would occur also with their vacuoles, organelles that accumulate citric acid, which are secreted when the active citrate transport system is induced. This induction happens when there is limitation of intracellular nitrogen, followed by high concentrations of intracellular ammonium ion and energy. According to Anastassiadis et al. [34], the increase in the intracellular concentration of ammonium ion in C. oleophila can be explained by proteolysis that occurs as a result of intracellular nitrogen limitation, which would lead to the extracellular exhaustion. This limitation would cause conversion of glucose in citrate, in order to obtain energy.

2.3.3. Effect of Temperature, pH and Oxygenation on the Production of Citric Acid. The fermentation temperature is a variable that affects directly in the production of citric acid. To determine the optimum temperature, it is usual to accept a variation of only ±1°C, to ensure an effective process [44]. The pH of the growth medium has an important influence on the production of citric acid since it affects the metabolism of Y. lipolytica. Changes in pH, along with the time of cultivation, are influenced mainly by the microorganism used [42], by the technique employed and by the nature of the substances produced. In the production of organic acids, such as citric acid, there is a decrease in the pH of the medium [41, 42].

Karasu-Yalcin et al. [41] achieved high yield of citric acid from strains of *Y. lipolytica* at an optimum temperature of 30°C, while Crolla and Kennedy [44] reported that the optimum temperature for both citric acid production and biomass formation by *C. lipolytica* was 26–30°C. In Karasu-Yalcin et al. [41] studies, the *Y. lipolytica* (NBRC 1658 and a domestic strain 57) strains showed optimum production of citric acid at initial pH values of 7.0 and 5.2, respectively. On the other hand, Kamzolova et al. [48] reported that the pH of the medium influenced both the production of citric acid and isocitric acid. *Y. lipolytica* produced similar amounts of these acids at pH value of 4.5. However, at higher pH (6.0), the yeast produced higher amounts of isocitric acid. This occurred because the citric acid transport across the membrane is stimulated at low pH values, while the isocitric acid transport is independent of the pH of the medium [53].

As is known, oxygen is essential for aerobic bioprocess. Microbial growth in a reactor depends on the oxygen transfer rate, which is widely used to study the behavior of microorganisms. An increase in the availability of dissolved oxygen in the culture often results in improving the yield of secondary metabolites [19]. Kamzolova et al. [5] reported that need of oxygen by *Y. lipolytica* N1 for growth and for citric acid production, depended on the iron concentration in medium. The effect of iron and oxygen concentrations affected the functioning of the electron transport chain in the mitochondria of yeast. According to the authors, when it was applied a relatively low pressure (20%) and furnished a high concentration of iron (3.5 mg/L), yield of citric acid production increased (120 g/L).

2.3.4. Inductors and Inhibitors of Citric Acid Production.
Attention should be given also to trace elements, which should be strictly controlled. Bivalent metal ions, such as zinc, manganese, iron, copper, and magnesium, affect the production of citric acid [42]. Iron salts are also essential, because they activate the production of acetyl coenzyme A, precursor of citric acid. However, excess iron activates the production of aconitase, enzyme that catalyzes the isomerization of citrate to isocitrate, directing the reactions to formation of isocitric acid, therefore reducing production of citric acid [44]. Finogenova et al. [45] described that, for the production of citric and isocitric acids by mutant strain of *Y. lipolytica* N1 grown in ethanol, high concentrations of zinc and iron were necessary. On cultivation conditions where zinc concentration was limited, cell growth was low and there was no production of citric acid. When zinc was added in the medium using nitrogen-limited conditions, production of citric and isocitric acids increased considerably. Similar behavior was observed in relation to the effect of iron concentration on the production of citric acid. Under nitrogen-limiting conditions, an increase in the intracellular iron content from 0.13 to 2.5 mg/g resulted in increased production of citric acid. But, with the increase in iron concentration from 0.25 to 0.48%, there was observed a

decrease in the production of citric acid. Soccol et al. [42] also claim that low levels of phosphate in the medium have a positive effect on yield of citric acid since phosphate would act on the enzymatic regulation of acid production.

2.4. Accumulation of Lipids.
Oleaginous microorganisms have the ability to transform organic acids into acetyl-CoA, an intermediate that is used for the biosynthesis of lipids [54]. The lipid accumulation primarily depends on the physiology of the microorganism, nutrient limitation and environmental conditions such as temperature and pH. These factors also affect the production of other secondary metabolites, such as ethanol and citrate [12, 50, 55]. Beyond *Y. lipolytica*, the mainly oleaginous microorganisms are of the genera *Candida, Cryptococcus, Rhizopus,* and *Trichosporon,* and lipid profile produced differences between the species [25, 56]. On average, these yeasts accumulate lipids in a quantity corresponding to 40% of their biomass. In conditions in which there is limitation of nutrients, the accumulation of lipids can reach values that exceed 70% of their biomass.

Y. lipolytica is one of most interesting oil-producing yeasts. Wild or genetically modified strains of yeast species have been reported as being capable of producing large amounts of intracellular lipids, which are stocked in the lipid bodies, during t growth in various types of hydrophobic materials [16]. In spite of accumulating fewer lipids than some other oleaginous yeast species, *Y. lipolytica* is the only one able to accumulate large amounts of linoleic acid, representing more than 50% of the fatty acids accumulated by the yeast [12].

2.4.1. Lipid Synthesis.
The intracellular oil accumulation is a consequence of yeast metabolism imbalance. When all the nutrients are present in suitable amounts in the culture medium, there is new cell synthesis, that is, microbial growth. But when the microorganism is deliberately deprived from any essential nutrient, lipogenesis induction occurs, namely, production and storage of oil [55, 57].

Lipids can be stored inside the cell by two different routes: (1) *de novo* synthesis, which involves the production of fatty acid precursors, such as acetyl and malonyl-CoA and their integration in lipid storage biosynthesis and (2) via accumulation ex novo, which involves the capture of fatty acids, oils and triglycerides of the growth medium, followed by their accumulation within the cell. This requires the hydrolysis of hydrophobic substrates on the outside of the cell, fatty acid transport to the interior and reassimilation as triglycerides and esters, followed by their accumulation in lipid bodies [12].

When there is nutrient limitation, some metabolic pathways are repressed due to the synthesis of proteins and nucleic acids, while others are induced (synthesis of fatty acids and triglycerides). During the growth phase, nitrogen is essential for the synthesis of proteins and nucleic acids, required for cell proliferation. When nitrogen is limited, this process is slowed and the growth rate declines rapidly. The excess carbon is then piped to the synthesis of lipids, which leads to an accumulation of triglycerides in lipid bodies. During this phase of stocking, precursors (acetyl-CoA, malonyl-CoA, and glycerol) and energy (ATP and NADPH) are necessary for lipids synthesis [12, 50, 55]. This process is known as *de novo* synthesis, which occurs in oleaginous microorganisms. The process of lipids accumulation inside cells of oleaginous microorganisms is also related to production of citric acid, according to Figure 3.

The reactions of the citric acid cycle begin with the condensation reaction by an enzyme which catalyzes the acetyl CoA in the cycle, with the formation of citric acid (or citrate). Then, with the action of the enzyme aconitase, citric acid is acid-isomerized. Soon after, there is a reaction of oxidative decarboxylation of isocitrate that is catalyzed by the enzyme isocitrate dehydrogenase and isocitrate oxidized and decarboxylated to α-ketoglutarate. This reaction is reversible; that is, depending on the situation, formation of α-ketoglutarate or reductive carboxylation of α-ketoglutarate, forming isocitrate. This enzyme, isocitrate dehydrogenase occurs, is affected by adenosine monophosphate (AMP), a positive stimulator of the enzyme, which has the role of regulating dehydrogenase activity. With high concentration of ATP, the activity of isocitrate dehydrogenase decreases. Upon ATP consumption, there is an increase in AMP concentrations, which stimulates the rate of isocitrate oxidation. Citrate and isocitrate concentrations are increased when the activity of the enzyme isocitrate dehydrogenase decreases. This happens because the equilibrium constant of aconitase reaction greatly favors citric acid accumulation [58].

With nitrogen limitation, occurs activation of AMP deaminase that fills the cell lacking of ammonium ion. As a result, the concentration of mitochondrial AMP decreases, causing the fall of activity of citrate dehydrogenase. The citric acid cycle is then blocked at the level of isocitrate, which accumulates and is balanced with the citrate action of aconitase. The citrate goes into the cytosol and is cleaved by ATP citrate lyase (ACL), giving rise to acetyl-CoA and oxaloacetate, which is the precursor of fatty acids biosynthesis. The acetyl-CoA excess is therefore the key element for the *new* synthesis of lipids in oleaginous microorganisms [25, 58].

When nonoleaginous microorganisms are submitted to nutrient limitation, cell growth also tends to end, but the carbon present in the medium is converted into several other polysaccharides, such as glycogen, glucans and mananes [12].

New extra- and intracellular fatty acids, previously absent in the substrate, can be produced by oleaginous yeast fermentation under certain conditions. An important application of new fats with polyunsaturated fatty acids (PUFA) is as food or nutritional supplements. However, the most important application of these metabolites consists in the production of lipids with high added value, such as in the production of exotic high-value fats such as Shea butter [59].

Lipids produced by *Y. lipolytica* represent an attractive source of edible oils [60] and have been considered as an alternative source for the production of PUFA, cocoa butter substitutes (CBS), and structured lipids. Papanikolaou et al. [61] studied the production of lipids by *Y. lipolytica* detecting stearic, oleic, linoleic, and palmitic acids. In all cases, the microorganism demonstrated ability to increase the concentration of stearic acid, even if this fatty acid was not present in high concentrations in the substrate. This ability allowed the synthesis of an interesting profile of lipids with high percentages of stearic, palmitic and oleic acids, and composition similar to cocoa butter.

The microorganisms begin to accumulate lipids when there are restrictions, principally of nitrogen, in the form of ammonium ion in culture medium and excess carbon source such as glucose [12, 50, 55, 62]. The fatty acids profiles produced, the amount, the productivity, and the efficiency of the conversion are also influenced by several factors during the fermentation process, such as the type of substrate used, choice of limiting nutrient, temperature, pH and aeration [50, 61, 63, 64]. The carbon source is the main factor that influences the lipid composition of the oils produced by yeasts [63, 65].

Athenstaedt et al. [64] cultivated *Y. lipolytica* in a medium whose unique carbon source was glucose and replaced it by oleic acid. This resulted in the change of lipid composition, as well as increased lipid production.

Currently, the cost of microbial oil production is greater than that of plant and animal oils. However, there are features to lower the cost of the production process of this type of oil. Using alternative carbon sources and as a result, increasing accessibility, is an alternative [54], since this factor corresponds to 80% of the total cost of biodiesel produced from microbial lipids, when using glucose as carbon source [57, 66].

The use of fermentable carbohydrates such as glucose is needed, but these carbohydrates can be obtained also by the hydrolysis of cornstarch or similar raw materials. Another alternative of carbon sources is the molasses of sugar cane and beet sugar, rich in sucrose, but not all oil-producing yeasts can metabolize them. The oleaginous yeasts are known for using these two types of sugars simultaneously [57]. When glycerol is added to dextrose or xylose in culture medium, greater amounts of unsaturated fatty acids are produced while glycerol activates the expression of different enzymes used in the synthesis of these acids [65].

The most commonly used nitrogen sources are yeast extract, peptone, nitrate, and ammonium sulfate [16, 60, 61, 67] but combination of nitrogen sources is also commonly used [11, 49]. The critical nitrogen concentration in the medium to induce lipid accumulation in *Y. lipolytica* is 10^{-3} mol/L [68]. It is important that nitrogen concentration not exceed this value, to avoid production of other metabolites (citric acid), therefore affecting production and accumulation of lipids [12, 69].

In addition to nitrogen, other nutrients may be limited to induce the production of lipids, such as metal ions magnesium, iron, zinc or phosphorous [12, 50].

The pH and temperature are parameters that must be controlled in the growth medium for lipid production, once that interfere in the production and storage of oil inside cells. In the literature, the most pH values used for the cultivation of *Y. lipolytica* with purpose to produce oil are included in the range between 5.0 and 7.0 [11, 49, 54, 60, 61]. Papanikolaou et al. [60] observed mainly cell growth using initial pH 6.0. In

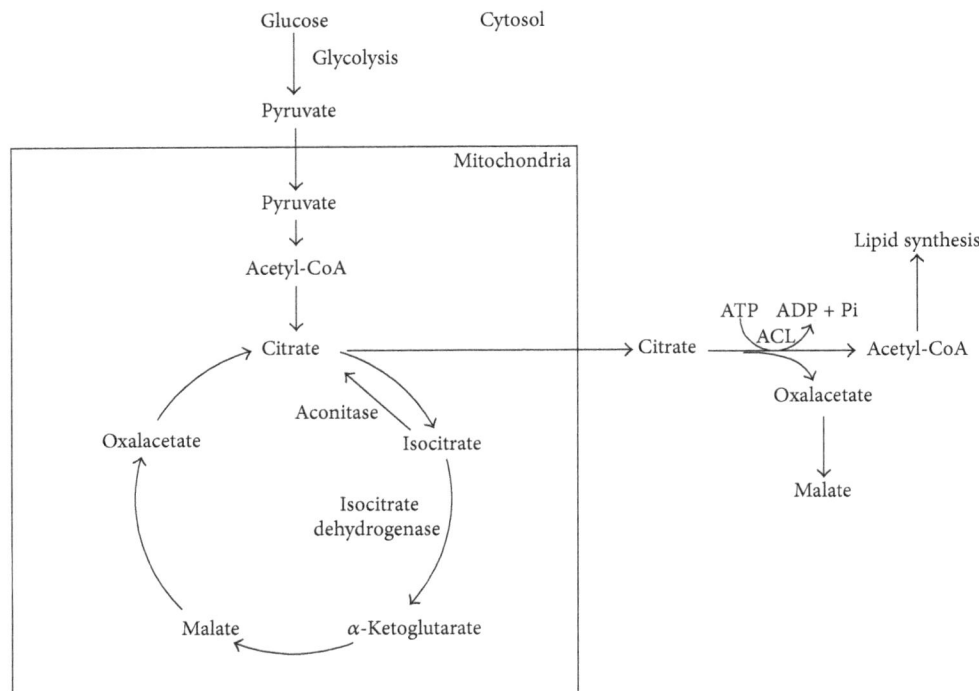

FIGURE 3: Synthesis of lipids by citrate and excess nitrogen limitation. Scheme of the major metabolic pathways for lipid synthesis in *Y. lipolytica*. The glucose undergoes glycolysis and enters in the mitochondria as pyruvate to be used in the tricarboxylic acid cycle. Excess acetyl-CoA is transported from the mitochondria to the cytoplasm in the form of citrate. The cytosolic acetyl-CoA is the precursor for the synthesis of lipids in the lipid bodies. Adapted from Rossi et al. [55] and Tai and G. Stephanopoulos [10].

relation to the temperature, it is reported that the optimum temperature for lipase production by *Y. lipolytica* is in the range between 28 and 30°C [11, 49, 54, 60, 61]. As these yeasts are aerobic, they need to be grown in bioreactors with agitation and aeration, usually between 100 and 250 m³ [57].

2.4.2. Carbon and Nitrogen Ratio in the Growth Medium for Accumulation of Lipids. The excess carbon and nitrogen deprivation are essential factors for the induction of lipogenesis in oleaginous microorganisms. The use of media with the proper C/N ratio is essential to maximize the production of lipids. In a batch process, converting lipid carbons will depend on the duration of the growth phase that depends on the C/N ratio in the fermentation medium [55].

In some oleaginous fungus, excess of carbon source in the medium leads to over production of organic acids, such as pyruvic acid and several other products, to the detriment of lipids accumulation [55]. In the case of *Y. lipolytica*, in the conversion of glucose into lipids, at high initial C/N ratio (80–120 C mol/mol N), cell growth can be followed by citric acid production, leading to low lipid accumulation [12].

Fontanille et al. [54] cultivated *Y. lipolytica* on glucose and glycerol, keeping the ratio C/N equal 62 (80 g/L of C and 3 g/L of N). After 48 h of fermentation, they obtained about 16 g/L of lipids, achieving a 20% conversion yield of carbon into lipids. Najjar et al. [11] also cultivated *Y. lipolytica*, using two different culture media, the first one containing

glucose and olive oil (YPDO) and the other containing only olive oil (YPO). In both media, the nitrogen sources used were bacteriological peptone and yeast extract. In the medium YPDO, glucose was the main source of carbon and energy responsible for cell growth. Using YPO, the free fatty acids and triglycerides stocked were consumed and used to increase the concentration of biomass.

The C/N ratio used by Easterling et al. [65] was 10 : 1 for the production and accumulation of lipids by yeast *Rhodotorula glutinis*. The authors aimed at studying the influence of different carbon sources (dextrose, fructose and glycerol, and xylose) on lipid production by *R. glutinis* and observed that the amount of accumulated fat was increased by using glycerol more than using dextrose and glucose. With glycerol as a carbon source, the lipid production increased by of average 13%, whereas using dextrose or xylose as a substrate, the lipid accumulation decreased by about 9% for both substrates. The efficiency of the conversion of glucose into fat increased from 0.25 to 0.40 (mol/mol glucose lipids), when the C/N ratio was increased from 150 to 350 (mol/mol glucose nitrogen) for the yeast *R. glutinis*. However, when the ratio C/N exceeded 350 g C/g N, due to severe nitrogen deficiency, there was a rapid decrease in cell viability before they reached lipid accumulation stage [12].

2.5. Lipase Production. Lipases are among the most interesting primary metabolites produced by *Yarrowia*, mainly because of the various possibilities for industrial applications.

Lipases of yeasts do not present region selectivity in relation to the position of fatty acids on glycerol molecule and have potential application in the synthesis of esters of short chains, which can be used as flavoring compounds [13, 70–73]. *Y. lipolytica* is able to produce both intracellular and extra lipases [74, 75]. Several factors influence the synthesis of lipase by microorganisms. The main factors are carbon and nitrogen sources, presence of inductors, stimulators, or inhibitors, agents affecting the oil/water interface, incubation temperature, and pH of the culture medium and inoculum [71, 76–78].

Lipase production reaches a peak when the culture enters the stationary phase of growth [73]. Lipase activity declines rapidly after reaching the maximum level, because of proteolysis, catalyzed by proteases that are formed during the phase of cell growth and released with the cell lyse [79].

Composition of the culture medium, in particular in the incorporation of different lipid substances, may result in the production of isozymes [79]. Isozymes are enzymes that occur in more than one form, differing in their structures, but that catalyze the same reaction. Enzymes can be inductive loads, that is, produced by microorganisms in the presence of an inductor, which may be the substrate or product of their hydrolysis [80, 81].

Yeasts are able to respond to stimulus and are constantly monitoring and adapting themselves to the medium in which they are grown. In the presence of oleic acid, yeast cells induce the expression of a set of enzymes needed for the reactions of β-oxidation of fatty acids and proteins that are involved in the expansion of the peroxisomal compartment and participating in the respiratory chain, for energy production [87]. Even under these conditions, there is a direct regulation of the genes that express lipase, by specific genes to oleic acid, identified as SOA genes (specific for oleic acid). These genes encode proteins that control gene expression LIP-2, responsible for the extracellular lipase by *Y. lipolytica* [14].

The literature shows that free fatty acids from triglycerides hydrolysis, mainly oleic acid, are better lipase inducers of the triglycerides and this is valid not only for *Y. lipolytica* but also for other yeasts. Other fatty acids are also reported as inducing the production of lipase [71, 75, 77, 88].

The complex culture media, containing protein hydrolysates such as peptones, yeast and malt extracts are used in fermentation processes but are very expensive. An alternative to decrease the costs would again be the use of agroindustrial wastes, inspite of being required to conduct nutrient supplementation using such culturing materials [89, 90].

2.5.1. The Effect of Carbon Source on Production of Lipase. The carbon source present in the culture medium can stimulate or inhibit the synthesis of lipase. Some fats or oils, as well as carbohydrates, organic acids, glycerol, and other alcohols and fatty acids have been used as inductors of lipase production by *Yarrowia* [71, 82]. The lipase activity is usually detected in growth medium containing lipid materials such as olive oil, soybean oil, tributyrin and oleic acid, which suggests that the production of the enzyme is induced by these substrates [83–85]. On the other hand, unrelated substrates to fats and oils such as carbohydrates, provide good cell growth but are not good for synthesis of lipase [79, 86].

In the study by Obradors et al. [76], oleic acid showed efficient biomass yield and lipase production in relation to other fatty acids with different sizes of carbonic chain. Tan et al. [82] reported that vegetable oils containing oleic and linoleic acid were the best for the biosynthesis of lipase from *Candida* species. Najjar et al. [11] also observed that the olive oil induced the production of lipase by *Y. lipolytica*, when it was used as a sole source of carbon. But when olive oil was added along with the glucose in the culture medium, glucose acted as an inhibitor of lipase production. Low production continued until glucose was fully consumed, being detected late in reaching the peak of enzyme production. The olive oil is widely used as carbon source for production of lipase, since it contains about 70% oleic acid (C18:1). Montesinos et al. [77] reported that when olive oil is used as sole source of carbon, the microorganism follows a sequential mode. Olive oil is initially hydrolyzed by the residual lipase present in the inoculum. Then the microorganism consumes glycerol released as carbon source, but even without producing lipase. Finally, the free fatty acids are consumed, acting as inducers of the formation of a significant amount of lipase.

2.5.2. Choice of the Nitrogen Source for Production of Lipase. The nitrogen source is a factor that has shown great influence on the lipolytic enzymes. Both the organic and inorganic nitrogen have an important role in the synthesis of the enzyme [82, 89]. Among nitrogen compounds used for lipase production, the most common sources that have been hydrolyzed proteins, peptones, amino acids, yeast and malt extracts, urea, nitrate and ammonia salts, agroindustrial waste, such as corn and water soy flour [71, 77, 89]. The tryptone, obtained by the hydrolysis of casein by trypsin, and yeast extract increases the production of lipase and cell growth. Generally, nitrogen sources are widely used because they provide amino acids and vitamins are enzyme cofactors, which are essential for cellular physiology [91].

The selection of the most suitable nitrogen source depends on the microorganism used and association with other ingredients of the culture medium [77]. To verify the influence on the growth of *Y. lipolytica* lipase production, various sources of nitrogen (organic and mineral) were tested by Shirazi et al. [83]. The sources of mineral nitrogen did not demonstrate significant effect on the growth of yeast in the production of lipase. However, increased production of lipase in media containing certain organic nitrogen sources has not been observed. The largest production of lipase was obtained in the presence of tryptone N1, a casein hydrolysate, which is rich in amino acids and peptide-free, compared with other nitrogen-containing organic or mineral substrates. In this way, the use of tryptone N1 can be considered interesting technological point of view, although its cost is high [75].

Almeida et al. [91] also tested the effect of mineral and organic nitrogen sources in the production of lipase and *Candida viswanathii* growth. They noted that one of the mineral sources, the chloride, and ammonium sulfate provided good cell growth, but had no effect on lipase production. However, the ammonium nitrate influenced positively both the growth and production of lipase. The organic sources (tryptone, peptone and yeast extract) increased lipase activity and cell growth, the yeast extract is considered the best source of nitrogen.

Among all organic nitrogen sources tested, the acid hydrolysate, composed exclusively of free amino acids, but low in tryptophan, yielded low lipase production, while the other enzymatic hydrolysates of casein led to high yield of production. Fickers et al. [75] suggest that specific peptides in enzymatic hydrolysates can regulate the production of the enzyme with lipase activity.

2.5.3. Carbon and Nitrogen Ratio (C/N) and Other Nutrients for Lipase Production.
The production of microbial enzymes also depends on the carbon-nitrogen ratio [99]. A balanced culture medium can contain ten times more carbon than nitrogen. This ratio 10 : 1 ensures high protein content, while a higher ratio such as 50 : 1 favors the accumulation of alcohol, metabolites derived from acetate, and extracellular polysaccharides or lipids [100].

In general, carbon, hydrogen, oxygen, nitrogen, sulfur, phosphorus, magnesium and potassium are required in large quantities because they participate in almost all of the cellular substances. Some elements (sulfur, phosphorus, magnesium and potassium) should be provided in the form of salts [101].

The mineral elements (phosphorus, sulfur, potassium, calcium, magnesium, sodium, iron and chlorine) and a small amount of trace elements (manganese, copper, zinc, molybdenum, chromium, nickel, cobalt and boron), which play an important role as constituents of enzymes and coenzymes are usually needed in the culture media for production of enzymes for *Y. lipolytica* [102].

According to Tan et al. [82], the metal ions actually interfere with the synthesis of lipase. According to the authors, magnesium ions, potassium and sodium are beneficial for the biosynthesis of lipase, while the calcium inhibits because form complexes with the fatty acids, changing its solubility and behavior in the oil/water interface.

2.5.4. Water Activity.
The amount of water available in the culture media for *Yarrowia* interferes directly in cell growth and productivity. Many studies of lipolytic enzymes production by yeasts have been conducted in submerged crops, that is, in liquid cultures, with high water activity which is identified as submerged fermentation (SmF). The production of microbial enzymes can also be made by solid substrate fermentation process (SSF). A solid-state culture can be defined as the one in which there is microbial growth in solid material, with low amount of free water. The use of solid media allows a fast population growth of filamentous fungi, allowing the detection of specific enzymes [78, 103].

By-products generated in food processing, rich in fatty acids, triglycerides and/or sugars [78] have been used for the production of lipase by yeasts grown on semisolid culture medium. Using these raw materials, the total capital invested in the production of lipase is significantly lower in strong cultures than in submerged ones. However, one of the problems with this method is low oxygenation, which destabilizes the aerobic processes.

2.5.5. Effects of Temperature, pH and Oxygenation in the Production of Lipase.
The temperature affects the microbial growth parameters as the adaptation time (lag phase), the specific growth rate and total income, and influences the biosynthesis of primary and secondary metabolites [100]. Corzo and Revah [96] detected that the temperature was the factor that most influenced lipase production by *Y. lipolytica* 681. The optimum temperature for the detection of lipase activity was 29.5°C.

The hydrogen ion concentration in a culture medium can affect microbial growth indirectly, affecting nutrient availability or directly, acting on cellular surfaces [100].

It is commonly reported that lipase production by liquid substrate fermentation, using different microorganisms, is accomplished when lowering pH occurs at the end of cultivation [70, 96, 104, 105].

Lopes et al. [17], using a pressurized bioreactor for *Y. lipolytica* cultivation, found that the increased availability of oxygen caused induction of antioxidant enzyme superoxide dismutase, a defense mechanism of the cells against oxidative stress. The increased pressure has not affected the ability of lipase production, but using air pressure at 5 bar, there has been an increase of 96% of extracellular lipase activity.

In another study also involving *Y. lipolytica*, the authors added perfluorocarbon (PFC) to the growth medium, to increase the availability of oxygen to the yeast and thus increase productivity of lipase. Higher rates of growth of *Y. lipolytica* were found with increasing concentration of PFC and upon increasing agitation speed. Lipase production by yeast was increased 23 times with addition of 20% (v/v) of PFC to 250 rpm. In addition, it was demonstrated that the use of PFC along with glucose is more effective in the production of lipase than the conventional use of olive oil, precisely because of the increase in oxygen transfer [18]. Table 1 shows the types of substrates, nitrogen sources and fermentation more used for *Y. lipolytica* cultivation for production of citric acid, lipids, lipases and biomass.

3. Conclusions

There is an expectation around *Y. lipolytica*, which is considered as a promising and efficient microorganism to be used by biotechnological industry. Its major application stands out due to its rich and varied biosynthetic potential, being capable of producing various types of metabolites. Additionally, simple alterations in the fermentative process can be accomplished in order to modulate metabolites production by this species. The versatility of the metabolites produced, along with the easiness of scale increasing, is raising numerous proposals for industrial applications of *Y. lipolytica* in the most diverse segments, mainly in the sectors of energy, foods and pharmaceutical industries that move a great volume of the world economy. To make the production of these metabolites industrially viable, it is

TABLE 1: Substrates, nitrogen sources, type of fermentation used to produce citric acid (CA), lipids, lipase, and biomass by *Y. lipolytica*.

Substrate	Nitrogen source	Fermentation	Product	Reference
Glucose	Ammonium nitrate and yeast extract	SmF	CA	Antonucci et al. (2001) [40]
Ethanol	Ammonium sulphate and hydrolyzed yeast	SmF	CA	Arzumanov et al. (2000) [38]
n-Paraffin	Iron nitrate	SmF	CA	Crolla and Kennedy (2001)] [44
Ethanol	Ammonium sulfate	SmF	CA	Finogenova et al. (2002) [45]
Glycerol	Yeast extract	SmF	CA	Imandi et al. (2007) [47]
Pineapple residue	Yeast extract	SSF	CA	Imandi et al. (2008) [92]
Sunflower oil	Ammonium sulfate and yeast extract	SmF	CA	Kamzolova et al., (2008) [48]
Sucrose, glucose, and glycerol	Ammonium chloride	SmF	CA	Lazar et al. (2011) [93]
Glycerol	Ammonium sulfate	SmF	CA	Levinson et al. (2007) [46]
Glycerol	Ammonium sulfate and yeast extract	SmF	CA	Papanikolaou et al. (2002) [60]; Makri et al. (2010) [52]
Glucose	Ammonium chloride and yeast extract	SmF	CA	Karasu-Yalcin et al. (2010) [41]
Glucose and stearin	Ammonium sulfate and yeast extract	SmF	CA and lipids	Papanikolaou et al. (2006) [94]
Methanol	Peptone and yeast and malt extracts	SmF	Lipids	Rupčić et al. (1996) [63]
Glucose and olive oil	Peptone and yeast extract	SmF	Lipids and lipase	Najjar et al. (2011) [11]
Industrial fats (stearin)	Peptone and yeast extract	SmF	Lipids	Papanikolaou et al. (2001) [61]
Stearin	Ammonium sulfate and yeast extract	SmF	Lipids, lipase, and biomass	Papanikolaou et al. (2007) [16]
Coconut fat	Ammonium sulfate	SSF	Saturated fatty acid	Parfene et al. (2013) [95]
Sugar cane bagasse	Peptone	SmF	Lipid	Tsigie et al. (2011) [67]
Olive oil, corn oil, and glucose	Urea	SmF	Lipase	Corzo and Revah (1999) [96]
Olive oil	Urea	SmF	CA, lipase, and biomass	Darvishi et al. (2009) [97]
Olive oil	Peptone and yeast extract	SmF	Lipase	Deive et al. (2003) [84]
Olive oil	Peptone	SmF	Lipase	Gonçalves et al. (2013) [98]

CA: citric acid; SmF: submerged fermentation; SSF: solid substrate fermentation.

important to obtain detailed knowledge of suitable growing conditions of *Y. lipolytica*. Investments in the optimization of the metabolic production, as well as in the operationalization of the fermentative processes and the improvement of the producing strains, by mutation or by selecting new strains, should soon result in novel marketed products from this microorganism.

Conflict of Interests

The authors declare that there is no conflict of interests regarding the publication of this article.

Acknowledgments

The authors thank the Brazilian Funding Agencies FAPEMIG, CNPq, and CAPES for financial support and scholarships.

References

[1] A. L. Demain, "Pharmaceutically active secondary metabolites of microorganisms," *Applied Microbiology and Biotechnology*, vol. 52, no. 4, pp. 455–463, 1999.

[2] F. A. G. Torres, J. L. D. Marco, M. J. Poças-Fonseca, and M. S. S. Felipe, "O uso de leveduras e fungos filamentosos para produção heteróloga de enzimas," in *Enzimas em Biotecnologia: Produção, Aplicações e Mercado*, E. P. S. Bon, M. A. Ferrara, M. L. Corvo et al., Eds., pp. 65–69, Interciência, Rio de Janeiro, Brazil, 2008.

[3] J. F. T. Spencer, A. L. Ragout de Spencer, and C. Laluce, "Non-conventional yeasts," *Applied Microbiology and Biotechnology*, vol. 58, no. 2, pp. 147–156, 2002.

[4] J.-M. Beckerich, A. Boisramé-Baudevin, and C. Gaillardin, "*Yarrowia lipolytica*: a model organism for protein secretion studies," *International Microbiology*, vol. 1, no. 2, pp. 123–130, 1998.

[5] S. V. Kamzolova, N. V. Shishkanova, I. G. Morgunov, and T. V. Finogenova, "Oxygen requirements for growth and citric acid production of *Yarrowia lipolytica*," *FEMS Yeast Research*, vol. 3, no. 2, pp. 217–222, 2003.

[6] C. Madzak, C. Gaillardin, and J.-M. Beckerich, "Heterologous protein expression and secretion in the non-conventional yeast *Yarrowia lipolytica*: a review," *Journal of Biotechnology*, vol. 109, no. 1-2, pp. 63–81, 2004.

[7] M. Aguedo, N. Gomes, E. E. Garcia et al., "Decalactone production by *Yarrowia lipolytica* under increased O_2 transfer rates," *Biotechnology Letters*, vol. 27, no. 20, pp. 1617–1621, 2005.

[8] G. Barth and C. Gaillardin, "Physiology and genetics of the dimorphic fungus *Yarrowia lipolytica*," *FEMS Microbiology Reviews*, vol. 19, no. 4, pp. 219–237, 1997.

[9] J. M. Nicaud, "*Yarrowia lipolytica*," *Yeast*, vol. 29, pp. 409–418, 2012.

[10] M. Tai and G. Stephanopoulos, "Engineering the push and pull of lipid biosynthesis in oleaginous yeast *Yarrowia lipolytica* for biofuel production," *Metabolic Engineering*, vol. 15, pp. 1–9, 2013.

[11] A. Najjar, S. Robert, C. Guérin, M. Violet-Asther, and F. Carrière, "Quantitative study of lipase secretion, extracellular lipolysis, and lipid storage in the yeast *Yarrowia lipolytica* grown in the presence of olive oil: analogies with lipolysis in humans,"

Applied Microbiology and Biotechnology, vol. 89, no. 6, pp. 1947–1962, 2011.

[12] A. Beopoulos, J. Cescut, R. Haddouche, J.-L. Uribelarrea, C. Molina-Jouve, and J.-M. Nicaud, "*Yarrowia lipolytica* as a model for bio-oil production," *Progress in Lipid Research*, vol. 48, no. 6, pp. 375–387, 2009.

[13] M. A. Z. Coelho, P. F. F. Amaral, and I. Belo, "lipolytica: an industrial workhorse," in *Current Research, Technology and Education Topics in Applied Microbiology and Microbial Biotechnology*, pp. 930–944, 2010.

[14] T. Desfougères, R. Haddouche, F. Fudalej, C. Neuvéglise, and J.-M. Nicaud, "SOA genes encode proteins controlling lipase expression in response to triacylglycerol utilization in the yeast *Yarrowia lipolytica*: RESEARCH ARTICLE," *FEMS Yeast Research*, vol. 10, no. 1, pp. 93–103, 2010.

[15] P. Fickers, P.-H. Benetti, Y. Waché et al., "Hydrophobic substrate utilisation by the yeast *Yarrowia lipolytica*, and its potential applications," *FEMS Yeast Research*, vol. 5, no. 6-7, pp. 527–543, 2005.

[16] S. Papanikolaou, I. Chevalot, M. Galiotou-Panayotou, M. Komaitis, I. Marc, and G. Aggelis, "Industrial derivative of tallow: a promising renewable substrate for microbial lipid, single-cell protein and lipase production by *Yarrowia lipolytica*," *Electronic Journal of Biotechnology*, vol. 10, no. 3, pp. 425–435, 2007.

[17] M. Lopes, N. Gomes, M. Mota, and I. Belo, "*Yarrowia lipolytica* growth under increased air pressure: influence on enzyme production," *Applied Biochemistry and Biotechnology*, vol. 159, no. 1, pp. 46–53, 2009.

[18] P. F. F. Amaral, M. H. M. Rocha-Leão, I. M. Marrucho, J. A. P. Coutinho, and M. A. Z. Coelho, "Improving lipase production using a perfluorocarbon as oxygen carrier," *Journal of Chemical Technology and Biotechnology*, vol. 81, no. 8, pp. 1368–1374, 2006.

[19] S. Suresh, V. C. Srivastava, and I. M. Mishra, "Techniques for oxygen transfer measurement in bioreactors: a review," *Journal of Chemical Technology and Biotechnology*, vol. 84, no. 8, pp. 1091–1103, 2009.

[20] R. Boopathy, "Factors limiting bioremediation technologies," *Bioresource Technology*, vol. 74, no. 1, pp. 63–67, 2000.

[21] M. Vidali, "Bioremediation. An overview," *Pure and Applied Chemistry*, vol. 73, no. 7, pp. 1163–1172, 2001.

[22] E. Z. Ron and E. Rosenberg, "Biosurfactants and oil bioremediation," *Current Opinion in Biotechnology*, vol. 13, no. 3, pp. 249–252, 2002.

[23] C. Scioli and L. Vollaro, "The use of *Yarrowia lipolytica* to reduce pollution in olive mill wastewaters," *Water Research*, vol. 31, no. 10, pp. 2520–2524, 1997.

[24] A. V. Bankar, A. R. Kumar, and S. S. Zinjarde, "Environmental and industrial applications of *Yarrowia lipolytica*," *Applied Microbiology and Biotechnology*, vol. 84, no. 5, pp. 847–865, 2009.

[25] A. Beopoulos, T. Chardot, and J.-M. Nicaud, "*Yarrowia lipolytica*: a model and a tool to understand the mechanisms implicated in lipid accumulation," *Biochimie*, vol. 91, no. 6, pp. 692–696, 2009.

[26] N. Christofi and I. B. Ivshina, "Microbial surfactants and their use in field studies of soil remediation," *Journal of Applied Microbiology*, vol. 93, no. 6, pp. 915–929, 2002.

[27] I. M. Banat, "Biosurfactants production and possible uses in microbial enhanced oil recovery and oil pollution remiedlation: a review," *Bioresource Technology*, vol. 51, no. 1, pp. 1–12, 1995.

[28] M. C. Cirigliano and G. M. Carman, "Isolation of a bioemul-sifier from *Candida lipolytica*," *Applied and Environmental Microbiology*, vol. 48, no. 4, pp. 747–750, 1984.

[29] L. A. Sarubbo, M. D. C. Marçal, M. L. C. Neves, M. D. P. C. Silva, L. F. Porto, and G. M. Campos-Takaki, "Bioemulsifier production in batch culture using glucose as carbon source by *Candida lipolytica*," *Applied Biochemistry and Biotechnology A*, vol. 95, no. 1, pp. 59–67, 2001.

[30] A. Braga, N. Gomes, and I. Belo, "Lipase induction in *Yarrowia lipolytica* for castor oil hydrolysis and its effect on γ-decalactone production," *Journal of the American Oil Chemists' Society*, vol. 89, pp. 1041–1047, 2012.

[31] M. Aguedo, M. H. Ly, I. Belo, J. A. Teixeira, J.-M. Belin, and Y. Waché, "The use of enzymes and microorganisms for the production of aroma compounds from lipids," *Food Technology and Biotechnology*, vol. 42, no. 4, pp. 327–336, 2004.

[32] Y. Pagot, A. le Clainche, J.-M. Nicaud, Y. Wache, and J.-M. Belin, "Peroxisomal β-oxidation activities and γ-decalactone produc-tion by the yeast *Yarrowia lipolytica*," *Applied Microbiology and Biotechnology*, vol. 49, no. 3, pp. 295–300, 1998.

[33] J. Schrader, M. M. W. Etschmann, D. Sell, J.-M. Hilmer, and J. Rabenhorst, "Applied biocatalysis for the synthesis of natural flavour compounds—current industrial processes and future prospects," *Biotechnology Letters*, vol. 26, no. 6, pp. 463–472, 2004.

[34] S. Anastassiadis, A. Aivasidis, and C. Wandrey, "Citric acid production by *Candida* strains under intracellular nitrogen limitation," *Applied Microbiology and Biotechnology*, vol. 60, no. 1-2, pp. 81–87, 2002.

[35] M. Sauer, D. Porro, D. Mattanovich, and P. Branduardi, "Micro-bial production of organic acids: expanding the markets," *Trends in Biotechnology*, vol. 26, no. 2, pp. 100–108, 2008.

[36] S. Papanikolaou and G. Aggelis, "Biotechnological valorization of biodiesel derived glycerol waste through production of single cell oil and citric acid by *Yarrowia lipolytica*," *Lipid Technology*, vol. 21, no. 4, pp. 83–87, 2009.

[37] W. Rymowicz, A. Rywińska, B. Zarowska, and P. Juszczyk, "Citric acid production from raw glycerol by acetate mutants of *Yarrowia lipolytica*," *Chemical Papers*, vol. 60, no. 5, pp. 391–394, 2006.

[38] T. E. Arzumanov, I. A. Sidorov, N. V. Shishkanova, and T. V. Finogenova, "Mathematical modeling of citric acid production by repeated batch culture," *Enzyme and Microbial Technology*, vol. 26, no. 9-10, pp. 826–833, 2000.

[39] M. Papagianni, "Fungal morphology and metabolite produc-tion in submerged mycelial processes," *Biotechnology Advances*, vol. 22, no. 3, pp. 189–259, 2004.

[40] S. Antonucci, M. Bravi, R. Bubbico, A. di Michele, and N. Verdone, "Selectivity in citric acid production by *Yarrowia lipolytica*," *Enzyme and Microbial Technology*, vol. 28, no. 2-3, pp. 189–195, 2001.

[41] S. Karasu-Yalcin, M. T. Bozdemir, and Z. Y. Ozbas, "Effects of different fermentation conditions on growth and citric acid production kinetics of two *Yarrowia lipolytica* strains," *Chemical and Biochemical Engineering Quarterly*, vol. 24, no. 3, pp. 347–360, 2010.

[42] C. R. Soccol, L. P. S. Vandenberghe, C. Rodrigues, and A. Pandey, "New perspectives for citric acid production and application," *Food Technology and Biotechnology*, vol. 44, no. 2, pp. 141–149, 2006.

[43] S. K. Yalcin, M. T. Bozdemir, and Z. Y. Ozbas, "Citric acid pro-duction by yeasts: fermentation conditions, process optimiza-tion and strain improvement," in *Current Research, Technology and Education Topics in Applied Microbiology and Microbial Biotechnology*, A. Mendez-Vilas, Ed., pp. 1374–1382, Formatex, Badajoz, Spain, 2010.

[44] A. Crolla and K. J. Kennedy, "Optimization of citric acid production from *Candida lipolytica* Y-1095 using *n*-paraffin," *Journal of Biotechnology*, vol. 89, no. 1, pp. 27–40, 2001.

[45] T. V. Finogenova, S. V. Kamzolova, E. G. Dedyukhina et al., "Biosynthesis of citric and isocitric acids from ethanol by mutant *Yarrowia lipolytica* N 1 under continuous cultivation," *Applied Microbiology and Biotechnology*, vol. 59, no. 4-5, pp. 493–500, 2002.

[46] W. E. Levinson, C. P. Kurtzman, and T. M. Kuo, "Characteri-zation of *Yarrowia lipolytica* and related species for citric acid production from glycerol," *Enzyme and Microbial Technology*, vol. 41, no. 3, pp. 292–295, 2007.

[47] S. B. Imandi, V. R. Bandaru, S. R. Somalanka, and H. R. Garapati, "Optimization of medium constituents for the pro-duction of citric acid from byproduct glycerol using Doehlert experimental design," *Enzyme and Microbial Technology*, vol. 40, no. 5, pp. 1367–1372, 2007.

[48] S. V. Kamzolova, T. V. Finogenova, and I. G. Morgunov, "Microbiological production of citric and isocitric acids from sunflower oil," *Food Technology and Biotechnology*, vol. 46, no. 1, pp. 51–59, 2008.

[49] D. Sarris, M. Galiotou-Panayotou, A. A. Koutinas, M. Komaitis, and S. Papanikolaou, "Citric acid, biomass and cellular lipid production by *Yarrowia lipolytica* strains cultivated on olive mill wastewater-based media," *Journal of Chemical Technology and Biotechnology*, vol. 86, no. 11, pp. 1439–1448, 2011.

[50] S. D. Dyal and S. S. Narine, "Implications for the use of *Mortierella* fungi in the industrial production of essential fatty acids," *Food Research International*, vol. 38, no. 4, pp. 445–467, 2005.

[51] A. Rywińska, W. Rymowicz, B. Zarowska, and M. Wojtatowicz, "Biosynthesis of citric acid from glycerol by acetate mutants of *Yarrowia lipolytica* in fed-batch fermentation," *Food Technology and Biotechnology*, vol. 47, no. 1, pp. 1–6, 2009.

[52] A. Makri, S. Fakas, and G. Aggelis, "Metabolic activities of biotechnological interest in *Yarrowia lipolytica* grown on glyc-erol in repeated batch cultures," *Bioresource Technology*, vol. 101, no. 7, pp. 2351–2358, 2010.

[53] S. Anastassiadis, C. Wandrey, and H.-J. Rehm, "Continuous citric acid fermentation by *Candida oleophila* under nitrogen limitation at constant C/N ratio," *World Journal of Microbiology and Biotechnology*, vol. 21, no. 5, pp. 695–705, 2005.

[54] P. Fontanille, V. Kumar, G. Christophe, R. Nouaille, and C. Larroche, "Bioconversion of volatile fatty acids into lipids by the oleaginous yeast *Yarrowia lipolytica*," *Bioresource Technology*, vol. 114, pp. 443–449, 2012.

[55] M. Rossi, A. Amaretti, S. Raimondi, and A. Leonardi, "Getting lipids for biodiesel production from oleaginous fungi," in *Feedstocks and Processing Technologies*, M. Stoytcheva, Ed., In Tech, 2011, http://www.intechopen.com/books/biodi-esel-feedstocks-and-processing-technologies/getting-lipids-for-biodiesel-production-from-oleaginous-fungi.

[56] J. P. Wynn and C. Ratledge, "Oils from microorganisms," in *Bailey's Industrial Oil and Fat Products*, Fereidoon Shaihidi, 2005.

[57] C. Ratledge and Z. Cohen, "Microbial and algal oils: do they have a future for biodiesel or as commodity oils?" *Lipid Technology*, vol. 20, pp. 155–160, 2008.

[58] C. Ratledge, "Regulation of lipid accumulation in oleaginous micro-organisms," *Biochemical Society Transactions*, vol. 30, no. 6, pp. 1047–1050, 2002.

[59] S. Papanikolaou and G. Aggelis, "Lipids of oleaginous yeasts. Part I: biochemistry of single cell oil production," *European Journal of Lipid Science and Technology*, vol. 113, no. 8, pp. 1031–1051, 2011.

[60] S. Papanikolaou, I. Chevalot, M. Komaitis, I. Marc, and G. Aggelis, "Single cell oil production by *Yarrowia lipolytica* growing on an industrial derivative of animal fat in batch cultures," *Applied Microbiology and Biotechnology*, vol. 58, no. 3, pp. 308–312, 2002.

[61] S. Papanikolaou, I. Chevalot, M. Komaitis, G. Aggelis, and I. Marc, "Kinetic profile of the cellular lipid composition in an oleaginous *Yarrowia lipolytica* capable of producing a cocoa-butter substitute from industrial fats," *Antonie van Leeuwenhoek*, vol. 80, no. 3-4, pp. 215–224, 2001.

[62] C. Ratledge and Z. Cohen, "Microbial and algal oils: do they have a future for biodiesel or as commodity oils?" *Lipid Technology*, vol. 20, pp. 155–160, 2008.

[63] J. Rupčić, B. Blagović, and V. Marić, "Cell lipids of the *Candida lipolytica* yeast grown on methanol," *Journal of Chromatography A*, vol. 755, no. 1, pp. 75–80, 1996.

[64] K. Athenstaedt, P. Jolivet, C. Boulard et al., "Lipid particle composition of the yeast *Yarrowia lipolytica* depends on the carbon source," *Proteomics*, vol. 6, no. 5, pp. 1450–1459, 2006.

[65] E. R. Easterling, W. T. French, R. Hernandez, and M. Licha, "The effect of glycerol as a sole and secondary substrate on the growth and fatty acid composition of *Rhodotorula glutinis*," *Bioresource Technology*, vol. 100, no. 1, pp. 356–361, 2009.

[66] Q. Fei, H. N. Chang, L. Shang, and J.-D. Choi, "Exploring low-cost carbon sources for microbial lipids production by fed-batch cultivation of *Cryptococcus albidus*," *Biotechnology and Bioprocess Engineering*, vol. 16, no. 3, pp. 482–487, 2011.

[67] Y. A. Tsigie, C.-Y. Wang, C.-T. Truong, and Y.-H. Ju, "Lipid production from *Yarrowia lipolytica* Po1g grown in sugarcane bagasse hydrolysate," *Bioresource Technology*, vol. 102, no. 19, pp. 9216–9222, 2011.

[68] J. Cescut, *Accumulation d'acylglycérols par des espèces levuriennes à usage carburant aéronautique: physiologie et performances de procédés*, Université de Toulouse, Toulouse, France, 2009.

[69] L. M. Granger, P. Perlot, G. Goma, and A. Pareilleux, "Efficiency of fatty acid synthesis by oleaginous yeasts: prediction of yield and fatty acid cell content from consumed C/N ratio by a simple method," *Biotechnology and Bioengineering*, vol. 42, no. 10, pp. 1151–1156, 1993.

[70] K. H. Tan and C. O. Gill, "Effect of culture conditions on batch growth of Saccharomycopsis lipolytica on olive oil," *Applied Microbiology and Biotechnology*, vol. 20, no. 3, pp. 201–206, 1984.

[71] W. Hadeball, "Production of lipase by *Yarrowia lipolytica* I. Lipases from yeasts (Review)," *Acta Biotechnologica*, vol. 11, no. 2, pp. 159–167, 1991.

[72] D. J. Glover, R. K. McEwen, C. R. Thomas, and T. W. Young, "pH-regulated expression of the acid and alkaline extracellular proteases of *Yarrowia lipolytica*," *Microbiology*, vol. 143, no. 9, pp. 3045–3054, 1997.

[73] M. Zarevúcka, Z. Kejík, D. Saman, Z. Wimmer, and K. Demnevorá, "Enantioselective properties of induced lipases from *Geotrichum*," *Enzyme and Microbial Technology*, vol. 37, pp. 481–486, 2005.

[74] M. E. Guerzoni, R. Lanciotti, L. Vannini et al., "Variability of the lipolytic activity in *Yarrowia lipolytica* and its dependence on environmental conditions," *International Journal of Food Microbiology*, vol. 69, no. 1-2, pp. 79–89, 2001.

[75] P. Fickers, J. M. Nicaud, C. Gaillardin, J. Destain, and P. Thonart, "Carbon and nitrogen sources modulate lipase production in the yeast *Yarrowia lipolytica*," *Journal of Applied Microbiology*, vol. 96, no. 4, pp. 742–749, 2004.

[76] N. Obradors, J. L. Montesinos, F. Valero, F. J. Lafuente, and C. Sola, "Effects of different fatty acids in lipase production by *Candida rucosa*," *Biotechnology Letters*, vol. 15, no. 4, pp. 357–360, 1993.

[77] J. L. Montesinos, N. Obradors, M. A. Gordillo, F. Valero, J. Lafuente, and C. Solà, "Effect of nitrogen sources in batch and continuous cultures to lipase production by *Candida rugosa*," *Applied Biochemistry and Biotechnology A*, vol. 59, no. 1, pp. 25–37, 1996.

[78] A. Domínguez, M. Costas, M. A. Longo, and A. Sanromán, "A novel application of solid state culture: production of lipases by *Yarrowia lipolytica*," *Biotechnology Letters*, vol. 25, no. 15, pp. 1225–1229, 2003.

[79] E. Dalmau, J. L. Montesinos, M. Lotti, and C. Casas, "Effect of different carbon sources on lipase production by *Candida rugosa*," *Enzyme and Microbial Technology*, vol. 26, no. 9-10, pp. 657–663, 2000.

[80] R. Sharma, Y. Chisti, and U. C. Banerjee, "Production, purification, characterization, and applications of lipases," *Biotechnology Advances*, vol. 19, no. 8, pp. 627–662, 2001.

[81] M. Roveda, M. Hemkemeier, and L. M. Crolla, "Avaliação da produção de lipases por diferentes cepas de microrganismos isolados em efluentes de laticínios por fermentação submersa," *Ciência e Tecnologia de Alimentos*, vol. 30, pp. 126–131, 2010.

[82] T. Tan, M. Zhang, B. Wang, C. Ying, and L. Deng, "Screening of high lipase producing *Candida* sp. and production of lipase by fermentation," *Process Biochemistry*, vol. 39, no. 4, pp. 459–465, 2003.

[83] S. H. Shirazi, S. R. Rahman, and M. M. Rahman, "Short communication: production of extracellular lipases by *Saccharomyces cerevisiae*," *World Journal of Microbiology and Biotechnology*, vol. 14, no. 4, pp. 595–597, 1998.

[84] F. J. Deive, M. Costas, and M. A. Longo, "Production of a thermostable extracellular lipase by *Kluyveromyces marxianus*," *Biotechnology Letters*, vol. 25, no. 17, pp. 1403–1406, 2003.

[85] L. Freitas, T. Bueno, V. H. Perez, J. C. Santos, and H. F. de Castro, "Enzymatic hydrolysis of soybean oil using lipase from different sources to yield concentrated of polyunsaturated fatty acids," *World Journal of Microbiology and Biotechnology*, vol. 23, no. 12, pp. 1725–1731, 2007.

[86] M. Lotti, S. Monticelli, J. Luis Montesinos, S. Brocca, F. Valero, and J. Lafuente, "Physiological control on the expression and secretion of *Candida rugosa* lipase," *Chemistry and Physics of Lipids*, vol. 93, no. 1-2, pp. 143–148, 1998.

[87] A. Gurvitz and H. Rottensteiner, "The biochemistry of oleate induction: transcriptional upregulation and peroxisome proliferation," *Biochimica et Biophysica Acta*, vol. 1763, no. 12, pp. 1392–1402, 2006.

[88] J. Destain, D. Roblain, and P. Thonart, "Improvement of lipase production from *Yarrowia lipolytica*," *Biotechnology Letters*, vol. 19, no. 2, pp. 105–107, 1997.

[89] H. Treichel, D. de Oliveira, M. A. Mazutti, M. di Luccio, and J. V. Oliveira, "A review on microbial lipases production," *Food and Bioprocess Technology*, vol. 3, no. 2, pp. 182–196, 2010.

[90] A. Salihu, M. Z. Alam, M. I. AbdulKarim, and H. M. Salleh, "Lipase production: an insight in the utilization of renewable agricultural residues," *Resources, Conservation and Recycling*, vol. 58, pp. 36–44, 2012.

[91] A. F. Almeida, S. M. Taulk-Tomisielo, and E. C. Carmona, "Influence of carbon and nitrogen sources on lipase production by a newly isolated *Candida viswanathii* strain," *Annals of Microbiology*, vol. 63, no. 4, pp. 1225–1234, 2012.

[92] S. B. Imandi, V. V. R. Bandaru, S. R. Somalanka, S. R. Bandaru, and H. R. Garapati, "Application of statistical experimental designs for the optimization of medium constituents for the production of citric acid from pineapple waste," *Bioresource Technology*, vol. 99, no. 10, pp. 4445–4450, 2008.

[93] Z. Lazar, E. Walczak, and M. Robak, "Simultaneous production of citric acid and invertase by *Yarrowia lipolytica* SUC$^+$ transformants," *Bioresource Technology*, vol. 102, no. 13, pp. 6982–6989, 2011.

[94] S. Papanikolaou, M. Galiotou-Panayotou, I. Chevalot, M. Komaitis, I. Marc, and G. Aggelis, "Influence of glucose and saturated free-fatty acid mixtures on citric acid and lipid production by *Yarrowia lipolytica*," *Current Microbiology*, vol. 52, no. 2, pp. 134–142, 2006.

[95] G. Parfene, V. Horincar, A. K. Tyagi, A. Malik, and G. Bahrim, "Production of medium chain saturated fatty acids with enhanced antimicrobial activity from crude coconut fat by solid state cultivation of *Yarrowia lipolytica*," *Food Chemistry*, vol. 136, pp. 1345–1349, 2013.

[96] G. Corzo and S. Revah, "Production and characteristics of the lipase from *Yarrowia lipolytica* 681," *Bioresource Technology*, vol. 70, no. 2, pp. 173–180, 1999.

[97] F. Darvishi, I. Nahvi, H. Zarkesh-Esfahani, and F. Momenbeik, "Effect of plant oils upon lipase and citric acid production in *Yarrowia lipolytica* yeast," *Journal of Biomedicine and Biotechnology*, vol. 2009, Article ID 562943, 7 pages, 2009.

[98] F. A. G. Gonçalves, G. Colen, and J. A. Takahashi, "Optimization of cultivation conditions for extracellular lipase production by *Yarrowia lipolytica* using response surface method," *African Journal of Biotechnology*, vol. 12, pp. 2270–2278, 2013.

[99] A. Hiol, M. D. Jonzo, N. Rugani, D. Druet, L. Sarda, and L. C. Comeau, "Purification and characterization of an extracellular lipase from a thermophilic *Rhizopus oryzae* strain isolated from palm fruit," *Enzyme and Microbial Technology*, vol. 26, no. 5-6, pp. 421–430, 2000.

[100] M. Carlile and S. C. Watkinson, *The Fungi*, Academic Press, London, UK, 1997.

[101] G. Colen, *Isolamento e seleção de fungos filamentosos produtores de lipases [Ph.D. thesis]*, Faculdade de Farmácia da UFMG, Universidade Federal de Minas Gerais, Belo Horizonte, Brazil, 2006.

[102] D. H. Jennings, *The Physiology of Fungal Nutrition*, Cambridge University Press, Cambridge, UK, 1995.

[103] M. H. Alves, G. M. Campos-Takaki, A. L. Figueiredo Porto, and A. I. Milanez, "Screening of *Mucor spp.* for the production of amylase, lipase, polygalacturonase and protease," *Brazilian Journal of Microbiology*, vol. 33, no. 4, pp. 325–330, 2002.

[104] R. Bussamara, A. M. Fuentefria, E. S. D. Oliveira et al., "Isolation of a lipase-secreting yeast for enzyme production in a pilot-plant scale batch fermentation," *Bioresource Technology*, vol. 101, no. 1, pp. 268–275, 2010.

[105] K. N. Sathish Yadav, M. G. Adsul, K. B. Bastawde, D. D. Jadhav, H. V. Thulasiram, and D. V. Gokhale, "Differential induction, purification and characterization of cold active lipase from *Yarrowia lipolytica* NCIM 3639," *Bioresource Technology*, vol. 102, no. 22, pp. 10663–10670, 2011.

The Effectiveness of Anti-*R. equi* Hyperimmune Plasma against *R. equi* Challenge in Thoroughbred Arabian Foals of Mares Vaccinated with *R. equi* Vaccine

Osman Erganis,[1] **Zafer Sayin,**[1] **Hasan Huseyin Hadimli,**[1] **Asli Sakmanoglu,**[1] **Yasemin Pinarkara,**[2] **Ozgur Ozdemir,**[3] **and Mehmet Maden**[4]

[1] *Department of Microbiology, Faculty of Veterinary Medicine, University of Selcuk, 42075 Konya, Turkey*
[2] *Program of Food Technology, Sarayonu Vocational School, University of Selcuk, 42075 Konya, Turkey*
[3] *Department of Pathology, Faculty of Veterinary Medicine, University of Selcuk, 42075 Konya, Turkey*
[4] *Department of Internal Medicine, Faculty of Veterinary Medicine, University of Selcuk, 42075 Konya, Turkey*

Correspondence should be addressed to Zafer Sayin; zafersayin@gmail.com

Academic Editors: T. Betakova, V. Fedorenko, Y. Mu, Z. Shi, and H. Zaraket

This study aimed to determine the effectiveness of a pregnant mare immunization of a *Rhodococcus equi* (*R. equi*) vaccine candidate containing a water-based nanoparticle mineral oil adjuvanted (Montanide IMS 3012) inactive bacterin and virulence-associated protein A (VapA), as well as the administration of anti-*R. equi* hyperimmune (HI) plasma against *R. equi* challenge in the mares' foals. The efficacy of passive immunizations (colostral passive immunity by mare vaccination and artificial passive immunity by HI plasma administration) was evaluated based on clinical signs, complete blood count, blood gas analysis, serological response (ELISA), interleukin-4 (IL-4) and interferon gamma (IFN-γ), total cell count of the bronchoalveolar lavage fluids (BALF) samples, reisolation rate of *R. equi* from BALF samples (CFU/mL), lung samples (CFU/gr), and lesion scores of the organs and tissue according to pathological findings after necropsy in the foals. The vaccination of pregnant mares and HI plasma administration in the foals reduced the severity of *R. equi* pneumonia and lesion scores of the organs and tissue by 3.54-fold compared to the control foals. This study thus indicates that immunization of pregnant mares with *R. equi* vaccine candidate and administration of HI plasma in mares' foals effectively protect foals against *R. equi* challenge.

1. Introduction

Rhodococcus equi (*R. equi*) is a Gram-positive, nonmotile, obligate aerobe, intracellular microorganism. Virulent *R. equi* causes pyogranulomatous bronchopneumonia in young foals aged from 1 to 6 months [1]. Young foals may also develop extrapulmonary disease, such as septic arthritis, osteomyelitis, ulcerative enterocolitis, mesenteric lymphadenopathy, neonatal diarrhea, and sudden death. *R. equi* is additionally considered as an opportunistic pathogen of immunosuppressed people, especially AIDS patients [2]. *R. equi* was initially isolated from pulmonary lesions of foals by Magnusson in 1923 [3]. *R. equi* bacterium is present in soil and horse feces. Foals are thought to become infected when, within the first few days of life, they ingest or breathe in soil, dust, or fecal particles harboring the bacteria [2, 4]. Inhalation of aerosolized virulent *R. equi* from the environment and intracellular replication within alveolar macrophages is essential components of pathogenesis of *R. equi* pneumonia in foals [5]. Virulence in foals is associated with the presence of 80–90 kb plasmids that encode the 15–17 kDa lipoprotein "virulence-associated protein A" (VapA) [6]. The disease is endemic on some farms and sporadic on other farms, but nonexistent on most farms. Recent epidemiologic studies indicate that the difference in the disease's prevalence on farms directly relates to differences in foal population density,

The Effectiveness of Anti-R. equi Hyperimmune Plasma against R. equi Challenge in Thoroughbred Arabian
Foals of Mares Vaccinated with R. equi Vaccine

149

farm management, and environmental factors, such as temperature, dust, soil pH, and the number of virulent *R. equi* organisms in the soil [7].

Several antimicrobial agents are active against *R. equi* in vitro. However, since *R. equi* is a facultative intracellular pathogen that survives, replicates in macrophages, and causes pyogranulomatous lesions, many of these agents are ineffective in vivo [8]. *R. equi* pneumonia significantly impacts the equine industry by posing financial losses since foals that recover from the disease are less likely to race as adults. The cost of therapy and occasional death of foals also pose financial risks. Furthermore, treatment with long-term antibiotics does not guarantee full recovery [9].

Due to the epitheliochorial placentation of equines, foals must obtain all maternally derived antibodies from the ingestion of colostrum [10]. Ingestion of colostrum from hyperimmunized mares was found to be associated with protection against *R. equi* in foals normally hypogammaglobulinemic at birth [11, 12]. Foals become infected approximately when maternal antibody concentrations wane [13]. Immunization of mares has been suggested by several researchers to prevent *R. equi* infection in foals [11, 12, 14–17]. Traditional hyperimmune plasma therapy is currently the only proven method for prevention of *R. equi* in foals, especially those exhibiting passive antibody transfer failure [11, 12, 15].

Due to the presence of the maternal antibody and the immaturity of foals' immune system, vaccination of neonate presents different results [18–20], yet none of the control strategies to protect horses from *R. equi* infection have proven successful. Several vaccines have been investigated for the prevention of *R. equi*, though none have been developed for widespread vaccination [21].

This study thus aimed to determine the effectiveness of a pregnant mare immunization with a *R. equi* vaccine candidate and the administration of anti-*R. equi* hyperimmune plasma against *R. equi* challenge in these mares' foals.

2. Materials and Methods

2.1. Immunization of Mares. Four pregnant thoroughbred Arabian mares were vaccinated three times at months 8, 9, and 10 during pregnancy. Vaccination was performed intramuscularly with the *R. equi* vaccine candidate containing a water-based nanoparticle mineral oil adjuvanted (IMS 3012, SEPPIC, Paris, France) inactive antigen and VapA. Four mares not vaccinated formed the control group. Serum samples were collected from each mare at birth to test the presence of an anti-*R. equi*-specific antibody by ELISA. ELISA was carried out according to Takai et al. [22].

Nine healthy Arabian mares were selected for the production of anti-*R. equi* hyperimmune plasma. After proving to be free of equine infectious anemia (EIA), dourine, glanders, African horse sickness, and *S. abortus equi*, the mares were hyperimmunized 5 days apart with four doses of inactive *R. equi*. After 10 to 15 days, mares were vaccinated 21 days apart with 3 doses of *R. equi* vaccine candidate [23, 24]. After 15 to 20 days following the most previous

immunization, serum samples were obtained from the mares and tested by ELISA for anti-*R. equi* antibody titers [22]. Horses having anti-*R. equi* antibody titers ≥1/12800 by ELISA were selected as plasma donors. Donor horses were bled, and the hyperimmune plasma was separated from the blood cells by plasmapheresis (PCS2, Haemonetics, Braintree, MA, USA). The plasma samples were packed in 200 mL sterile bottles in a BSL 2 cabinet and stored at 4°C. Sterility tests for aerobic, anaerobic bacteria, mycoplasma, and mycotic agents as well as mouse safety tests were performed, after which the hyperimmune plasma samples were used. Donor horses were subsequently vaccinated at intervals of 50 to 60 days and tested 10 to 15 days later, and if the titers were again satisfactory, they were again bled.

2.2. Challenge. To determine the effectiveness of a pregnant mare immunization using a *R. equi* vaccine candidate and HI plasma activity against *R. equi* infection in foals, 4 weeks old mares which born four vaccinated and four unvaccinated mares challenged the 2 mL of 1.0×10^5 CFU pathogen *R. equi* by intercostal injection in the lobe of the left lung [25, 26]. Before receiving the challenge, foals were kept together with their dams approximately 3 weeks after birth to ingest a sufficient amount of colostrum. Two days before the challenge, 150 mL of HI plasma was administered to each foal of the vaccinated mares by intravenous infusion and 50 mL by subcutaneous infusion at days 1, 5, 9, 13, and 17 after the challenge. HI plasma was not given to the foals of unvaccinated mares.

2.3. Laboratory Tests. Blood samples were obtained from the challenged foals to determine the presence of anti-*R. equi* specific antibodies using a ELISA on the challenge day (day 0) and on days 10 and 20, to measure the interleukin-4 (IL-4) and interferon gamma (IFN-γ) concentrations on days 0, 1, 10, 20, and 30 and for complete blood count and blood gas analysis on days 0, 1, 5, 10, 14, 20, and 30 [25]. BALF samples were taken by passing a nasotracheal tube and in fusing 20 mL of sterile saline solution to bacterial culture and measure total cell count (TCC) on days 0, 1, 5, 10, 14, 20, and 30 [25, 26]. BALF samples were collected according to the method described by Mansmann and Knight [27] and Higuchi et al. [28]. The IL-4 and IFN-γ levels were measured using the Horse IL-4 ELISA kit (product code CSB-E14223Hs [96 T], Cusabio, Wuhan, Hubei, China) and Equine IFN-γ ELISA kit (ALP) (product code 3117-1A-6, MabTech, Thomastown, VIC, Australia) according to the manufacturer's instructions. The optimal dilutions of the serum samples, conjugate, substrate, and concentration of the coating antigen were standardized in our laboratory.

2.4. Clinical Examination. Fever (F), respiratory rate (R), pulse rate (P), cough, bronchial sounds, nasal discharge, and mucous membranes of foals were examined on days 0, 1, 5, 10, 14, 20, and 30. Clinical and respiratory system signs were defined as regular (0), mild (1), moderate (2), and severe (3) and normal (0), congested (1), and cyanotic (2) for mucous membranes.

TABLE 1: Macroscopic lesions scores in lungs and organs after necropsy.

Organs	No lesion (0)	Postmortem Lesion Scores			
		Mild (1)	Moderate (2)	Severe (3)	Very severe (4)
Lung	Without lesions	With 2 to 3 small pyogranulomas or large abscess formation limited to only one lobe of the lung and focal consolidation	With four to five pyogranulomas limited to two lobes of the lung multifocal consolidation and large abscess formation	With a large number of differently sized pyogranulomas limited to three lobes or between the right and left lobes of the lung	With military or differently sized pyogranulomas spread over all lobes of the lung
Lymph node	Without lesions	With slight growth	With mild growth	Overgrown and with limited pus foci of the cross-sectional surface	Overgrown and purulent
Heart	Without lesions	With slight paleness	With a wide area of paleness in the epicardium and endocardium or pyogranuloma	With a wide area of paleness in the cross-sectional area or a few pyogranulomas	With a wide area of paleness in the cross-sectional area or several pyogranulomas
Liver	Without lesions	Slightly enlarged and congested	Enlarged and mottled appearance or with pyogranuloma	With numerous pyogranulomas in the cross-sectional area and under the capsule	With pyogranulomas spread over the entire surface and the cross-sectional area
Kidney	Without lesions	Slightly enlarged	Enlarged kidneys and mottled appearance	Enlarged kidneys and with pyogranuloma in the cross-sectional area and infarction area	Enlarged kidneys and with numerous pyogranulomas in the cross-sectional area and infarction area
Spleen	Without lesions	Slightly enlarged	Overgrown and with pyogranuloma in the cross-sectional area	Overgrown and with a few pyogranulomas in the cross-sectional area	Overgrown and with numerous pyogranulomas in the cross-sectional area

2.5. Postmortem Examination. Foals were euthanized with an intravenous injection of a mixture of 100 mg of suxamethonium chloride and 22.5 mg sodium chloride (Lysthenon 2%, Fako, Turkey) 30 days after the challenge, and necropsy was performed. Macroscopic lesions in lungs and organs were scored according to Table 1.

Samples from these organs were fixed in 10% formalin and embedded in paraffin wax blocks. Block sections of 5 μm thickness were stained with Luna [29]. All sections were evaluated under a light microscope.

2.6. Statistical Analysis. All statistical analyses were performed using the Student's *t*-test.

3. Results

Two foals (numbers 5 and 7) of unvaccinated mares died before the challenge, and *R. equi*, *Streptococcus* sp., and *Corynebacterium* sp. were isolated from the lungs of these dead foals. In the place of the dead foals, two unvaccinated mares' foals were included in the study. One of these foals died on day 7 and another on day 13 after the challenge.

Anti-*R. equi* antibody titer measured in vaccinated mares was higher than in the unvaccinated mares at birth by ELISA. In the foals of vaccinated mares, anti-*R. equi* antibody titer was determined to be 1/1600 of the maximum after

the administration of hyperimmune plasma, and 1/200 titer was also determined in control foals (Table 2).

The IL-4 concentration was measured to have a mean of <16 pg/mL at days 0, 10, and 20 and a mean of 80 pg/mL at day 30 in foals of the vaccinated group, as well as a mean of 32 pg/mL in control foals. IFN-γ concentration increased on day 10 compared to the challenge day in both vaccine and control groups. By day 20, the control group remained stable, while the vaccine group exhibited doubling. By day 30, it increased in both groups but was measured to be 4.1-fold more in the vaccine group than in the control group (Table 3).

The reisolation rate of *R. equi* from BALF samples was determined to have increased by day 10, decreased by day 20, and increased again by day 30 in the vaccinated and HI plasma-administered group. An increase was observed in the control group compared to day 1 (Table 4, Figure 1).

A decrease in the concentration of pO_2 and SO_2 in the vaccine group was observed on day 5. The concentration of pCO_2 and tCO_2 increased, while the concentration of SO_2 decreased on day 10. A decrease in the concentration of pO_2 and SO_2 in the control group was observed on day 30 ($P < 0.05$) (Table 6).

An insignificant ($P > 0.05$) increase was observed in WBC concentration in both the control and vaccine group during the study period. Significant ($P < 0.05$) differences were determined in percentages of LYM, MON,

The Effectiveness of Anti-R. equi Hyperimmune Plasma against R. equi Challenge in Thoroughbred Arabian
Foals of Mares Vaccinated with R. equi Vaccine

151

TABLE 2: Anti-*R. equi* antibody ELISA titer of vaccinated and unvaccinated mares and challenged foals.

Mare group	Mare number	ELISA titer at birth	Foal group	Foal number	ELISA titer after challenge		
					Day 0	Day 10	Day 20
Vaccinated	1	1/12800	Vaccine + HI plasma	1	1/800	1/1600	1/1600
	2	1/6400		2	1/100	1/400	1/800
	3	1/3200		3	1/800	1/1600	1/1600
	4	1/6400		4	1/400	1/800	1/800
Unvaccinated	5	Negative	Control	5	0	1/200	*
	6	1/200		6	0	Negative	1/200
	7	Negative		7	0	*	*
	8	Negative		8	0	Negative	1/200

*Because foals died, ELISA titer was not measured on these days.

TABLE 3: IL-4 (pg/mL) and IFN-γ (ng/mL) concentration in challenged foals.

Foal group	Foal number	Day 0 IL-4/IFN-γ	Day 10 IL-4/IFN-γ	Day 20 IL-4/IFN-γ	Day 30 IL-4/IFN-γ
Vaccine + HI plasma	1	<16/<0.01	<16/0.01	<16/0.02	**128/0.05**
	2	<16/<0.005	<16/0.005	<16/0.01	**128/0.05**
	3	<16/<0.005	<16/0.005	<16/0.01	32/0.05
	4	<16/<0.005	<16/0.005	<16/0.01	32/0.05
	Mean	**<16/<0.0075**	**<16/0.0075**	**<16/0.015**	**80/0.05**
Control	5	<16/<0.005	<16/0.005	*	*
	6	<16/<0.005	<16/0.005	<16/0.001	32/0.005
	7	<16/<0.005	*	*	*
	8	<16/<0.005	<16/0.005	<16/0.01	32/0.02
	Mean	**<16/<0.005**	**<16/0.005**	**<16/0.0055**	**32/0.012**

*Because the foals died, IL-4 and IFN-γ concentration were not measured on these days.

and GRAN concentration. The RBC concentration decreased significantly ($P < 0.05$) on days 14 and 30 (Table 6).

TCC of BALF and lung scores had increased in both groups by day 14 (Table 5). These increases were statistically significant ($P < 0.05$) in the control group, yet insignificant ($P > 0.05$) in the vaccine group.

The clinical and lung auscultation findings indicate that clinical signs commenced on day 5 and increased until day 10 in both groups. In the vaccine group, however, it began to decrease after day 10 and because of the sepsis two foals died on days 7 and 15 in the control group.

The mean of total lesion scores of the organs and tissue was determined to be 78 in control group and 22 in the vaccine group. According to pathological findings, the severity of *R. equi* pneumonia and lesion scores of the organs and tissue was observed 3.54-fold less in the vaccinated and HI plasma-administered foals compared to the control foals (Table 7, Figure 2).

4. Discussion

Cell-mediated immunity is thought to play an important role in eliminating the facultative intracellular pathogen from foals, yet humoral immunity seems to be critically involved in the early protection in young foals. Foals are the most susceptible to the effects of virulent organisms when maternal antibody levels wane [30, 31]. The passive transfer of immunity plays a critical role in the foals' resistance to a variety of infectious agents. Due to the epitheliochorial placentation of equines, foals must obtain all of their maternally derived antibodies by ingesting colostrum [10]. The lowest circulating antibody titers in foals appear from 1 to 6 months via the combined effects of waning maternally derived antibodies and low endogenous antibody production [32]. As a result, foals are susceptible to *R. equi* pneumonia during this period. Due to age-dependent susceptibility to *R. equi*, foals need to develop anti-*R. equi* immunity shortly after birth [25].

Since *R. equi* lives within macrophages, it resists many common antibiotics, and antibiotics-based therapy is prolonged, expensive, possibly associated with adverse effects, and inconsistently successful [33].

Studies investigating the active immunization of mares as a means of enhancing the passive transfer of virulent *R. equi* antibodies in colostrum and protecting foals from *R. equi* pneumonia have yielded mixed results. Solo vaccination of mares has not proven protective against *R. equi* pneumonia in foals, despite a significant increase in a colostral-specific antibody [14, 34]. Martens et al. [34], Madigan et al. [35], and Varga et al. [36] did not observe protection in foals against

TABLE 4: Reisolation rate of *R. equi* from BALF samples (CFU/mL) and lung samples (CFU/gr) after necropsy in foals.

Foal group	Foal number	Day 0	Day 1	Day 10	Day 20	Day 30	Necropsy
Vaccine + HI plasma	1	0	65000	135	1400	140	50
	2	0	1000	55	1220	0	21000
	3	0	1300	120	12700	0	0
	4	0	2600	195	90000	170000	**
	Mean	**0**	**17475**	**126**	**26330**	**42535**	
Control	5	0	19000	800	died	—	**
	6	0	4100	3600	1300000	1100000	320000
	7	0	5500	Died	—	—	**
	8	0	83000	640000	2300	100000	3250000
	Mean	**0**	**27900**	**214666**	**651150**	**600000**	

** Too many to count.

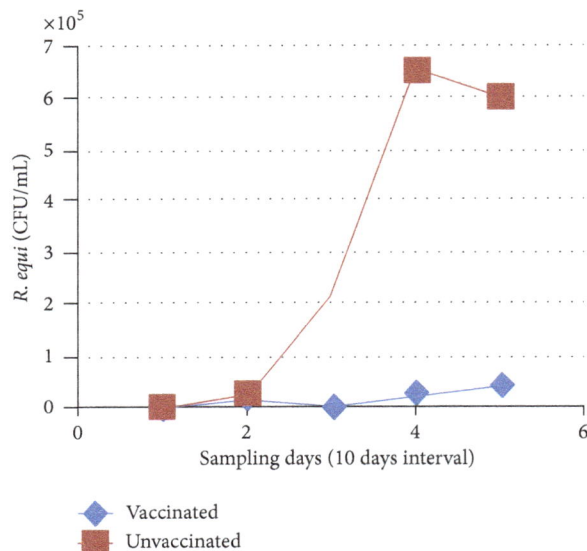

FIGURE 1: Reisolation rate of *R. equi* from foals' BALF samples (CFU/mL).

R. equi pneumonia after mare vaccination. Moreover, Hines et al. [31] reported that immunoglobulin in mares may not be efficiently transferred via colostrum. However, according to other researchers, passive antibody transfer from ingested colostrum was found to be associated with protection against *R. equi* in foals normally hypogammaglobulinemic at birth [11, 12, 14, 15]. Immunization of pregnant mares with virulent *R. equi* and VapA protein antigen associated with a water-based nanoparticle adjuvant as a candidate vaccine developed a higher serum IgG and opsonic activity, which resulted in passive antibody-mediated protection of foals [14]. Muscatello [17] reported that the protective effect was associated with an increase in the opsonic capacity of polymorphonuclear leucocytes against virulent *R. equi* in foals from immunized mares.

Martens et al. [37] were the first to show the immunoprophylactic capacity of specific hyperimmune plasma in an experimental model of *R. equi* pneumonia in foals. Other researchers have reported a reduction in foal morbidity and mortality due to *R. equi* as a result of HI plasma administration [14, 35, 37–39]. However, several other studies report no protective effects of HI plasma [11, 12, 40].

According to our results, the immunization of pregnant mares with a *R. equi* vaccine candidate and the administration of the anti-*R. equi* HI plasma in vaccinated mares' foals proved to have protective effects during experimental *R. equi* infection. The clinical signs of pneumonia were significantly delayed, and the reisolation rate of *R. equi* from BALF samples decreased. The natural mortality rate due to *R. equi* infection was less than 50% in these foals compared to the controls. The severity of *R. equi* pneumonia and lesion scores of the organs and tissue determined 3.54-fold less than control foals.

The protective components of HI plasma are not completely known. Antibodies to Vap proteins, specifically VapA, appear to be crucially important [17]. It has been reported that there is no correlation between the total serum IgG levels and concentration of the specific anti-*R. equi* antibody [34, 35]. The phagocytic ability of foal neutrophils has been found comparable to adults, but the lymphocyte stimulation response alone did not influence the course of *R. equi* infection, while the opsonic ability of foal serum was found to be a limited factor for phagocytosis from the ages of 1 to 6 weeks [41, 42]. Phagocytic activity of foal neutrophils was found to improve when mixed with adult serum or plasma [43], which may be related to unknown, nonspecific immune factors provided by HI plasma and normal adult equine plasma that are absent from colostrum, such as fibronectin, complement, and cytokines [12, 40, 44]. The effectiveness of HI plasma is likely to be affected by the dosage, timing of administration, innate immune system competence, management conditions, and number of virulent bacterin in the environment [45].

Cytokines IFN-γ and IL-4 are major macrophage and neutrophil-activating factors, as well as upregulated microbial killing pathways [46]. It is reported that newborn foals had a deficiency of IFN-γ/IL-4 and levels not reaching adult status until approximately 4 months of age [47]. Reduced IFN-γ and IL-4 expression have a limited killing capacity of phagocytes in young foals [47, 48]. In our study,

The Effectiveness of Anti-R. equi Hyperimmune Plasma against R. equi Challenge in Thoroughbred Arabian
Foals of Mares Vaccinated with R. equi Vaccine

153

TABLE 5: Total cell count (TCC) of BALF samples, clinical signs, respiratory system signs, mucous membrane scores, and temperature of foals.

	Foal group	Day 0	Day 1	Day 5	Day 10	Day 14	Day 20	Day 30
BALF-TCC	Control	300.00 ± 57.74	300.00 ± 57.74	300.00 ± 57.74	300.00 ± 57.74	466.67 ± 266.67	8800.00 ± 8400.00	1500.00 ± 1100.00
	Vaccinated	300.00 ± 57.74	300.00 ± 57.74	300.00 ± 57.74	300.00 ± 57.74	2600.00 ± 1116.54	1000.00 ± 258.20	1950.00 ± 1040.43
Clinical score	Control	0.00 ± 0.00	0.00 ± 0.00	0.00 ± 0.00	0.00 ± 0.00	1.00 ± 1.00	0.00 ± 0.00	0.00 ± 0.00
	Vaccinated	0.00 ± 0.00	0.00 ± 0.00	0.00 ± 0.00	0.00 ± 0.00	0.00 ± 0.00	0.00 ± 0.00	0.00 ± 0.00
Respiratory system score	Control	1.00 ± 0.00	0.00 ± 0.00	0.00 ± 0.00 B	1.00 ± 0.58	0.67 ± 0.33 B	0.00 ± 0.00 B	0.00 ± 0.00 B
	Vaccinated	1.00 ± 0.00	0.00 ± 0.00	1.25 ± 0.48 A	2.00 ± 0.00	0.275 ± 0.25 A	2.75 ± 0.25 A	2.75 ± 0.25 A
Mucous membrane score	Control	0.00 ± 0.00	0.00 ± 0.00	0.00 ± 0.00	0.00 ± 0.00	0.33 ± 0.33	0.00 ± 0.00	0.00 ± 0.00
	Vaccinated	0.00 ± 0.00	0.00 ± 0.00	0.00 ± 0.00	0.00 ± 0.00	0.00 ± 0.00	0.00 ± 0.00	0.00 ± 0.00
Temperature	Control	38.3 ± 0.12	38.43 ± 0.11	38.05 ± 0.23	27.90 ± 9.31	37.87 ± 0.24	38.20 ± 0.20	38.45 ± 0.05
	Vaccinated	38.13 ± 0.13	38.33 ± 0.22	38.50 ± 0.11	38.45 ± 0.06	38.05 ± 0.16	38.60 ± 0.07	38.60 ± 0.07

TABLE 6: Complete blood count and blood gas analysis results of foals.

Foal group		Day 0	Day 1	Day 5	Day 10	Day 14	Day 20	Day 30
pH (7.34–7.43)	Control	7.45 ± 0.02	7.41 ± 0.02	7.43 ± 0.02	7.35 ± 0.03	7.42 ± 0.01	7.45 ± 0.02	7.40 ± 0.02
	Vaccinated	7.43 ± 0.01	7.42 ± 0.01	7.41 ± 0.01	7.41 ± 0.01	7.41 ± 0.01	7.39 ± 0.01	7.42 ± 0.01
pCO2 (38–48 mmHg)	Control	37.50 ± 1.19	42.25 ± 0.75	45.75 ± 1.93	52.75 ± 2.78 A	47.33 ± 2.73	40.50 ± 1.50	47.00 ± 3.00
	Vaccinated	40.00 ± 1.78	42.00 ± 1.47	42.00 ± 0.41	40.25 ± 1.25 B	42.00 ± 2.74	45.00 ± 3.14	40.50 ± 1.26
pO2 (37–56 mmHg)	Control	35.00 ± 2.12	35.00 ± 2.12	36.75 ± 1.55 A	33.00 ± 1.35	32.00 ± 2.08	38.00 ± 4.00	34.00 ± 1.01 B
	Vaccinated	37.25 ± 3.71	33.50 ± 2.63	29.00 ± 1.91 B	36.25 ± 1.75	53.75 ± 21.46	33.75 ± 2.29	41.25 ± 1.38 A
tCO2	Control	26.98 ± 1.26	28.03 ± 1.21	31.53 ± 1.81	30.53 ± 0.61 A	31.87 ± 1.18	29.05 ± 0.55	30.20 ± 0.90
	Vaccinated	27.85 ± 0.45	28.35 ± 0.73	28.05 ± 0.29	26.75 ± 0.84 B	27.95 ± 1.47	54.15 ± 27.30	27.65 ± 0.57
SO2 (>60 mmHg)	Control	70.00 ± 2.48	67.00 ± 3.37	71.50 ± 1.32 A	71.50 ± 1.32 A	62.67 ± 4.67	74.50 ± 4.50	65.00 ± 1.00 B
	Vaccinated	71.25 ± 6.70	64.75 ± 4.87	55.50 ± 4.29 B	55.50 ± 4.29 B	71.50 ± 9.50	53.90 ± 7.42	77.75 ± 1.38 A
WBC (5–12 ×10³)	Control	7.08 ± 1.01	7.68 ± 1.31	7.32 ± 1.16	7.02 ± 1.12	5.65 ± 1.66	4.64 ± 0.37	10.53 ± 2.61
	Vaccinated	6.66 ± 0.87	5.86 ± 0.33	8.82 ± 1.29	8.86 ± 0.87	8.60 ± 0.66	26.90 ± 15.77	10.33 ± 1.57
LYM (% 20–40)	Control	15.60 ± 2.46 B	15.70 ± 0.90 B	22.65 ± 2.41	22.80 ± 1.86 B	29.13 ± 11.52	36.85 ± 14.75	17.50 ± 10.00
	Vaccinated	35.60 ± 3.98 A	30.65 ± 2.23 A	34.63 ± 6.43	36.75 ± 4.88 A	37.83 ± 4.26	27.11 ± 6.51	31.85 ± 3.40
MON (% 2–8)	Control	1.40 ± 0.12 B	1.68 ± 0.11	3.55 ± 0.13	3.43 ± 0.77	3.67 ± 1.22	3.50 ± 0.90	1.60 ± 0.70 A
	Vaccinated	2.75 ± 0.45 A	2.73 ± 0.52	3.00 ± 0.48	3.05 ± 0.46	3.58 ± 0.33	8.00 ± 5.90	2.53 ± 0.46
GRAN (% 50–70)	Control	83.00 ± 2.55 A	82.63 ± 0.92 A	73.80 ± 2.46	73.78 ± 2.32	67.20 ± 12.72	59.65 ± 15.65	80.90 ± 10.70
	Vaccinated	61.65 ± 3.70 B	64.75 ± 4.87 B	62.38 ± 6.69	60.20 ± 5.10	58.60 ± 4.42	49.43 ± 16.35	65.63 ± 3.78
RBC (7–13 ×10⁶)	Control	8.87 ± 0.48	9.28 ± 0.89	9.27 ± 0.39	9.10 ± 0.71	8.10 ± 0.64 B	9.01 ± 1.16	8.05 ± 0.24 B
	Vaccinated	9.65 ± 0.32	11.30 ± 1.02	10.22 ± 0.34	10.03 ± 0.22	9.89 ± 0.26 A	25.16 ± 15.82	9.98 ± 0.27 A
MCV (35–60 fl)	Control	32.63 ± 0.29 A	32.83 ± 0.24 A	32.78 ± 0.71 A	34.48 ± 0.47 A	33.23 ± 0.91	32.70 ± 0.90	32.00 ± 1.90
	Vaccinated	29.43 ± 1.14 B	29.95 ± 0.76 B	29.50 ± 0.93 B	29.30 ± 0.99 B	30.85 ± 0.92	26.24 ± 5.43	30.50 ± 0.68
HCT (% 32–53)	Control	28.85 ± 1.39	30.43 ± 3.04	30.30 ± 1.10	31.38 ± 2.66	26.77 ± 1.39	29.30 ± 3.00	25.65 ± 0.75 B
	Vaccinated	28.43 ± 1.98	33.70 ± 2.90	30.10 ± 1.11	29.35 ± 1.42	30.43 ± 0.92	29.18 ± 0.71	30.33 ± 0.50 A
MCHC (30–42 g/dL)	Control	39.75 ± 0.43	36.18 ± 2.50	39.63 ± 0.26	36.78 ± 0.86	37.43 ± 0.45	38.85 ± 0.15	40.50 ± 1.20
	Vaccinated	39.50 ± 1.67	40.60 ± 0.94	43.65 ± 3.07	41.73 ± 1.15	37.93 ± 0.61	38.40 ± 0.93	38.70 ± 1.09
RDW (8–12)	Control	14.45 ± 0.10	14.60 ± 0.47 B	14.30 ± 0.27 B	14.10 ± 0.43 B	14.13 ± 0.29	15.05 ± 0.85	16.65 ± 0.65
	Vaccinated	15.40 ± 0.54	16.00 ± 0.31 A	15.85 ± 0.37 A	16.18 ± 0.55 A	16.30 ± 0.83	16.70 ± 1.00	17.15 ± 0.84
HB (11–17 g/dL)	Control	11.48 ± 0.47	10.83 ± 0.59	12.03 ± 0.51	11.53 ± 0.86	10.03 ± 0.64	11.40 ± 1.20	10.40 ± 0.00
	Vaccinated	11.18 ± 0.55	13.68 ± 1.12	12.40 ± 0.44	12.23 ± 0.43	11.55 ± 0.39	11.23 ± 0.44	11.75 ± 0.48
THR (100–400 m/mm³)	Control	278.50 ± 87.36	330.50 ± 87.36	287.00 ± 44.53	238.50 ± 29.83 B	171.33 ± 30.69	240.00 ± 37.00	241.50 ± 8.50
	Vaccinated	248.50 ± 30.01	317.00 ± 40.26	410.00 ± 49.83	384.50 ± 9.56 A	244.00 ± 19.20	341.50 ± 26.38	279.00 ± 25.65

The Effectiveness of Anti-R. equi Hyperimmune Plasma against R. equi Challenge in Thoroughbred Arabian Foals of Mares Vaccinated with R. equi Vaccine

155

TABLE 7: Lesion scores of the organs and tissue according to pathological findings.

Foal group	Foals number	Lung	Bronchial LN	Mediastinal LN	Heart	Liver	Kidney	Spleen	Cecum LN	Total foal score	Total group score	Mean group score
Vaccine + HI plasma	1	1	1	0	0	0	0	0	2	4		
	2	0	2	2	0	0	0	1	1	6	22	5.5
	3	1	1	1	0	0	0	0	0	3		
	4	3	3	2	0	0	0	0	1	9		
Control	5	4	4	4	4	4	4	4	4	32		
	6	3	1	1	2	0	0	0	1	8	78	19.5
	7	4	4	4	4	4	4	4	4	32		
	8	3	1	1	0	0	0	0	1	6		

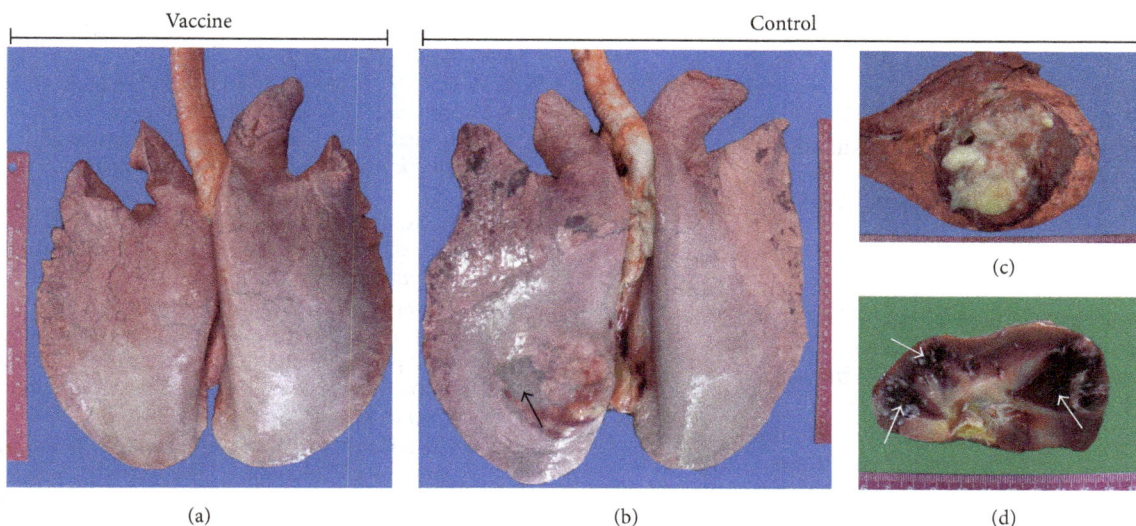

FIGURE 2: (a) Normal lung in a foal of vaccine group, (b) pyogranuloma formation in the *R. equi* inoculated left lung of a control foal (arrow), (c) pyogranuloma in lung, and (d) infarction areas in kidney of control foals (arrows).

the IFN-γ and IL-4 concentration measured 4.1-fold and 2.5-fold more in the foals of vaccinated mares and HI plasma-administered, respectively, compared to the control foals.

According to changes in blood gases of foals, it was observed that lung ventilation in the challenge group had been affected by day 5, intensified after day 10, and continued to increase during the study period.

The important decrease ($P < 0.05$) of the RBC concentration on days 14 and 30 and significant differences ($P < 0.05$) of the MCV and RDW values in the vaccine and control group were evaluated as microcytic/normocytic-normochromic-regenerative anemia which was interpreted as a response to emerging infectious.

Given the decrease in the number in the controls, BALF-TCC parameters were detected as statistically insignificant ($P > 0.05$).

According to clinical scores and laboratory findings, the effect of infection in the vaccine group began on day 5, increased from day 9 to 14, and was constant from days 20 to 30. The effect of infection was similar in the control group, though the clinical findings increased from days 7 to 15, and two foals died of sepsis during this latter period. These findings show that resistance to infection was low in the control group.

As yet there is no licensed vaccine for the prevention of *R. equi*. However, on January 27, 2011, Intervet/Schering-Plough Animal Health announced that a vaccine against *Rhodococcus equi* infection in foals had entered the final stages of development [49].

5. Conclusions

Our results indicate that the immunization of pregnant mares with a water-based nanoparticle mineral oil adjuvanted (IMS 3012) inactive bacterin and VapA and the administration of HI plasma in foals of these mares effectively protect foals against *R. equi* challenge. Foals are born into the *R. equi* contaminated environment due to mares carrying the *R. equi* in their intestines. *R. equi* infection can be controlled by both the mares' vaccination and anti-*R. equi*-HI plasma administration in foals of such dams.

Conflict of Interests

The authors declare that there is no conflict of interests regarding the publication of this paper.

Acknowledgments

This study was supported by the Scientific and Technological Research Council of Turkey (Project no. 108 G 030). This research has been approved (date 15/02/2008 and 2008/010) by the Ethics Committee of the Faculty of Veterinary Medicine at the University of Selcuk in Konya, Turkey.

References

[1] J. F. Prescott, "Epidomiology of *Rhodococcus equi* infection in horses," *Veterinary Microbiology*, vol. 14, no. 3, pp. 211–214, 1987.

[2] W. G. Meijer and J. F. Prescott, "*Rhodococcus equi*," *Veterinary Research*, vol. 35, no. 4, pp. 383–396, 2004.

[3] H. Magnusson, "Spezifische Infektiose Pneumonie beim Fohlen. Ein neuer Eitererreger beim Pferde," *Archiv fur Wissenschaftliche und Praktische Tierheilkunde*, vol. 50, pp. 22–38, 1923.

[4] J. B. Woolcock, M. D. Mutimer, and A. M. T. Farmer, "Epidemiology of Corynebacterium equi in horses," *Research in Veterinary Science*, vol. 28, no. 1, pp. 87–90, 1980.

[5] G. Muscatello, "*Rhodococcus equi* pneumonia in the foal. Part 1: pathogenesis and epidemiology," *Veterinary Journal*, vol. 192, no. 1, pp. 20–26, 2012.

[6] S. Takai, "Epidemiology of *Rhodococcus equi* infections: a review," *Veterinary Microbiology*, vol. 56, no. 3-4, pp. 167–176, 1997.

[7] R. J. Martens, S. Takai, N. D. Cohen, M. K. Chaffin, H. Liu, and S. Lingsweiler, "Prevalence and virulence of Rhodococcus equi in sick foals and soil of horse-breeding farms in Texas," in *Proceedings of the 45th Annual Convention of the American Association of Equine Practitioners (AAEP)*, vol. 45, pp. 53–55, 1999.

[8] S. Giguere, "Update on *R. equi* infections in foals," in *Proceedings of the 11th Congress on Equine Medicine and Surgery*, pp. 136–141, Geneva, Switzerland, 2009.

[9] D. M. Ainsworth, S. W. Eicker, A. E. Yeagar et al., "Associations between physical examination, laboratory, and radiographic findings and outcome and subsequent racing performance of foals with *Rhodococcus equi* infection: 115 cases (1984–1992)," *Journal of the American Veterinary Medical Association*, vol. 213, no. 4, pp. 510–515, 1998.

[10] C. W. Kohn, D. Knight, W. Hueston, R. Jacobs, and S. M. Reed, "Colostral and serum IgG, IgA, and IgM concentrations in Standardbred mares and their foals at parturition," *Journal of the American Veterinary Medical Association*, vol. 195, no. 1, pp. 64–68, 1989.

[11] J. R. Hurley and A. P. Begg, "Failure of hyperimmune plasma to prevent pneumonia caused by *Rhodococcus equi* in foals," *Australian Veterinary Journal*, vol. 72, no. 11, pp. 418–420, 1995.

[12] S. Giguère, J. M. Gaskin, C. Miller, and J. L. Bowman, "Evaluation of a commercially available hyperimmune plasma product for prevention of naturally acquired pneumonia caused by *Rhodococcus equi* in foals," *Journal of the American Veterinary Medical Association*, vol. 220, no. 1, pp. 59–63, 2002.

[13] S. S. Caston, S. R. McClure, R. J. Martens et al., "Effect of hyperimmune plasma on the severity of pneumonia caused by *Rhodococcus equi* in experimentally infected foals," *Veterinary Therapeutics*, vol. 7, no. 4, pp. 361–375, 2006.

[14] T. Becú, G. Polledo, and J. M. Gaskin, "Immunoprophylaxis of *Rhodococcus equi* pneumonia in foals," *Veterinary Microbiology*, vol. 56, no. 3-4, pp. 193–204, 1997.

[15] J. E. Butler, "Immunoglobulin diversity, B-cell and antibody repertoire development in large farm animals," *OIE Revue Scientifique et Technique*, vol. 17, no. 1, pp. 43–70, 1998.

[16] J. Cauchard, C. Sevin, J.-J. Ballet, and S. Taouji, "Foal IgG and opsonizing anti-*Rhodococcus equi* antibodies after immunization of pregnant mares with a protective VapA candidate vaccine," *Veterinary Microbiology*, vol. 104, no. 1-2, pp. 73–81, 2004.

[17] G. Muscatello, "*Rhodococcus equi* pneumonia in the foal. Part 2: diagnostics, treatment and disease management," *Veterinary Journal*, vol. 192, no. 1, pp. 27–33, 2012.

[18] H. A. Gans, A. M. Arvin, J. Galinus, L. Logan, R. Dehovitz, and Y. Maldonado, "Deficiency of the humoral immune response to measles vaccine in infants immunized at age 6 months," *Journal of the American Medical Association*, vol. 280, no. 6, pp. 527–532, 1998.

[19] H. A. Gans, Y. Maldonado, L. L. Yasukawa et al., "IL-12, IFN-γ, and T cell proliferation to measles in immunized infants," *Journal of Immunology*, vol. 162, no. 9, pp. 5569–5575, 1999.

[20] K. E. Hooper-Mcgrevy, B. N. Wilkie, and J. F. Prescott, "Virulence-associated protein-specific serum immunoglobulin G-isotype expression in young foals protected against *Rhodococcus equi* pneumonia by oral immunization with virulent *R. equi*," *Vaccine*, vol. 23, no. 50, pp. 5760–5767, 2005.

[21] R. van der Geize, A. W. F. Grommen, G. I. Hessels, A. A. C. Jacobs, and L. Dijkhuizen, "The steroid catabolic pathway of the intracellular pathogen *Rhodococcus equi* is important for pathogenesis and a target for vaccine development," *PLoS Pathogens*, vol. 7, no. 8, Article ID e1002181, 2011.

[22] S. Takai, S. Kawazu, and S. Tsubaki, "Enzyme-linked immunosorbent assay for diagnosis of *Corynebacterium* (Rhodococcus) equi infection in foals," *American Journal of Veterinary Research*, vol. 46, no. 10, pp. 2166–2170, 1985.

[23] O. Erganiş and E. İstanbulluoğlu, *İmmünoloji*, Mimoza Basım Yayım A.Ş. Yayın, Konya, Turkey, 1993.

[24] G. C. Hovard and M. R. Kaser, *Making and Using Antibodies: A Practical Handbook*, CRC Pres Taylor & Francis Group, Boca Raton, Fla, USA, 2007.

[25] A. M. Lopez, H. G. G. Townsend, A. L. Allen, and M. K. Hondalus, "Safety and immunogenicity of a live-attenuated auxotrophic candidate vaccine against the intracellular pathogen *Rhodococcus equi*," *Vaccine*, vol. 26, no. 7, pp. 998–1009, 2008.

[26] T. Phumoonna, *Further development of vaccines and diagnostic tests for Rhodococcus equi infection in foals [Ph.D. thesis]*, Charles Sturt University, Bathurst, Australia, 2005.

[27] R. A. Mansmann and H. D. Knight, "Transtracheal aspiration in the horse," *Journal of the American Veterinary Medical Association*, vol. 160, no. 11, pp. 1527–1529, 1972.

[28] T. Higuchi, S. Hashikura, C. Gojo et al., "Clinical evaluation of the serodiagnostic value of enzymelinked immunosorbent assay for *Rhodococcus equi* infection in foals," *Equine Veterinary Journal*, vol. 29, no. 4, pp. 274–278, 1997.

[29] L. G. Luna, *Manuals of Histologic Staining Methods of the Armed Forces Institute of Pathology*, McGraw-Hill Book Company, New York, NY, USA, 1968.

The Effectiveness of Anti-R. equi Hyperimmune Plasma against R. equi Challenge in Thoroughbred Arabian
Foals of Mares Vaccinated with R. equi Vaccine

157

[30] R. B. Brandon, R. Hill, and R. P. Wilson, "Development of hyperimmune equine plasma to virulence associated protein A (VapA) of *Rhodococcus equi*," Plasvacc White Paper, 2009.

[31] S. A. Hines, S. T. Kanaly, B. A. Byrne, and G. H. Palmer, "Immunity to *Rhodococcus equi*," *Veterinary Microbiology*, vol. 56, no. 3-4, pp. 177–185, 1997.

[32] S. K. Hietala, A. A. Ardans, and A. Sansome, "Detection of *Corynebacterium* equi-specific antibody in horses by enzyme-linked immunosorbent assay," *American Journal of Veterinary Research*, vol. 46, no. 1, pp. 13–15, 1985.

[33] J. Cauchard, S. Taouji, C. Sevin et al., "Immunogenicity of synthetic *Rhodococcus equi* virulence-associated protein peptides in neonate foals," *International Journal of Medical Microbiology*, vol. 296, no. 6, pp. 389–396, 2006.

[34] R. J. Martens, J. G. Martens, and R. A. Fiske, "Failure of passive immunisation by colostrum from immunised mares to protect foals against *Rhodococcus equi* pneumonia," *Equine Veterinary Journal*, vol. 23, pp. 19–22, 1991.

[35] J. E. Madigan, S. Hietala, and N. Muller, "Protection against naturally acquired *Rhodococcus equi* pneumonia in foals by administration of hyperimmune plasma," *Journal of Reproduction and Fertility*, vol. 44, pp. 571–578, 1991.

[36] J. Varga, L. Fodor, M. Rusvai, I. Soós, and L. Makrai, "Prevention of *Rhodococcus equi* pneumonia of foals using two different inactivated vaccines," *Veterinary Microbiology*, vol. 56, no. 3-4, pp. 205–212, 1997.

[37] R. J. Martens, J. G. Martens, R. A. Fiske, and S. K. Hietala, "*Rhodococcus equi* foal pneumonia: protective effects of immune plasma in experimentally infected foals," *Equine Veterinary Journal*, vol. 21, no. 4, pp. 249–255, 1989.

[38] T. Higuchi, T. Arakawa, S. Hashikura, T. Inui, H. Senba, and S. Takai, "Effect of prophylactic administration of hyperimmune plasma to prevent *Rhodococcus equi* infection on foals from endemically affected farms," *Journal of Veterinary Medicine B*, vol. 46, no. 9, pp. 641–648, 1999.

[39] S. S. Caston, S. R. McClure, R. J. Martens et al., "Epidemiology and prevention of *R. equi* pneumonia," in *Proceedings of the 10th International Congress of World Equine Veterinary Association*, pp. 343–346, Moscow, Russia, 2008.

[40] G. Perkins, A. Yeager, H. N. Erb, D. Nydam, T. Divers, and J. Bowman, "Survival of foals with experimentally induced *Rhodococcus equi* infection given either hyperimmune plasma containing *R. equi* antibody or normal equine plasma," *Veterinary Therapeutics*, vol. 3, no. 3, pp. 334–346, 2002.

[41] S. Demmers, A. Johannisson, G. Gröndahl, and M. Jensen-Waern, "Neutrophil functions and serum IgG in growing foals," *Equine Veterinary Journal*, vol. 33, no. 7, pp. 676–680, 2001.

[42] C. McTaggart, J. V. Yovich, J. Penhale, and S. L. Raidal, "A comparison of foal and adult horse neutrophil function using flow cytometric techniques," *Research in Veterinary Science*, vol. 71, no. 1, pp. 73–79, 2001.

[43] G. Gröndahl, S. Sternberg, M. Jensen-Waern, and A. Johannisson, "Opsonic capacity of foal serum for the two neonatal pathogens *Escherichia coli* and *Actinobacillus equuli*," *Equine Veterinary Journal*, vol. 33, no. 7, pp. 670–675, 2001.

[44] K. E. Hooper-McGrevy, S. Giguere, B. N. Wilkie, and J. F. Prescott, "Evaluation of equine immunoglobulin specific for *Rhodococcus equi* virulence-associated proteins A and C for use in protecting foals against *Rhodococcus equi*-induced pneumonia," *American Journal of Veterinary Research*, vol. 62, no. 8, pp. 1307–1313, 2001.

[45] T. R. M. Y. Dawson, D. W. Horohov, W. G. Meijer, and G. Muscatello, "Current understanding of the equine immune response to *Rhodococcus equi*. An immunological review of *R. equi* pneumonia," *Veterinary Immunology and Immunopathology*, vol. 135, no. 1-2, pp. 1–11, 2010.

[46] T. R. Mosmann and R. L. Coffman, "TH1 and TH2 cells: different patterns of lymphokine secretion lead to different functional properties," *Annual Review of Immunology*, vol. 7, pp. 145–173, 1989.

[47] C. C. Breathnach, T. Sturgill-Wright, J. L. Stiltner, A. A. Adams, D. P. Lunn, and D. W. Horohov, "Foals are interferon gamma-deficient at birth," *Veterinary Immunology and Immunopathology*, vol. 112, no. 3-4, pp. 199–209, 2006.

[48] C. Ryan, S. Giguère, J. Hagen, C. Hartnett, and A. E. Kalyuzhny, "Effect of age and mitogen on the frequency of interleukin-4 and interferon gamma secreting cells in foals and adult horses as assessed by an equine-specific ELISPOT assay," *Veterinary Immunology and Immunopathology*, vol. 133, no. 1, pp. 66–71, 2010.

[49] Anonymous, 2011, http://www.thoroughbredtimes.com/horse-health/2011/01/27/candidate-vaccine-against-rhodococcus-equi-infections-to-be-investigated.aspx.

18

Optimization of Large-Scale Culture Conditions for the Production of Cordycepin with *Cordyceps militaris* by Liquid Static Culture

Chao Kang,[1,2] **Ting-Chi Wen,**[1] **Ji-Chuan Kang,**[1] **Ze-Bing Meng,**[1] **Guang-Rong Li,**[1] **and Kevin D. Hyde**[3,4]

[1] *The Engineering and Research Center of Southwest Bio-Pharmaceutical Resources, Ministry of Education, Guizhou University, Guiyang, Guizhou 550025, China*

[2] *Institute of Biology, Guizhou Academy of Sciences, Guiyang, Guizhou 550009, China*

[3] *Institute of Excellence in Fungal Research, School of Science, Mae Fah Luang University, Chiang Rai 57100, Thailand*

[4] *Botany and Microbiology Department, College of Science, King Saud University, Riyadh 11442, Saudi Arabia*

Correspondence should be addressed to Ji-Chuan Kang; bcec.jckang@gzu.edu.cn

Academic Editor: Rajesh Jeewon

Cordycepin is one of the most important bioactive compounds produced by species of *Cordyceps sensu lato*, but it is hard to produce large amounts of this substance in industrial production. In this work, single factor design, Plackett-Burman design, and central composite design were employed to establish the key factors and identify optimal culture conditions which improved cordycepin production. Using these culture conditions, a maximum production of cordycepin was 2008.48 mg/L for 700 mL working volume in the 1000 mL glass jars and total content of cordycepin reached 1405.94 mg/bottle. This method provides an effective way for increasing the cordycepin production at a large scale. The strategies used in this study could have a wide application in other fermentation processes.

1. Introduction

Cordyceps militaris is an entomopathogenic fungus belonging to Ascomycota, Sordariomycetidae, Hypocreales, and Cordycipitaceae [1] and is one of the most important traditional Chinese medicinal mushrooms. *Cordyceps militaris* is the type species of *Cordyceps*, which internally parasitizes larva or pupa of lepidopteran insects and forms fruiting bodies on their insect hosts. *Cordyceps militaris* has long been recognized as a desirable alternative for natural *Ophiocordyceps sinensis* [2] as it has been given Chinese Licence number Z20030034/35. This is because the gathering of *Ophiocordyceps sinensis* is causing substantial reductions in populations [3]. *Cordyceps militaris* produces many bioactive compounds, including polysaccharides, cordycepin, adenosine, amino acid, organic selenium, ergosterol, sterols, cordycepic acid, superoxide dismutase (SOD), and multivitamins [4, 5].

Cordycepin (3′-deoxyadenosine), a nucleoside analog, was first isolated from *C. militaris* [6] and is one of the species most important biologically active metabolites. It has been regarded as a medicinal agent responsible for immunological regulation [7], anticancer [8], antifungus [9], antivirus [10], antileukemia [11, 12], and antihyperlipidemia [13] activities. Cordycepin is also a Phase I/II clinical stage drug candidate for treatment of refractory acute lymphoblastic leukemia (ALL) patients who express the enzyme terminal deoxynucleotidyl transferase (TdT) (http://www.ClinicalTrials.Gov verified by OncoVista, Inc., 2009).

In previous work, cordycepin has been synthesized by chemical [14, 15] and microbial fermentation using *C. militaris* [6] or *Aspergillus nidulans* [16, 17]. Solid-state fermentation [18, 19], submerged culture [4, 20–24], and surface liquid culture [25–27] have been used in microbial fermentation of cordycepin. Cordycepin obtained through chemistry

Optimization of Large-Scale Culture Conditions for the Production of Cordycepin with Cordyceps militaris
by Liquid Static Culture

159

pathways is hard to purify, and the cost is much higher than through biology fermentation. Thus a major need is to improve the biology methodology [28]. Fermentation time is too long and is difficult to achieve large scale production via solid-state fermentation [18, 19]. Productivity is generally low, the costs are high, and fermentation processes are easily contaminated in submerged culture in large fermenters [4, 20, 21, 29]. Productivity in surface culture techniques is higher as compared to other methods [29, 30] and the cost is lower [23]. New technologies, such as space mutation treatment and high-energy ion beam irradiation, have been used to obtain better Cordycepin producing, novel mutants of *C. militaris*. The resulting mutants were higher cordycepin produces, than the wild strain [30, 31]. Bu et al. [20] reported that the cordycepin in *C. militaris* was substantially increased by the elicitor of *Phytophthora* sp. Research result showed that glucose and yeast extract were effective media components for improved cordycepin production by *C. militaris* [32, 33]. There have been other studies using different culture conditions [21, 24, 25, 32], culture medium, and additives [4, 22–24, 26, 27] for the production of cordycepin via liquid culture. However, as far as we know, these reports studied cordycepin production in 250 mL or 500 mL Erlenmeyer flasks, and there have been no reports to improved cordycepin production using static liquid culture in 1000 mL glass jars. The latter process is a good way to scale up large scale cordycepin production from the laboratory to industry.

In this study, the effects of working volume, carbon sources, nitrogen sources, inorganic salts, growth factor, nucleoside analogue, and amino acid additions were studied in order to improve the cordycepin production by static liquid culture of *C. militaris* (strain CGMCC2459) in 1000 mL glass jars. The results suggested that the optimization medium conditions were helpful for improved large scale cordycepin production.

2. Materials and Methods

2.1. Microorganism and Seed Culture.
The isolate of *C. militaris* (strain CGMCC2459) used in the present study was collected from Mt. Qingcheng in Sichuan Province, China. The microorganism was maintained on potato dextrose agar (PDA) slants. Slants were incubated at 25°C for 7 days and then stored at 4°C. The seed culture was grown in a 250 mL flask containing 70 mL of basal medium (sucrose 20 g/L; peptone 20 g/L; KH_2PO_4 1 g/L; and $MgSO_4 \cdot 7H_2O$ 0.5 g/L) at 25°C on a rotary shaker incubator at 150 rev/min for 5 days [24].

2.2. Basal Medium and Static Culture of Glass Jars.
The basal medium composition for the fermentation was as follows: sucrose 20 g/L; peptone 20 g/L; KH_2PO_4 1 g/L; and $MgSO_4 \cdot 7H_2O$ 0.5 g/L. The pH was not adjusted, followed by autoclaving for 30 min on the 121°C. The static culture experiments were performed in 1000 mL glass jars (inner diameter 110 mm, height 150 mm) containing basal medium after inoculating with 10% (v/v; the biomass dry weight of seed culture is 54 mg/mL) of the seed culture. The culture was incubated at 25°C without moving for 35 days, and samples were collected at the end of the fermentation from the

glass jars for analyzing biomass dry weight and cordycepin production.

2.3. Static Culture Conditions.
The effects of factors affecting cell growth and the production of cordycepin by *C. militaris* were studied using a one-factor-at-a-time method for static culture. The effects of carbon sources on cordycepin production were studied by substituting carbon sources such as sucrose, lactose, soluble starch, and dextrin for glucose at 25°C for 35 days. Effects of nitrogen sources (yeast extract, beef extract, NH_4NO_3, $NaNO_3$, NH_4Cl, casein, and carbamide) and inorganic salts ($MgCl_2 \cdot 6H_2O$, $MgSO_4 \cdot 7H_2O$, KCl, $ZnSO_4$, $CaCl_2 \cdot 2H_2O$, $CaSO_4 \cdot 2H_2O$, $FeSO_4 \cdot 7H_2O$, and $K_2HPO_4 \cdot 3H_2O$) were also studied using static culture. Growth factors (Vitamin B_1 (VB_1), Vitamin B_6 (VB_6), Vitamin B_7 (VB_7), Vitamin B_{11} (VB_{11}), α-naphthylacetic acid (NAA), 3-Indoleacetlc acid (IAA), and 2,4-dichlorophenoxyacetic acid (2,4-D)) were supplemented for 10 mg/L in basal media. Nucleoside analogues (1 g/L) and amino acids (8 g/L) established in our previous study [23] as an initial concentration were separately added to the optimal concentration of carbon and nitrogen source, inorganic salts, and growth factors and cultivated at 25°C for 35 days. All experiments were carried out at triplicate, and mean of results is presented.

2.4. Analytical Methods.
Samples collected at 35 days from the glass jars were centrifuged at 2810 ×g for 20 min. The mycelium at the bottom of tubes was washed sufficiently with a large amount of distilled water and dried to a constant dry weight at 55°C.

For analysis of extracellular cordycepin, the resulting culture filtrate was obtained by centrifugation at 2810 ×g for 20 min. The supernatant was filtered through a 0.45 μm membrane and the filtrate was analyzed by HPLC (1100 series, Agilent Technology, USA). Accurate quantities of cordycepin (Sigma, USA) were dissolved in distilled water, to give various concentrations for calibration. The mobile phase was 10 mmol/L KH_2PO_4, which was dissolved in methanol/distilled water (6 : 94). Elution was performed at a flow rate of 1.0 mL/min with column temperature at 45°C and UV wavelength of 259 nm. Mean values were computed from triplicate samples.

2.5. Plackett-Burman Design.
The Plackett-Burman design, an effective technique for medium-component optimization [35, 36], was used to select factors that significantly influenced hydrogen production. Sucrose (X_1), peptone (X_2), $K_2HPO_4 \cdot 3H_2O$ (X_4), $MgSO_4 \cdot 7H_2O$ (X_5), and VB_1 (X_6) were investigated as key ingredients affecting cordycepin production. Based on the Plackett-Burman design, a 15-run was applied to evaluate eleven factors (including two virtual variables). Each factor was prepared in two levels: −1 for low level and +1 for high level. Table 1 illustrates the variables and their corresponding levels used in the experimental design. The values of two levels were set according to our preliminary experimental results. The Plackett-Burman design and the response value of cordycepin production are shown in Table 2.

Table 1: Range of different factors investigated with Plackett-Burman design.

Symbol	Variables	Experimental value	
		Low (−1)	High (+1)
X_1	Sucrose (g/L)	20	25
X_2	Peptone (g/L)	20	25
X_3	Virtual 1	−1	1
X_4	$K_2HPO_4 \cdot 3H_2O$ (g/L)	1	1.25
X_5	VB_1 (g/L)	10	12.5
X_6	$MgSO_4 \cdot 7H_2O$ (g/L)	1	1.25
X_7	Virtual 2	−1	1

2.6. Response Surface Methodology.

Response surface methodology using a central composite design was applied to batch cultures of *C. militaris*, for identifying the effects of process variables [35, 36]. In this study, the basic nutrient (carbon sources, nitrogen sources, inorganic salts, and growth factors) and additives (amino acid, nucleoside analogue) were studied for cordycepin production using static liquid culture. In the first test, a three-factor, five-level central composite design with 20 runs was employed. Tested variables (sucrose, $K_2HPO_4 \cdot 3H_2O$, and $MgSO_4 \cdot 7H_2O$) were denoted as X_1, X_4, and X_6, respectively, and each of them was assessed at five different levels, combining factorial points (−1, −1), axial points (−1.6818, +1.6818), and central point (0), as shown in Table 3. Based on the above results, another test, a three-factor, five-level central composite design with 20 runs was employed. Tested variables (amino acid, nucleoside analogue, and culture time) were denoted as A, B, and C, respectively, and each of them was assessed at five different levels, combining factorial points (−1, +1), axial points (−1.6818, +1.6818), and central point (0), as shown in Table 4.

2.7. Statistical Analysis.

Dry weight and cordycepin production are expressed as means ± SD. An analysis of variance (ANOVA) followed by Tukey's test was applied for multiple comparisons of significant analyses at $P < 0.05$. Statistical data analyses were performed in SPSS version 17.0 software packet. Design-Expert Version 8.0.5b software package (Stat-Ease Inc., Minneapolis, USA) was used for designing experiments as well as for regression and graphical analysis of the experimental data obtained.

3. Results and Discussion

3.1. Effects of Working Volume on the Biomass and Cordycepin Production.

Dissolved oxygen concentration is the key factor in the medium for cell growth and metabolite biosynthesis [21]. Dissolved oxygen does not only have an important function in the respiratory chain, but also in metabolite composition [37, 38]. A previous study showed that the highest cordycepin production and productivity were obtained at lower dissolved oxygen levels [21]. Masuda et al. [25] also reported that a lower medium depth was most efficient for cordycepin production in *C. militaris* by surface culture.

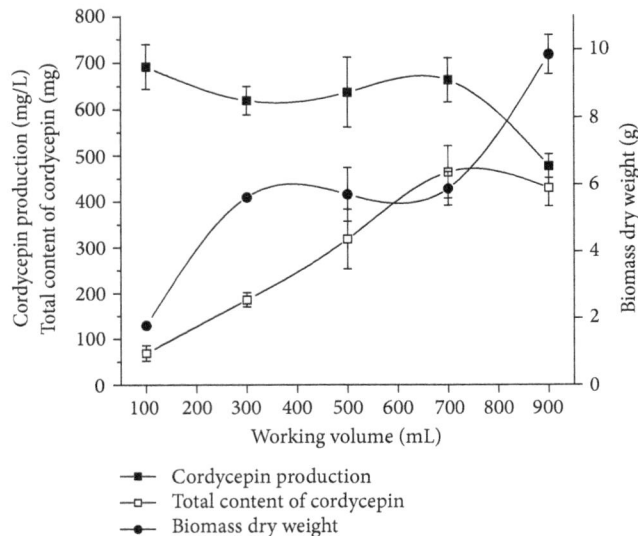

Figure 1: Effects of working volume on the production of cordycepin, total production of cordycepin, and biomass dry weight (total content of cordycepin (mg) = cordycepin production (mg/L) × working volume (mL)).

In this study, we tried to establish the most efficient working volume of medium for improved cordycepin production. Cultures of *C. militaris* were prepared at the working volumes of 100 to 900 mL (corresponding to a medium depth of 1.26 to 11.31 cm). As shown in Figure 1, cordycepin production reduced gradually with increasing working volume of the medium, from 100 to 700 mL. However, there was no significant difference in cordycepin production in different working volumes. Obviously, the highest working volume of 900 mL did not help in cordycepin production. The result indicates that there is an upper dissolved oxygen limit in the medium for cordycepin production [21]. Lower working volumes result in higher cordycepin productivity, with the highest peak (463.33 ± 56.72 mg of cordycepin) produced at using 700 mL of media. Changes in biomass values were small (between 300 and 700 mL) because of the restricted area and thickness of the mycelial mat. In order to obtain higher cordycepin production, the most effective medium amount was 700 mL (corresponding to an 8.8 cm medium depth) and used as the media volume for next experiment.

3.2. Effects of Carbon and Nitrogen Sources on Cordycepin Production.

To find a suitable carbon source for *C. militaris* cordycepin production we added various carbon sources at a concentration of 20 g/L to the sugar-free basal medium. Glucose was previously found to be an excellent precursor of cordycepin production [39]. However, as shown in Figure 2(a), sucrose and lactose proved to be better carbon sources for cordycepin production than glucose in this study. Cordycepin production reached 843.63 ± 66.70 mg/L of sucrose and 823.72 ± 85.64 mg/L of lactose, respectively. Therefore, sucrose was selected as the main carbon source in the remaining experiment.

Optimization of Large-Scale Culture Conditions for the Production of Cordycepin with Cordyceps militaris
by Liquid Static Culture

161

TABLE 2: Plackett-Burman design and response values.

Runs	Experimental value							Y (mg/L) Cordycepin production
	X_1	X_2	X_3	X_4	X_5	X_6	X_7	
1	1	−1	1	−1	−1	−1	1	812.36 ± 26.83
2	1	1	−1	1	−1	−1	−1	1395.18 ± 8.4
3	−1	1	1	−1	1	−1	−1	900.25 ± 10.29
4	1	−1	1	1	−1	1	−1	802.45 ± 45.43
5	1	1	−1	1	1	−1	1	1097.66 ± 25.57
6	1	1	1	−1	1	1	−1	845.87 ± 24.94
7	−1	1	1	1	−1	1	1	786.35 ± 7.61
8	−1	−1	1	1	1	−1	1	805.08 ± 29.1
9	−1	−1	−1	1	1	1	−1	920.48 ± 16.21
10	1	−1	−1	−1	1	1	1	694.01 ± 79.51
11	−1	1	−1	−1	−1	1	1	497.28 ± 4.44
12	−1	−1	−1	−1	−1	−1	−1	592.83 ± 16.13
13	0	0	0	0	0	0	0	1134.14 ± 2.59
14	0	0	0	0	0	0	0	1100.21 ± 0.08
15	0	0	0	0	0	0	0	1133.56 ± 1.85

TABLE 3: Factors and levels of central composite design for carbon sources and inorganic salts.

Symbol	Variables	Code level				
		−1.6818	−1	0	1	1.6818
X_1	Sucrose (g/L)	3.1821	10	20	30	36.8179
X_4	$K_2HPO_4 \cdot 3H_2O$ (g/L)	0.1591	0.5	1	1.5	1.8409
X_6	$MgSO_4 \cdot 7H_2O$ (g/L)	0.1591	0.5	1	1.5	1.8409

TABLE 4: Factors and levels of central composite design for amino acid, nucleoside analogue, and time.

Symbol	Variables	Code level				
		−1.6818	−1	0	1	1.6818
A	Hypoxanthine (g/L)	0.53	1	5	9	10.53
B	L-alanine (g/L)	5.27	8	12	16	18.72
C	Culture time (days)	−0.09	4	10	16	20.09

In previous work, nitrogen showed a regulating role important in cordycepin production and had two effects [40]. One effect was negative since, in excess, N promoted a faster mycelial growth and consequently diverted the source of carbon toward energy and biomass production. The other effect was positive because a moderate input contributed to the maintenance of citric acid productive biomass [40]. To investigate the effect of nitrogen sources on cordycepin production in *C. militaris*, various compounds containing nitrogen (inorganic and organic nitrogen) were added individually to nitrogen free basal medium at a concentration of 20 g/L. Among the 8 nitrogen sources tested, peptone, yeast extract, beef extract, casein, and NH_4NO_3 were favorable to the cordycepin production (Figure 2(b)). Organic nitrogen was advantageous to both growth and biosynthesis of metabolites. The result is consistent with the experimental data reported [18] and showed that maximum cordycepin production resulted when the peptone was used as a nitrogen source.

3.3. Effects of Inorganic Salt and Growth Factor on the Cordycepin Production.
Inorganic ion was one of the most important nutrition components of medium for the mycelial growth [41]. In order to investigate the effects of inorganic salt for the cordycepin production in *C. militaris*, we tested nine types (at 1 g/L) of inorganic salts (Figure 3(a)). Media with only 20 g/L glucose and 20 g/L peptone were used as the control. The

highest cordycepin production (1120.30 ± 105.28 mg/L) by *C. militaris* was observed in medium, when $K_2HPO_4 \cdot 3H_2O$ was used as an inorganic salt. KH_2PO_4, $MgSO_4 \cdot 7H_2O$, KCl, and $MgCl_2 \cdot 6H_2O$ were also useful inorganic salts. At last, $MgSO_4 \cdot 7H_2O$ and $K_2HPO_4 \cdot 3H_2O$ were recognized as favorable bioelements for production of cordycepin.

Growth factor is essential for growth response and metabolite production [42]. In order to find the optimal growth factor for cordycepin production, *C. militaris* was cultured in a basal medium with different vitamins and plant growth hormones in static liquid culture. Cordycepin production increased in media with added 10 mg/L of VB_1, NAA, and VB_{11} (Figure 3(b)). Maximum cordycepin production (1159.34 ± 109.01 mg/L) occurred when VB_1 was used as the growth factor.

3.4. Screening of Important Variables Using Plackett-Burman Design.
The data (Table 2) indicated wide variation in cordycepin production in the 15 tests. The data suggested that process optimization is important for improving the efficiency of cordycepin production. Analysis of the regression coefficients and t values of 7 factors (Table 5) showed that X_1, X_2, X_4, and X_5 had positive effects on cordycepin production. X_6 had negative effects. The variable affects with a confidence level above 95% are considered as significant factors. Based on these results, three factors (X_1, sucrose; X_4, $K_2HPO_4 \cdot 3H_2O$;

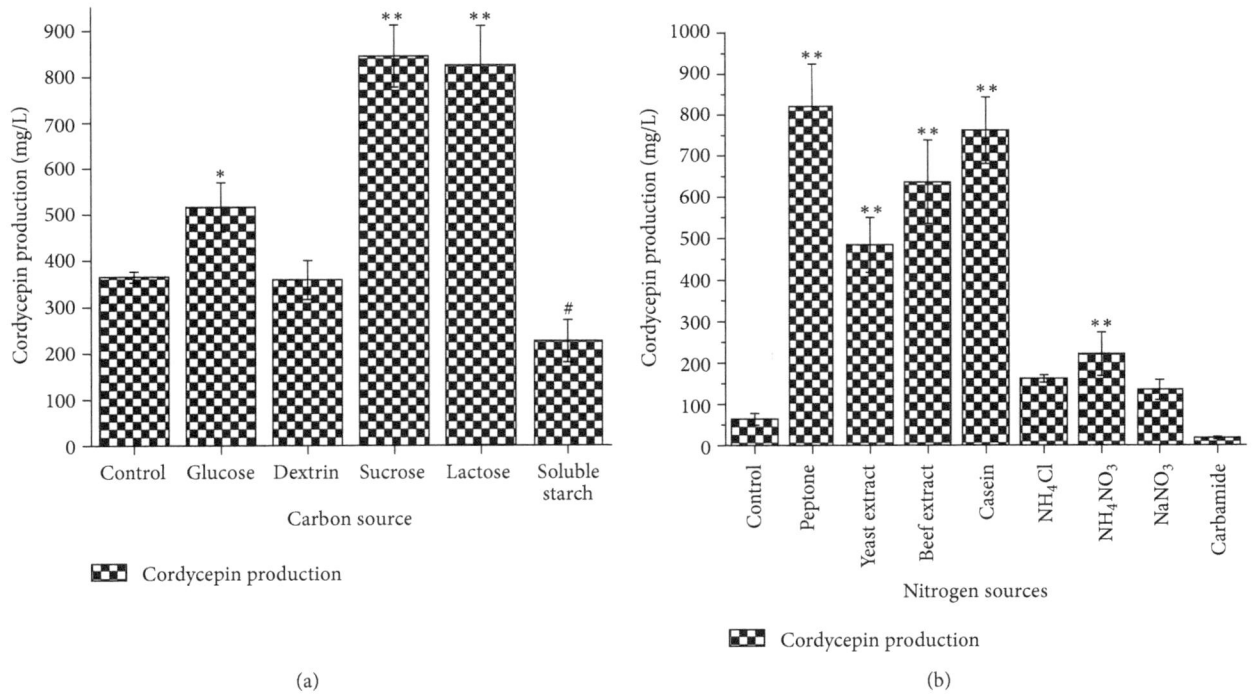

(a)

(b)

FIGURE 2: Effects of carbon sources and nitrogen sources on the production of cordycepin: carbon sources (a); nitrogen sources (b); *5% significance level (test group versus control group); **1% significance level (test group versus control group); $^{#}$5% significance level (control group versus test group).

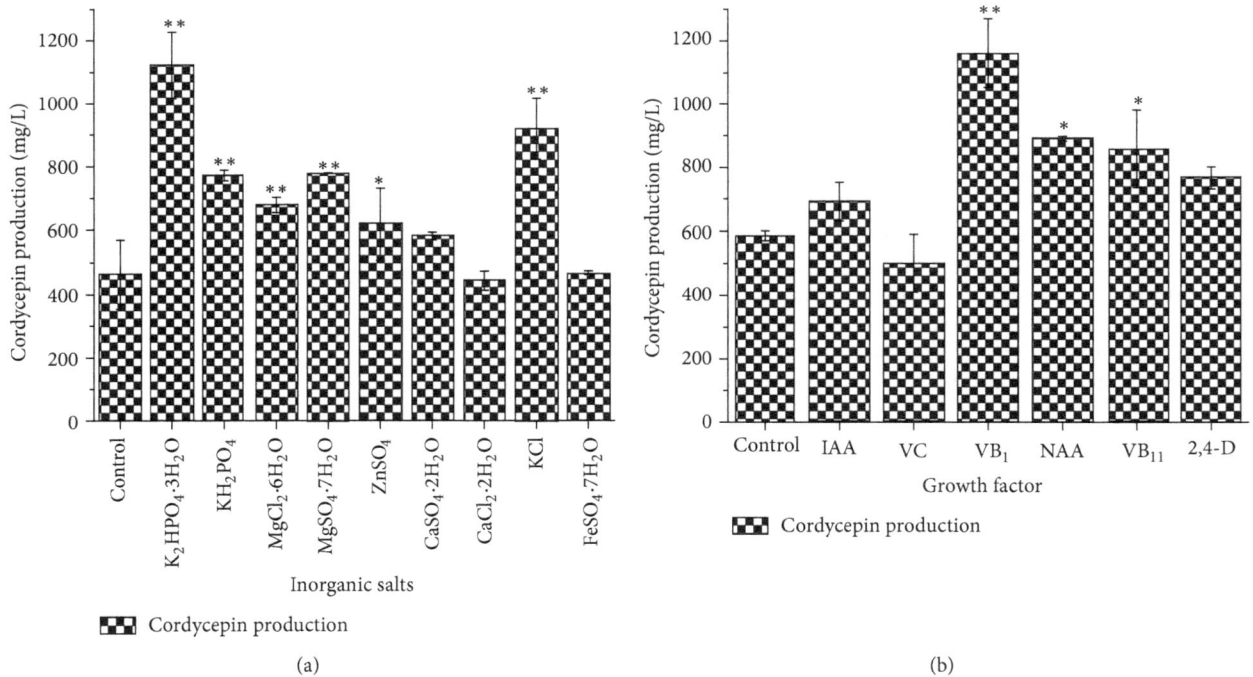

(a)

(b)

FIGURE 3: Effects of inorganic salt and growth factors on the production of cordycepin: inorganic salt (a); growth factors (b); *5% significance level (test group versus control group); **1% significance level (test group versus control group).

Optimization of Large-Scale Culture Conditions for the Production of Cordycepin with Cordyceps militaris by Liquid Static Culture

163

TABLE 5: Results of regression analysis of Plackett-Burman design.

| Symbol | Regression analysis | | | | |
	Effect	Coefficient	Standard error	T	P
		845.82	32.86	25.74	0.000**
X_1	190.88	95.44	32.86	2.90	0.027*
X_2	149.23	74.62	32.86	2.27	0.064
X_3	−40.85	−20.42	32.86	−0.62	0.557
X_4	244.10	122.05	32.86	3.71	0.010*
X_5	62.81	31.41	32.86	0.96	0.376
X_6	−176.15	−88.08	32.86	−2.68	0.037*
X_7	−127.39	−63.69	32.86	−1.94	0.101
Ct Pt		276.82	73.48	3.77	0.009**

*5% significance level; **1% significance level; X_1–X_7 are symbols shown in Table 1.

TABLE 6: Experimental design and responses of the central composite design for carbon sources and inorganic salts.

| Run | Variables Code | | | Y (mg/L) Cordycepin production | Run | Variables Code | | | Y (mg/L) Cordycepin production |
	X_1	X_4	X_6			X_1	X_4	X_6	
1	−1	−1	−1	1080.55 ± 109.69	11	0	0	0	1399.43 ± 124.44
2	1	−1	−1	1359.48 ± 12.61	12	0	0	0	1415.43 ± 42.13
3	−1	1	−1	1158.59 ± 12.15	13	0	0	0	1487.42 ± 16.38
4	1	1	−1	1289.90 ± 47.00	14	0	0	0	1409.43 ± 155.22
5	−1	−1	1	980.55 ± 34.72	15	−1.6818	0	0	876.91 ± 16.69
6	1	−1	1	1097.48 ± 52.42	16	1.6818	0	0	1290.00 ± 14.71
7	−1	1	1	987.95 ± 2.89	17	0	−1.6818	0	1110.57 ± 157.63
8	1	1	1	1117.38 ± 116.84	18	0	1.6818	0	1344.66 ± 65.54
9	0	0	0	1485.38 ± 12.19	19	0	0	−1.6818	958.39 ± 224.14
10	0	0	0	1333.48 ± 94.22	20	0	0	1.6818	958.00 ± 75.82

and X_6, MgSO$_4$·7H$_2$O) were considered as significant for cordycepin production by static liquid culture methodology.

3.5. Optimization by Response Surface Methodology for Carbon Sources and Inorganic Salts. In order to evaluate the influence of medium component on cordycepin production, sucrose, K$_2$HPO$_4$·3H$_2$O, and MgSO$_4$·7H$_2$O should be examined. The levels of variables for central composite design experiments were selected according to the above results of Plackett-Burman design. Table 6 shows the detailed experimental design and results. Regression analysis was performed to fit the response function (cordycepin production) with the experimental data. From the variables obtained (Table 6), the model is expressed by (1), which represents cordycepin production (Y_1) as a function of sucrose (X_1), K$_2$HPO$_4$·3H$_2$O (X_4), and MgSO$_4$·7H$_2$O (X_6) concentrations:

$$Y_1 = 1419.68 - 98.95X_1 + 31.45X_4 - 51.68X_6$$
$$- 16.89X_1X_4 - 20.48X_1X_6 + 2.36X_4X_6 \quad (1)$$
$$- 106.03X_1^2 - 55.06X_4^2 - 150.31X_6^2.$$

Results of F-test analysis of variance (ANOVA) showed that the regression was statistically significant at 95% and 99% confidence levels (Table 7). The "F value" of the model was 9.21, and the value of "Prob > F" < 0.01 indicated that the model was significant. In this case, linear terms of X_1 and quadratic terms of X_1^2, X_4^2, X_6^2 were significant of model terms for cordycepin production. The "*Lack of Fit F value*" of 0.0903 implied that the "*Lack of Fit*" was not significant relative to the pure error ($P > 0.05$). The Pred-R^2 of 0.3183 was not as close to the Adj-R^2 of 0.7954 as one might normally expect. The result suggested that some factors were not considered in the model. However, the "Adeq Precision" of 8.173 indicated that the model was adequate for prediction production of cordycepin.

The response surface plot obtained from (1) is shown in Figure 4. It is evident that cordycepin production reached its maximum at a combination of coded level (X_1, sucrose, level 0.47; X_4, K$_2$HPO$_4$·3H$_2$O, level 0.21; X_6, MgSO$_4$·7H$_2$O, level −0.20) when using canonical analysis of the Design-Expert Version 8.0.5b software package. The model predicted a maximum response of 1451.43 mg/L cordycepin production at levels of sucrose 24.7 g/L, K$_2$HPO$_4$·3H$_2$O 1.11 g/L, and MgSO$_4$·7H$_2$O 0.90 g/L as optimized medium components.

3.6. Effects of Nucleoside Analogue and Amino Acid on the Production of Cordycepin. Chassy and Suhadolnik [43] reported that adenine and adenosine were precursors for cordycepin synthesis. Amino acids were regarded as the best substance for improved cordycepin production [4, 23]. Based on these results, among 10 different kinds of nucleoside

TABLE 7: ANOVA for response surface quadratic polynomial model for carbon sources and inorganic salts.

Source	Sum of quares	df	Mean Square	F-value	P-value Prob $> F$
Model	$6.507E + 005$	9	72300.77	9.21	0.0009^{**}
X_1-X_1	$1.337E + 005$	1	$1.337E + 005$	17.03	0.0021^{**}
X_4-X_4	13504.30	1	13504.30	1.72	0.2190
X_6-X_6	36477.48	1	36477.48	4.65	0.0565
X_1X_4	2282.55	1	2282.55	0.29	0.6016
X_1X_6	3356.89	1	3356.89	0.43	0.5279
X_4X_6	44.38	1	44.38	$5.653E - 003$	0.9416
$X_1{}^2$	$1.620E + 005$	1	$1.620E + 005$	20.63	0.0011^{**}
$X_4{}^2$	43685.18	1	43685.18	5.56	0.0400^{**}
$X_6{}^2$	$3.256E + 005$	1	$3.256E + 005$	41.47	$<0.0001^{**}$
Residual	78513.73	10	7851.37		
Lack of Fit	61670.32	5	12334.06	3.66	0.0903
Pure Error	16843.41	5	3368.68		
Cor Total	$7.292E + 005$	19			

$R^2 = 0.8923$; CV $= 7.34\%$; Pred-$R^2 = 0.3183$; Adj-$R^2 = 0.7954$; Adeq Precision $= 8.173$; $^*5\%$ significance level; $^{**}1\%$ significance level.

TABLE 8: Experimental design and responses of the central composite design for amino acid, nucleoside analogue, and time.

Run	Variables Code A	B	C	Y (mg/L) Cordycepin production	Run	Variables Code A	B	C	Y (mg/L) Cordycepin production
1	−1	−1	−1	1383.01 ± 41.53	11	0	0	0	2041.25 ± 54.70
2	1	−1	−1	1422.52 ± 39.41	12	0	0	0	2020.97 ± 73.70
3	−1	1	−1	1216.88 ± 8.69	13	0	0	0	1998.18 ± 49.48
4	1	1	−1	1857.51 ± 164.86	14	0	0	0	2068.60 ± 72.79
5	−1	−1	1	1216.87 ± 253.38	15	−1.6818	0	0	1590.14 ± 222.14
6	1	−1	1	1111.18 ± 170.50	16	1.6818	0	0	1573.90 ± 776.16
7	−1	1	1	1536.05 ± 75.17	17	0	−1.6818	0	1636.44 ± 65.23
8	1	1	1	851.70 ± 17.01	18	0	1.6818	0	1527.98 ± 177.46
9	0	0	0	2073.27 ± 65.85	19	0	0	−1.6818	1211.14 ± 82.58
10	0	0	0	1743.09 ± 14.81	20	0	0	1.6818	676.97 ± 142.74

analogue were supplemented for 1 g/L in this study. As shown in Figure 5(a), cordycepin production increased obviously in the medium with hypoxanthine, thymine, and thymidine additives. The highest production of cordycepin was achieved, when hypoxanthine was used as the nucleoside analogue. Hypoxanthine's molecular structure is similar to purine bases found in cordycepin. Substituent on purine bases structure is –OH on hypoxanthine rather than –NH$_2$. The –OH should be replaced in metabolic pathways. In addition, among 14 different amino acids were tested for 8 g/L. As shown in Figure 5(b), L-alanine can improve cordycepin production. Previous research showed that adenine, adenosine, and glycine were good additives for increased cordycepin production [4, 23, 26, 27]. L-alanine may be an important nutritional element for *C. militaris* or component of cordycepin production. Hypoxanthine and L-alanine were the best additives to promote cordycepin production in this study.

3.7. Optimization by Response Surface Methodology for Amino Acid, Nucleoside Analogue, and Fermentation Time. Similarly, central composite design was also applied to study

the significant factors (hypoxanthine, L-alanine, and culture time) and their optimal levels. Figure 6 shows the morphological characteristics of *C. militaris* in 1000 mL glass jars after fermentation by static liquid fermentation. Table 8 shows the detailed experimental design and results. Regression analysis was performed to fit the response function (cordycepin production) with the experimental data. From the variables obtained (Table 9), the model was expressed by (2), which represented cordycepin production (Y_2) as a function of hypoxanthine (A), L-alanine (B), and culture time (C, time), concentrations:

$$Y_2 = 1991.13 - 10.05A + 10.70B - 151.02C$$
$$+ 2.81AB - 183.77AC - 26.14BC \quad (2)$$
$$- 146.09A^2 - 146.03B^2 - 371.65C^2.$$

Results of F-test analysis of variance (ANOVA) showed that the regression was statistically significant at 95% and 99% confidence level (Table 9). The "F value" of the model was 11.91, and the value of "Prob $> F$" < 0.01 indicated that the model was significant. In this case, linear terms

Optimization of Large-Scale Culture Conditions for the Production of Cordycepin with Cordyceps militaris by Liquid Static Culture

165

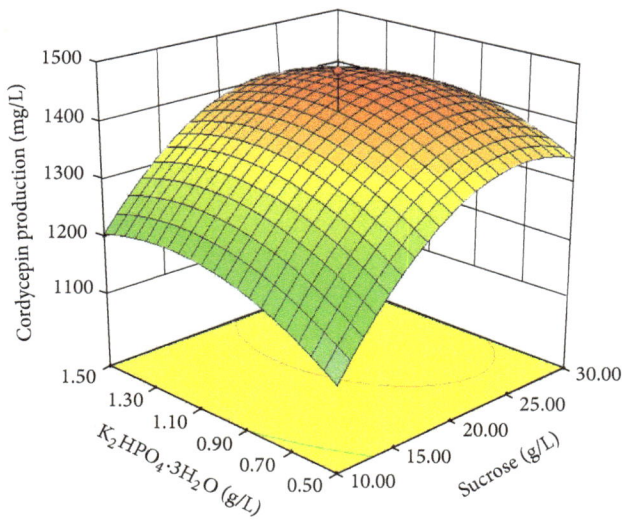

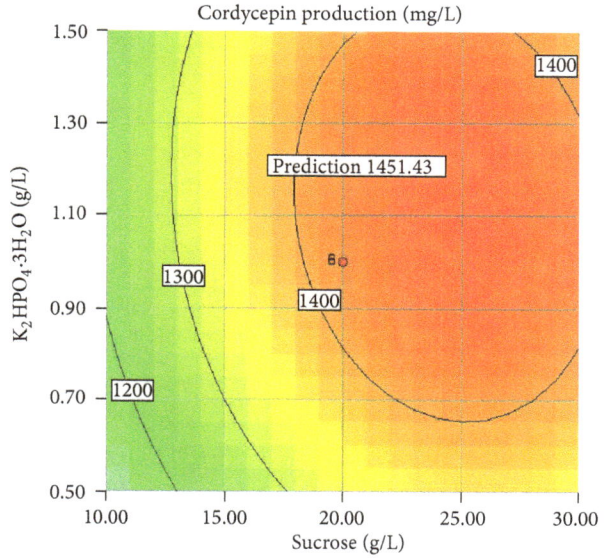

(a)

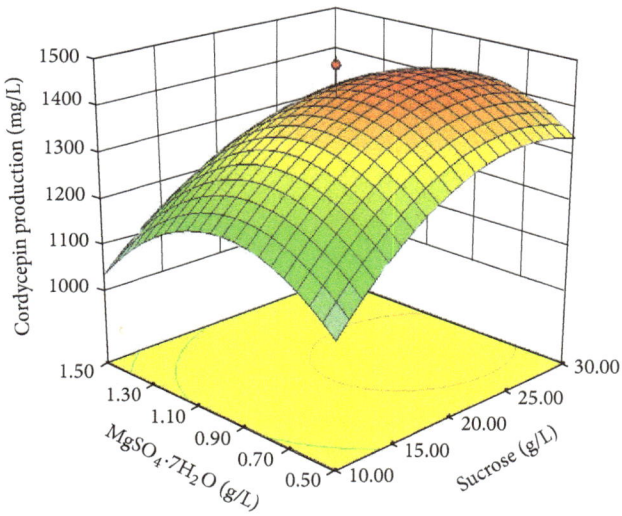

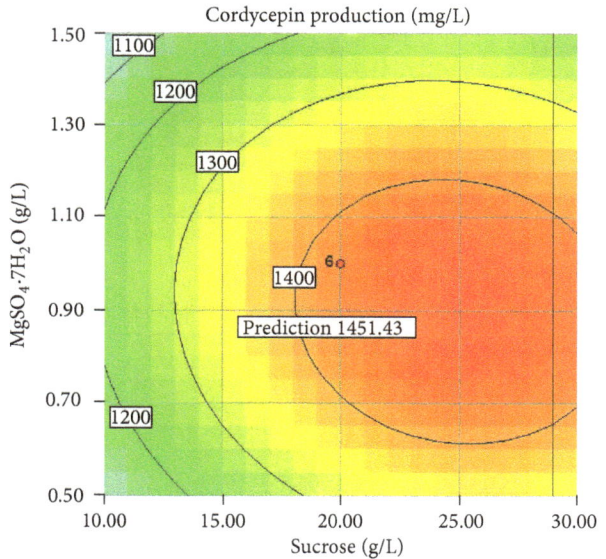

(b)

FIGURE 4: Continued.

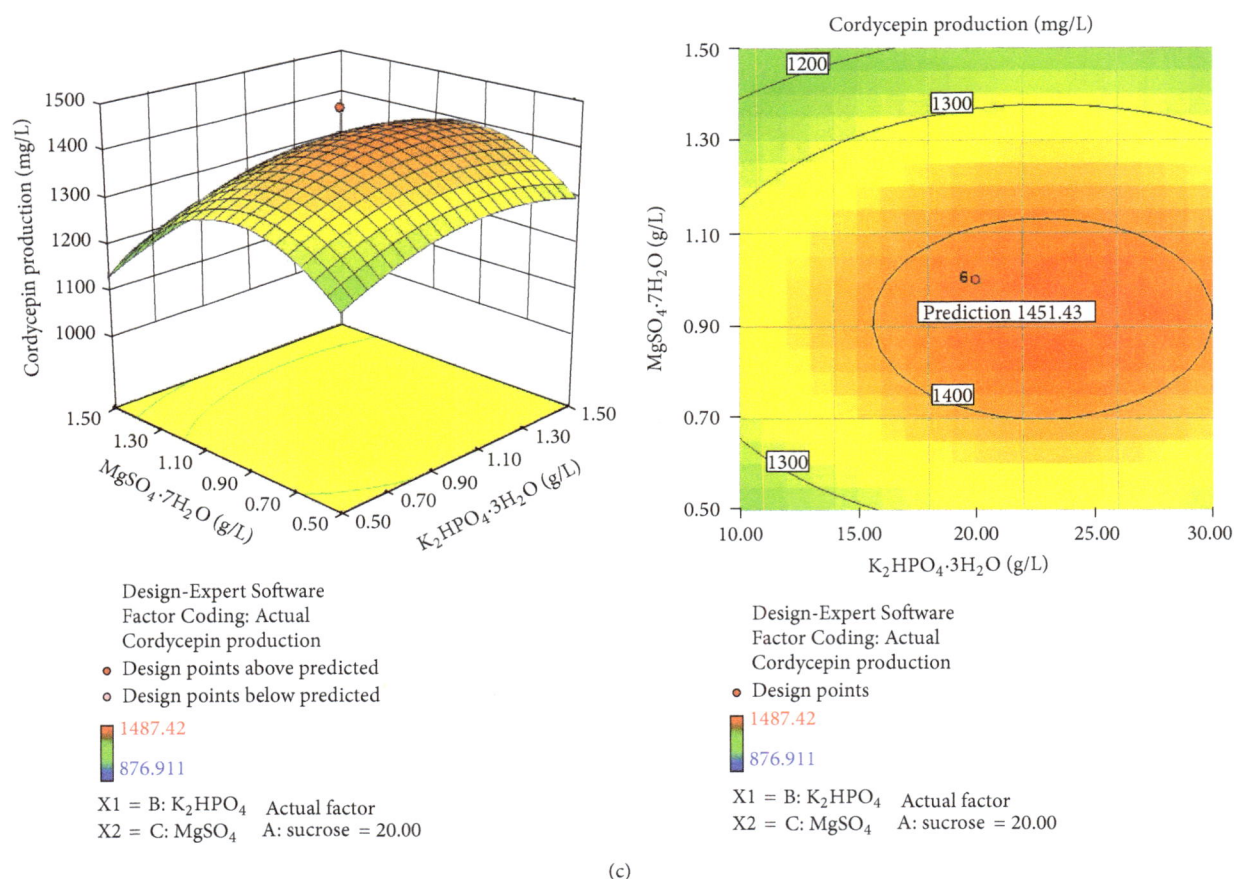

Design-Expert Software
Factor Coding: Actual
Cordycepin production
- Design points above predicted
- Design points below predicted

■ 1487.42
■ 876.911

X1 = B: K₂HPO₄   Actual factor
X2 = C: MgSO₄   A: sucrose = 20.00

Design-Expert Software
Factor Coding: Actual
Cordycepin production
- Design points

■ 1487.42
■ 876.911

X1 = B: K₂HPO₄   Actual factor
X2 = C: MgSO₄   A: sucrose = 20.00

(c)

FIGURE 4: Three-dimensional response surface plots and two-dimensional contour plots for cordycepin production by *C. militaris* (strain CGMCC2459) showing variable interactions of (a) sucrose and $K_2HPO_4\cdot3H_2O$; (b) sucrose and $MgSO_4\cdot7H_2O$; (c) $K_2HPO_4\cdot3H_2O$ and $MgSO_4\cdot7H_2O$.

TABLE 9: ANOVA for response surface quadratic polynomial model for amino acid, nucleoside analogue, and time.

Source	Sum of quares	df	Mean Square	F-value	P-value Prob > F
Model	$2.899E+006$	9	$3.221E+005$	11.91	0.0003**
A-A	1378.39	1	1378.39	0.051	0.8260
B-B	1564.06	1	1564.06	0.058	0.8148
C-C	$3.115E+005$	1	$3.115E+005$	11.51	0.0068*
AB	63.06	1	63.06	$2.331E-003$	0.9624
AC	$2.702E+005$	1	$2.702E+005$	9.99	0.0102*
BC	5468.49	1	5468.49	0.20	0.6626
A^2	$3.076E+005$	1	$3.076E+005$	11.37	0.0071**
B^2	$3.073E+005$	1	$3.073E+005$	11.36	0.0071**
C^2	$1.990E+006$	1	$1.990E+006$	73.58	<0.0001**
Residual	$2.705E+005$	10	27053.63		
Lack of Fit	$1.928E+005$	5	38561.98	2.48	0.1707
Pure Error	77726.41	5	15545.28		
Cor Total	$3.169E+006$	19			

$R^2 = 0.9146$; CV = 10.70%; Pred-R^2 = 0.4162; Adj-R^2 = 0.8378; Adeq Precision = 11.222; *5% significance level; **1% significance level.

Optimization of Large-Scale Culture Conditions for the Production of Cordycepin with Cordyceps militaris
by Liquid Static Culture

167

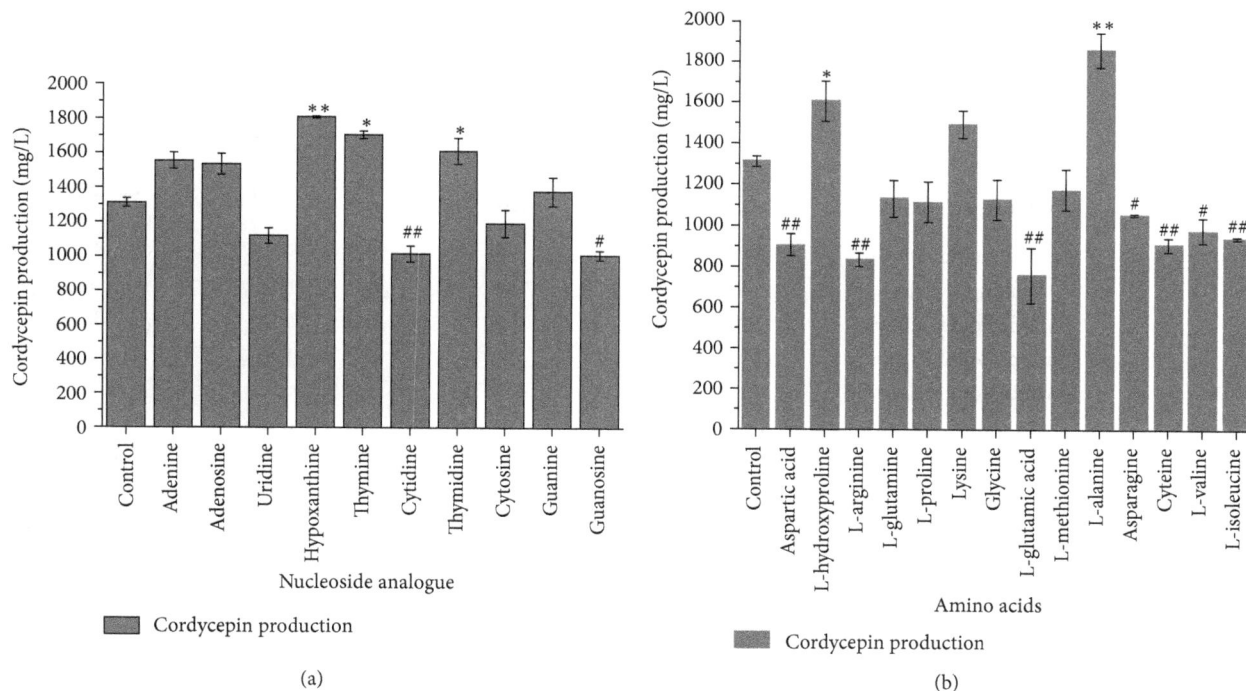

FIGURE 5: Effects of nucleoside analogue and amino acid on the production of cordycepin: nucleoside analogue (a); amino acid (b); *5% significance level (test group versus control group); **1% significance level (test group versus control group); #5% significance level (control group versus test group); ##1% significance level (control group versus test group).

of C; interactive terms of AC; and quadratic terms of A^2, B^2, and C^2 were significant in model terms for cordycepin production. The "*Lack of Fit F value*" of 0.1707 implied that the "*Lack of Fit*" was not significant relative to the pure error ($P > 0.05$). The Pred-R^2 of 0.4162 was not as close to the Adj-R^2 of 0.8378 as one might normally expect. The result suggested that some factors were not also considered in the model. However, the "Adeq Precision" of 11.222 indicated that the model was adequate for prediction production of cordycepin.

The response surface plot obtained from (2) is shown in Figure 7. It is evident that cordycepin production reached its maximum with a combination of coded level (A, hypoxanthine, level 0.11; B, L-alanine, level 0.06; C, time, level -0.23) by canonical analysis of the Design-Expert Version 8.0.5b software package. The model predicted a maximum response of 2008.48 mg/L cordycepin production at levels of hypoxanthine 5.45 g/L, L-alanine 12.23 g/L, and time 8.6 days (in the practical test 8 days) as optimized medium components.

In previous work, the orthogonal design method [44–46], Box-Behnken design [34, 47], and central composite design [31] were used to optimize culture conditions for cordycepin production by *Cordyceps* sp. These experimental designs have been successfully used to optimize medium for the mycelial growth and microbial metabolite production in liquid culture processes. In this study, static liquid culture conditions are optimized for the cordycepin production using response surface methodology and are an effective way to

enhance the productivity of cordycepin and biomass in *C. militaris*.

3.8. Verification Experiments and Batch Culture. Based on the results of response surface methodology, the optimized medium was prepared as follows: peptone 20 g/L; sucrose 24.7 g/L; $K_2HPO_4 \cdot 3H_2O$ 1.11 g/L; $MgSO_4 \cdot 7H_2O$ 0.90 g/L; VB_1 10 mg/L; hypoxanthine 5.45 g/L; and L-alanine 12.23 g/L. Five experiments were performed to confirm the above optimal culture requirements. The data were 2011.15 mg/L, 2000.69 mg/L, 1989.22 mg/L, 1969.6 mg/L, and 2061.37 mg/L, respectively. The average cordycepin production was 2006.41 ± 34.37 mg/L. The experimental values were particularly close to the predicted values (2008.48 mg/L). The result confirmed the model suited the predictive of hyperproduction of cordycepin by *C. militaris* in static liquid culture. Batch culture was carried for cordycepin production under optimized culture conditions (Figure 8).

3.9. In Vitro Cordycepin Production Using Liquid Culture in Other Studies. The highest report for cordycepin production was 14300 mg/L by Masuda et al. [29] (Table 10). In our experiment, cordycepin production at 2008.48 mg/L was lower. However, a maximum total content of cordycepin (1405.94 mg) was achieved in our study. This is a second higher report of cordycepin production in one single fermenter. The results showed that the culture conditions will provide an effective way for increasing cordycepin production.

TABLE 10: Cordycepin production in the medium by liquid culture in different studies.

No.	Methodology	Working volume of the medium v/v (mL/mL)	Cordycepin production (mg/L)	Total content of cordycepin in one bottle (mg)	References
1	Submerged culture	50/250	245.7	12.5	Mao et al., [32]
2	Submerged culture	50/250	420.5	21.03	Mao and Zhong [21]
3	Surface liquid culture	100/500	640	64	Masuda et al., [25]
4	Shaking + Static	100/250	2214.5	221.45	Shih et al., [34]
5	Surface liquid culture	100/500	2500	250	Masuda et al., [26]
6	Surface liquid culture	100/500	3100	310	Das et al., [30]
7	Surface liquid culture	100/500	8570	857	Das et al., [27]
8	Submerged culture	100/500	1644.21	164.42	Wen et al., [23]
9	Dark + Shaking	100/500	1015	101.5	Kang et al., [24]
10	Surface liquid culture	150/500	14300	2145	Masuda et al., [29]
11	**Static liquid culture**	**700/1000**	**2008.48**	**1405.94**	**In this study**

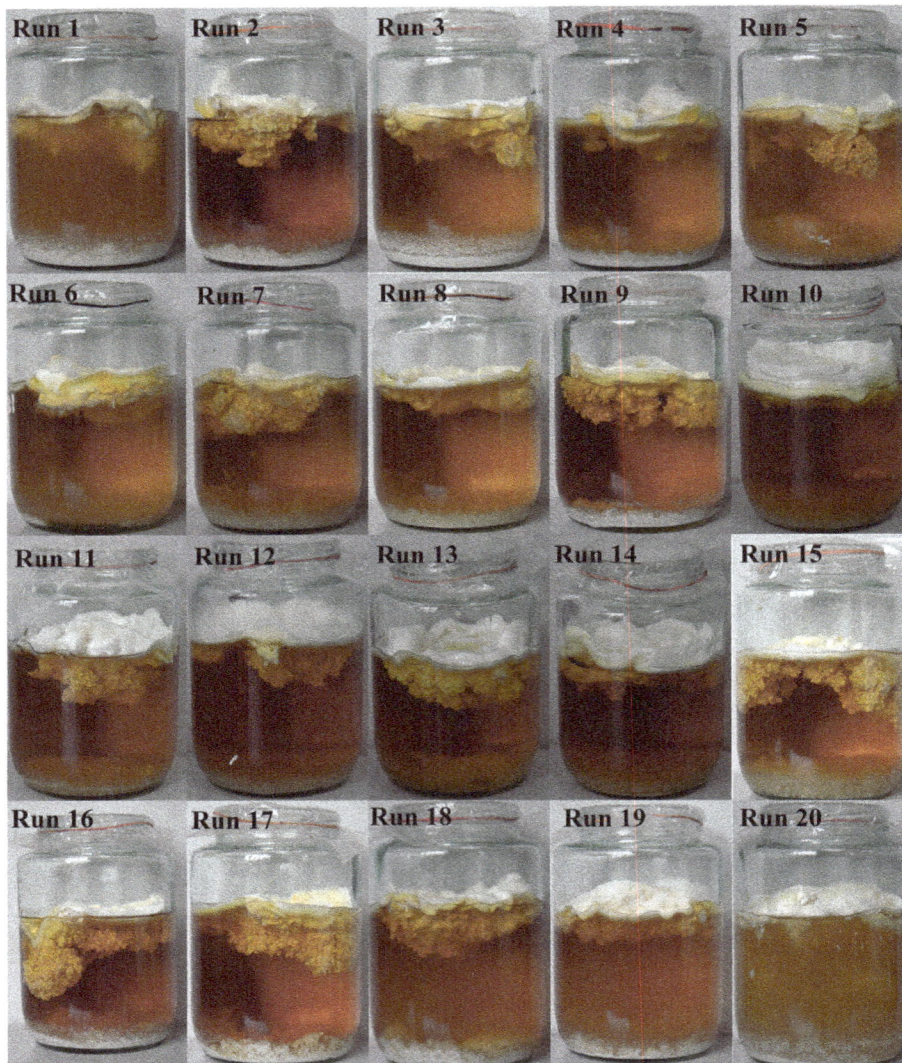

FIGURE 6: Morphology of *C. militaris* (strain CGMCC2459) in 700/1000 mL glass jars at the end of the fermentation process by response surface methodology: symbols in photos indicated 20 runs.

Optimization of Large-Scale Culture Conditions for the Production of Cordycepin with Cordyceps militaris by Liquid Static Culture

169

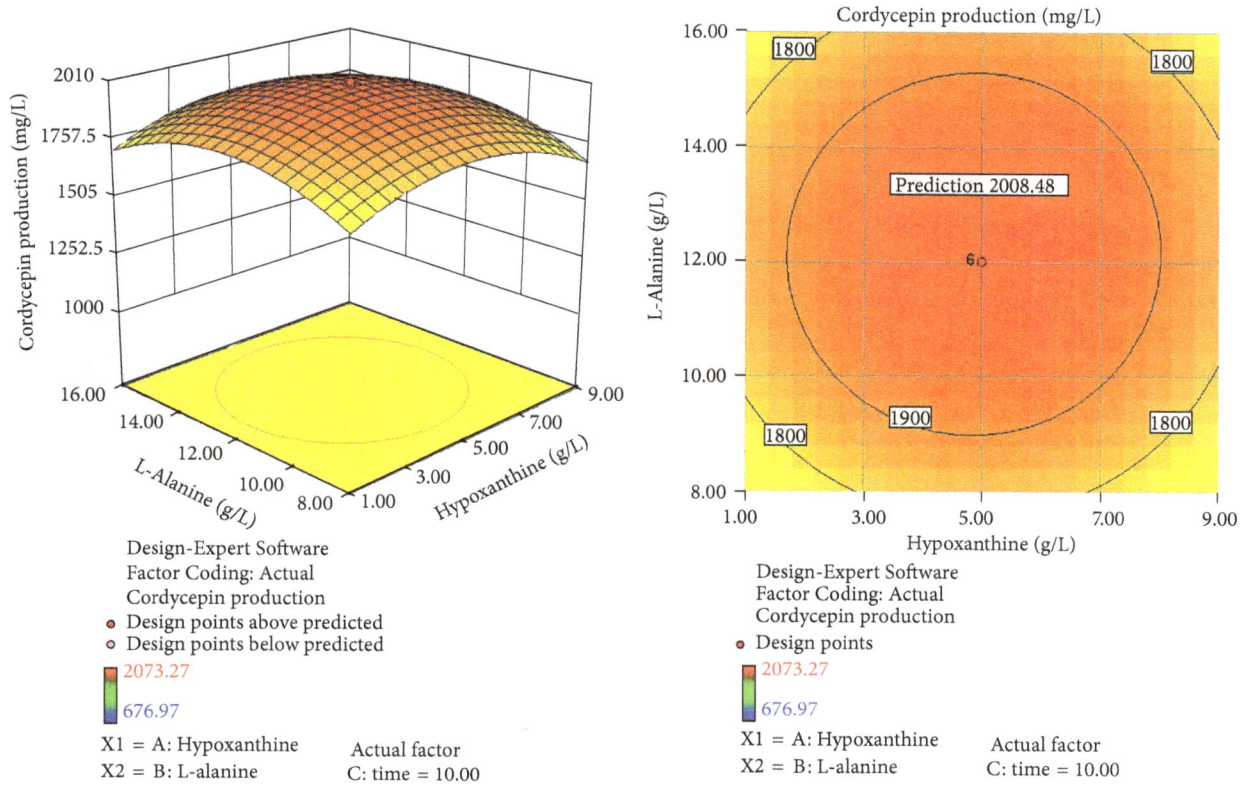

(a)

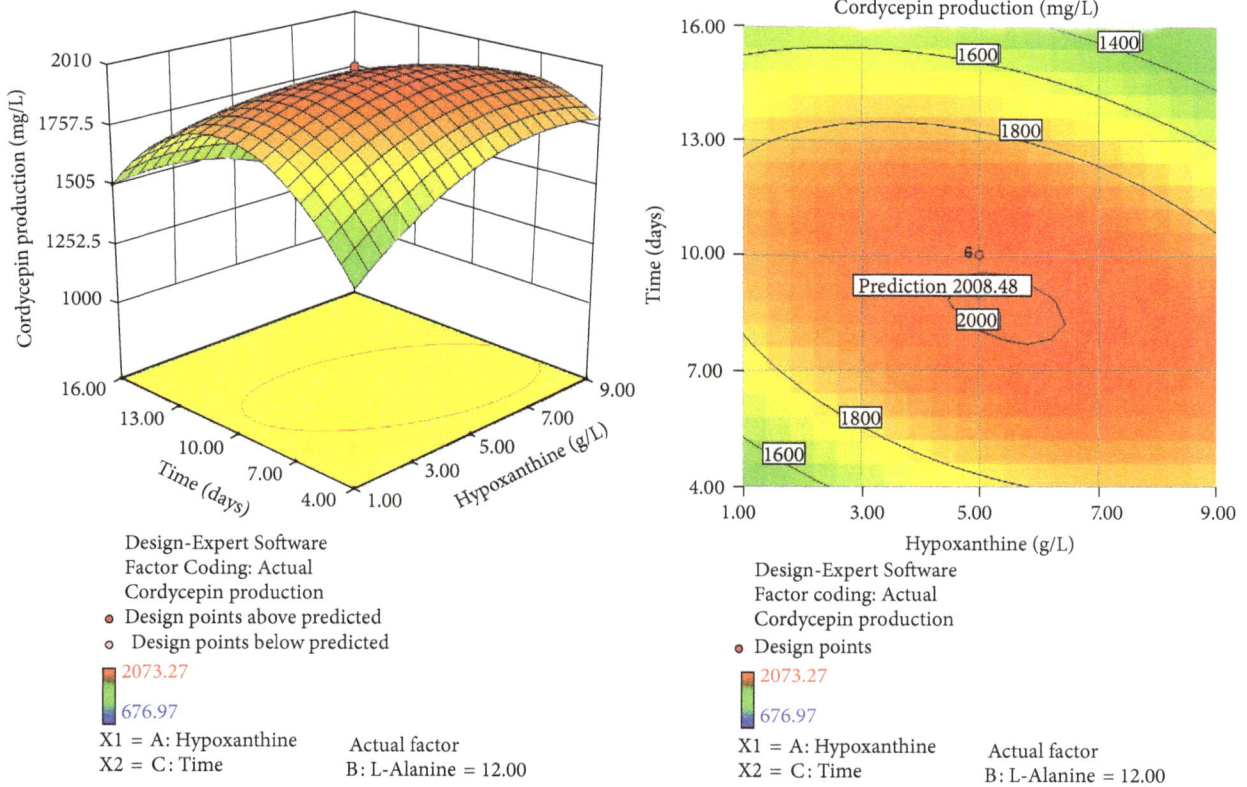

(b)

FIGURE 7: Continued.

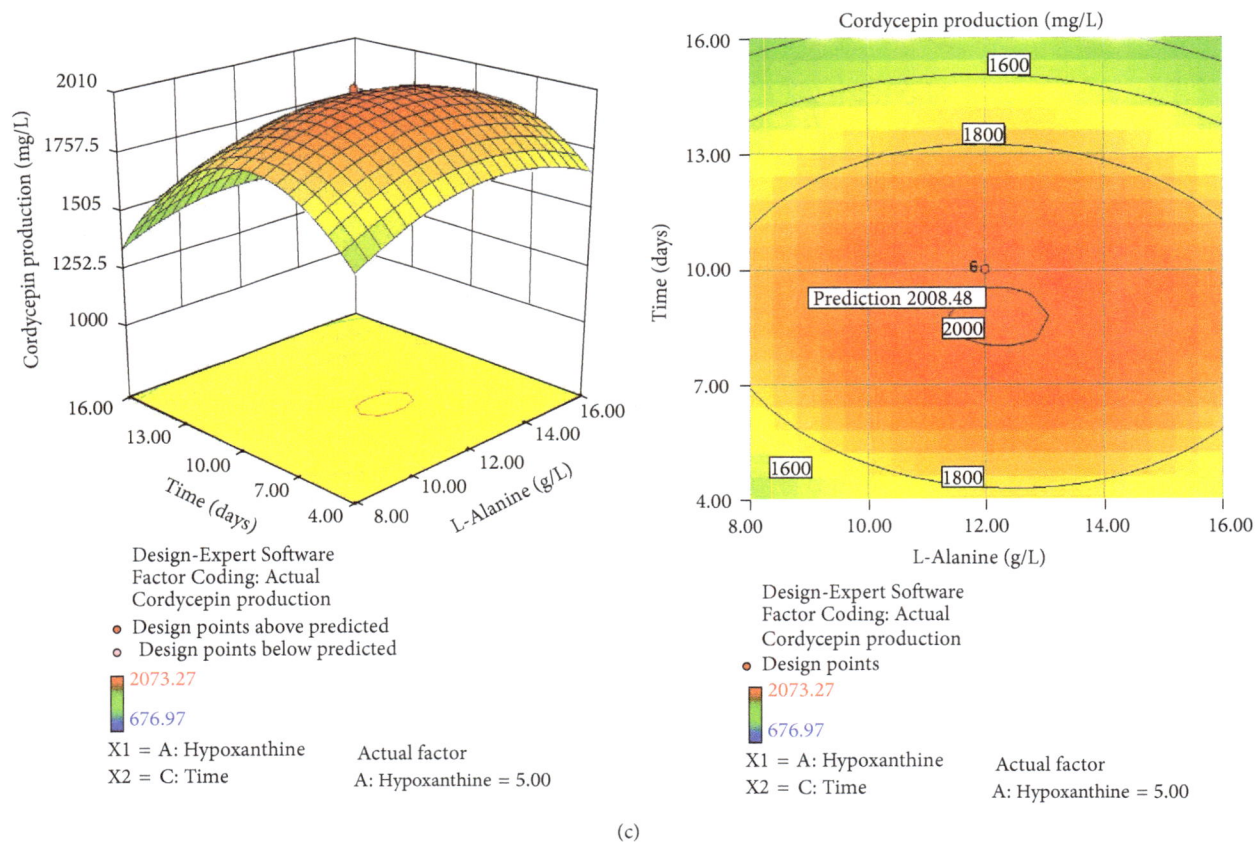

FIGURE 7: Three-dimensional response surface plots and two-dimensional contour plots for cordycepin production by *C. militaris* (strain CGMCC2459) showing variable interactions of (a) hypoxanthine and L-alanine; (b) hypoxanthine and time; (c) L-alanine and time.

FIGURE 8: Batch culture for cordycepin production under optimized culture conditions by static liquid culture using *C. militaris* (strain CGMCC2459).

4. Conclusion

In this work, single factor design, Plackett-Burman design, and central composite design were employed to establish the key factors and identify optimal culture conditions which improved cordycepin production by *C. militaris* CGMCC2459. Optimal media contained peptone 20 g/L; sucrose 24.7 g/L; $K_2HPO_4 \cdot 3H_2O$ 1.11 g/L; $MgSO_4 \cdot 7H_2O$ 0.90 g/L; VB_1 10 mg/L; hypoxanthine 5.45 g/L; and L-alanine 12.23 g/L. Hypoxanthine and L-alanine were added to the optimal medium at 8.6 days. Optimal incubation conditions were 25°C at an unaltered pH of 35 days. Using these culture conditions, a maximum production of cordycepin was 2008.48 mg/L for 700 mL working volume in the 1000 mL glass jars, and total content of cordycepin reached 1405.94 mg/bottle (700 mL/1000 mL). This method provides

Optimization of Large-Scale Culture Conditions for the Production of Cordycepin with Cordyceps militaris
by Liquid Static Culture

171

an effective way for increasing the cordycepin production at a large scale. The strategies used in this study could have a wide application in other fermentation process.

Conflict of Interests

The authors declare that there is no conflict of interests regarding the publication of this paper.

Authors' Contribution

Chao Kang and Ting-Chi Wen contributed equally to this work.

Acknowledgments

This work was supported by the Agricultural Science and Technology Foundation of Guizhou Province (no. (2011)3054), the National Natural Science Foundation of China (no. 31200016), and the Modernization of Traditional Chinese Medicine Program of Guizhou Province (no. (2012)5008).

References

[1] G.-H. Sung, N. L. Hywel-Jones, J.-M. Sung, J. J. Luangsa-ard, B. Shrestha, and J. W. Spatafora, "Phylogenetic classification of Cordyceps and the clavicipitaceous fungi," *Studies in Mycology*, vol. 57, pp. 5–59, 2007.

[2] J. T. Xu, *Medicinal Fungus in China*, Beijing Medical University, Peking Union Medical College Joint Press, Beijing, China, 1997 (Chinese).

[3] P. E. Mortimer, S. C. Karunarathna, Q. Li et al., "Prized edible Asian mushrooms: ecology, conservation and sustainability," *Fungal Diversity*, vol. 56, no. 1, pp. 31–47, 2012.

[4] T. C. Wen, B. X. Lei, J. C. Kang, G. R. Li, and J. He, "Enhanced production of mycelial culture using additives and cordycepin by submerged in Cordyceps militaris," *Food and Fermentation Industries*, vol. 35, no. 8, pp. 49–53, 2009 (Chinese).

[5] H. C. Li, P. Sun, and C. Q. Feng, "The research of cordycepin as an active component in Cordyceps," *Journal of Jinggangshan University (Natural Science)*, vol. 31, no. 2, pp. 93–96, 2010 (Chinese).

[6] K. G. Cunningham, W. Manson, F. S. Spring, and S. A. Hutchinson, "Cordycepin, a metabolic product isolated from cultures of Cordyceps militaris (Linn.) link," *Nature*, vol. 166, no. 4231, p. 949, 1950.

[7] D. D. de Silva, S. Rapior, F. Fons, A. H. Bahkali, and K. D. Hyde, "Medicinal mushrooms in supportive cancer therapies: an approach to anti-cancer effects and putative mechanisms of action," *Fungal Diversity*, vol. 55, pp. 1–35, 2012.

[8] N. Yoshikawa, S. Yamada, C. Takeuchi et al., "Cordycepin (3′-deoxyadenosine) inhibits the growth of B16-BL6 mouse melanoma cells through the stimulation of adenosine A3 receptor followed by glycogen synthase kinase-3β activation and cyclin D1 suppression," *Naunyn-Schmiedeberg's Archives of Pharmacology*, vol. 377, no. 4–6, pp. 591–595, 2008.

[9] A. M. Sugar and R. P. McCaffrey, "Antifungal activity of 3′-deoxyadenosine (cordycepin)," *Antimicrobial Agents and Chemotherapy*, vol. 42, no. 6, pp. 1424–1427, 1998.

[10] K. Hashimoto and B. Simizu, "Effect of cordycepin on the replication of western equine encephalitis virus," *Archives of Virology*, vol. 52, no. 4, pp. 341–345, 1976.

[11] E. N. Kodama, R. P. McCaffrey, K. Yusa, and H. Mitsuya, "Antileukemic activity and mechanism of action of cordycepin against terminal deoxynucleotidyl transferase-positive (TdT$^+$) leukemic cells," *Biochemical Pharmacology*, vol. 59, no. 3, pp. 273–281, 2000.

[12] D. D. de Silva, S. Rapior, E. Sudarman et al., "Bioactive metabolites from macrofungi: ethnopharmacology, biological activities and chemistry," *Fungal Diversity*, vol. 62, pp. 1–40, 2013.

[13] D. D. de Silva, S. Rapior, K. D. Hyde, and A. H. Bahkali, "Medicinal mushrooms in prevention and control of diabetes mellitus," *Fungal Diversity*, vol. 56, no. 1, pp. 1–29, 2012.

[14] A. Todd and T. L. V. Ulbricht, "Deoxynucleosides and related compounds—part IX: a synthesis of 3′-deoxyadenosine," *Journal of the Chemical Society*, pp. 3275–3277, 1960.

[15] D. W. Kwon, J.-H. Jeon, C. Kang, and Y. H. Kim, "New synthesis of 3′-deoxypurine nucleosides using samarium(III) iodide complex," *Tetrahedron Letters*, vol. 44, no. 43, pp. 7941–7943, 2003.

[16] E. A. Kaczka, E. L. Dulaney, C. O. Gitterman, H. B. Woodruff, and K. Folkers, "Isolation and inhibitory effects on KB cell cultures of 3′-deoxyadenosine from Aspergillus nidulans (Eidam) wint," *Biochemical and Biophysical Research Communications*, vol. 14, no. 5, pp. 452–455, 1964.

[17] H. X. Zhang, W. Wu, W. Chen, X. H. Gao, and L. S. Tang, "Comparative analysis of cordycepin and adenosine contents in fermentation supernatants between Aspergillus nidulans and Cordyceps militaris," *Acta Agriculturae Shanghai*, vol. 22, no. 2, pp. 28–31, 2006 (Chinese).

[18] T. C. Wen, J. C. Kang, B. X. Lei, G. R. Li, and J. He, "Effects of different solid culture condition on fruit body and cordycepin output of Cordyceps militaris," *Guizhou Agricultural Sciences*, vol. 36, no. 4, pp. 92–94, 2008 (Chinese).

[19] Z. H. Chen, H. Yu, W. B. Zeng, J. Y. Yang, J. Yuan, and Y. J. Chen, "Solid fermentation technique for cordycepin production by Cordyceps nigrella strain cnig-56," *Acta Edulis Fungi*, vol. 17, no. 1, pp. 80–82, 2010 (Chinese).

[20] L. Bu, Z. Y. Zhu, Z. Q. Liang, and A. Y. Liu, "Utilization of fungal elicitor in increasing cordycepin of Cordyceps militaris," *Mycosystema*, vol. 21, no. 2, pp. 252–256, 2002 (Chinese).

[21] X.-B. Mao and J.-J. Zhong, "Hyperproduction of cordycepin by two-stage dissolved oxygen control in submerged cultivation of medicinal mushroom Cordyceps militaris in bioreactors," *Biotechnology Progress*, vol. 20, no. 5, pp. 1408–1413, 2004.

[22] X. B. Mao and Y. Q. Tu, "Enhanced production of cordyceipn by fed batch cultivation using amino acid in Cordyceps militaris," *Chongqing Journal of Research on Chinese Drugs and Herbs*, no. 1, pp. 18–22, 2005 (Chinese).

[23] T. C. Wen, J. C. Kang, B. X. Lei, G. R. Li, and J. He, "Enhanced production of cordycepin by submerged culture using additives in Cordyceps militaris," *Food Science*, vol. 31, no. 5, pp. 175–179, 2010 (Chinese).

[24] C. Kang, T. C. Wen, J. C. Kang, Y. X. Qian, and B. X. Lei, "Effects of additives and different culture conditions on cordycepin production by the medicinal fungus Cordyceps militaris," *Mycosystem*, vol. 31, no. 3, pp. 389–397, 2012 (Chinese).

[25] M. Masuda, E. Urabe, A. Sakurai, and M. Sakakibara, "Production of cordycepin by surface culture using the medicinal mushroom *Cordyceps militaris*," *Enzyme and Microbial Technology*, vol. 39, no. 4, pp. 641–646, 2006.

[26] M. Masuda, E. Urabe, H. Honda, A. Sakurai, and M. Sakakibara, "Enhanced production of cordycepin by surface culture using the medicinal mushroom *Cordyceps militaris*," *Enzyme and Microbial Technology*, vol. 40, no. 5, pp. 1199–1205, 2007.

[27] S. K. Das, M. Masuda, A. Sakurai, and M. Sakakibara, "Effects of additives on cordycepin production using a *Cordyceps militaris* mutant induced by ion beam irradiation," *African Journal of Biotechnology*, vol. 8, no. 13, pp. 3041–3047, 2009.

[28] S. Aman, D. J. Anderson, T. J. Connolly, A. J. Crittall, and G. Ji, "From adenosine to 3'-deoxyadenosine: development and scale up," *Organic Process Research and Development*, vol. 4, no. 6, pp. 601–605, 2000.

[29] M. Masuda, S. K. Das, M. Hatashita, S. Fujihara, and A. Sakurai, "Efficient production of cordycepin by the *Cordyceps militaris* mutant G81-3 for practical use," *Process Biochemistry*, vol. 49, pp. 181–187, 2014.

[30] S. K. Das, M. Masuda, M. Hatashita, A. Sakurai, and M. Sakakibara, "A new approach for improving cordycepin productivity in surface liquid culture of *Cordyceps militaris* using high-energy ion beam irradiation," *Letters in Applied Microbiology*, vol. 47, no. 6, pp. 534–538, 2008.

[31] L. Wen, C. Zhang, M. Xia, and L. Weng, "Effects of *Cordyceps militaris* space mutation treatment on active constituent content," *Food Science*, vol. 29, no. 5, pp. 382–384, 2008.

[32] X.-B. Mao, T. Eksriwong, S. Chauvatcharin, and J.-J. Zhong, "Optimization of carbon source and carbon/nitrogen ratio for cordycepin production by submerged cultivation of medicinal mushroom *Cordyceps militaris*," *Process Biochemistry*, vol. 40, no. 5, pp. 1667–1672, 2005.

[33] S. K. Das, M. Masuda, M. Hatashita, A. Sakurai, and M. Sakakibara, "Optimization of culture medium for cordycepin production using *Cordyceps militaris* mutant obtained by ion beam irradiation," *Process Biochemistry*, vol. 45, no. 1, pp. 129–132, 2010.

[34] I.-L. Shih, K.-L. Tsai, and C. Hsieh, "Effects of culture conditions on the mycelial growth and bioactive metabolite production in submerged culture of *Cordyceps militaris*," *Biochemical Engineering Journal*, vol. 33, no. 3, pp. 193–201, 2007.

[35] W. Q. Guo, N. Q. Ren, X. J. Wang et al., "Optimization of submerged culture conditions for exo-biopolymer production by *Paecilomyces japonica*," *Bioresource Technology*, vol. 100, pp. 1192–1196, 2009.

[36] J. Zhou, X. Yu, C. Ding et al., "Optimization of phenol degradation by *Candida tropicalis* Z-04 using Plackett-Burman design and response surface methodology," *Journal of Environmental Sciences*, vol. 23, no. 1, pp. 22–30, 2011.

[37] D. F. Zhao, C. Y. Liang, X. Ren, and S. M. Sun, "Effect of dissolved oxygen on the biotransformation of tylosin to its acyl derivative," *Chinese Journal of Pharmaceutical*, vol. 37, no. 3, pp. 162–164, 2006 (Chinese).

[38] T. Xie, H. Y. Fang, G. B. Zhu, and G. J. Zhu, "Effect of dissolved oxygen on glycerol production by *Candida glycerinogenes*," *China Biotechnology*, vol. 28, no. 5, pp. 65–70, 2008 (Chinese).

[39] N. M. Kredich and A. J. Guarino, "Studies on the biosynthesis of cordycepin," *Biochimica et Biophysica Acta*, vol. 47, no. 3, pp. 529–534, 1961.

[40] J. Pintado, A. Torrado, M. P. González, and M. A. Murado, "Optimization of nutrient concentration for citric acid production by solid-state culture of *Aspergillus niger* on polyurethane foams," *Enzyme and Microbial Technology*, vol. 23, no. 1-2, pp. 149–156, 1998.

[41] J.-T. Bae, J. Sinha, J.-P. Park, C.-H. Song, and J.-W. Yun, "Optimization of submerged culture conditions for exo-biopolymer production by *Paecilomyces japonica*," *Journal of Microbiology and Biotechnology*, vol. 10, no. 4, pp. 482–487, 2000.

[42] K. Lei, Y. Ke, and N. Mao, "The optimization of the culture media for high production of cordycepin by *Cordyceps militaris* Fjnu-01," *Pharmaceutical Biotechnology*, vol. 18, no. 6, pp. 504–508, 2011 (Chinese).

[43] B. M. Chassy and R. J. Suhadolnik, "Nucleoside antibiotics IV. Metabolic fate of adenosine and cordycepin by *Cordyceps militaris* during cordycepin biosynthesis," *Biochimica et Biophysica Acta*, vol. 182, no. 2, pp. 307–315, 1969.

[44] J. H. Xiao, D. X. Chen, J. W. Liu et al., "Optimization of submerged culture requirements for the production of mycelial growth and exopolysaccharide by *Cordyceps jiangxiensis* JXPJ 0109," *Journal of Applied Microbiology*, vol. 96, no. 5, pp. 1105–1116, 2004.

[45] J. H. Xiao, D. X. Chen, Y. Xiao et al., "Optimization of submerged culture conditions for mycelial polysaccharide production in *Cordyceps pruinosa*," *Process Biochemistry*, vol. 39, no. 12, pp. 2241–2247, 2004.

[46] C.-H. Dong and Y.-J. Yao, "Nutritional requirements of mycelial growth of *Cordyceps sinensis* in submerged culture," *Journal of Applied Microbiology*, vol. 99, no. 3, pp. 483–492, 2005 (Chinese).

[47] C. Xie, G. Liu, Z. Gu, G. Fan, L. Zhang, and Y. Gu, "Effects of culture conditions on mycelium biomass and intracellular cordycepin production of *Cordyceps militaris* in natural medium," *Annals of Microbiology*, vol. 59, no. 2, pp. 293–299, 2009.

19

Growth Phase-Dependent Proteomes of the Malaysian Isolated *Lactococcus lactis* Dairy Strain M4 Using Label-Free Qualitative Shotgun Proteomics Analysis

Theresa Wan Chen Yap,[1] **Amir Rabu,**[1] **Farah Diba Abu Bakar,**[1] **Raha Abdul Rahim,**[2] **Nor Muhammad Mahadi,**[3] **Rosli Md. Illias,**[4] **and Abdul Munir Abdul Murad**[1]

[1] *School of Biosciences and Biotechnology, Faculty of Science and Technology, Universiti Kebangsaan Malaysia, 43600 Bangi, Selangor, Malaysia*
[2] *Faculty of Biotechnology and Biomolecular Sciences, Universiti Putra Malaysia (UPM), 43400 Serdang, Selangor, Malaysia*
[3] *Malaysia Genome Institute, Jalan Bangi, 43000 Kajang, Selangor, Malaysia*
[4] *Department of Bioprocess Engineering, Faculty of Chemical Engineering, Universiti Teknologi Malaysia, 81310 Skudai, Johor, Malaysia*

Correspondence should be addressed to Abdul Munir Abdul Murad; munir@ukm.my

Academic Editors: M. Fernández and M. Singh

Lactococcus lactis is the most studied mesophilic fermentative lactic acid bacterium. It is used extensively in the food industry and plays a pivotal role as a cell factory and also as vaccine delivery platforms. The proteome of the Malaysian isolated *L. lactis* M4 dairy strain, obtained from the milk of locally bred cows, was studied to elucidate the physiological changes occurring between the growth phases of this bacterium. In this study, ultraperformance liquid chromatography nanoflow electrospray ionization tandem mass spectrometry (UPLC- nano-ESI-MS^E) approach was used for qualitative proteomic analysis. A total of 100 and 121 proteins were identified from the midexponential and early stationary growth phases, respectively, of the *L. lactis* strain M4. During the exponential phase, the most important reaction was the generation of sufficient energy, whereas, in the early stationary phase, the metabolic energy pathways decreased and the biosynthesis of proteins became more important. Thus, the metabolism of the cells shifted from energy production in the exponential phase to the synthesis of macromolecules in the stationary phase. The resultant proteomes are essential in providing an improved view of the cellular machinery of *L. lactis* during the transition of growth phases and hence provide insight into various biotechnological applications.

1. Introduction

Lactic acid bacteria (LAB) are a heterogeneous group of Gram-positive bacteria that convert carbohydrates into lactic acid [1]. LAB are facultative anaerobic, nonspore forming rod- or coccus-shaped bacteria [2]. In nature, LAB are indigenous to food-related habitats such as plants and milk and occupy a niche on the mucosal surfaces of animals [3]. *Lactococcus lactis* is the most studied species among LAB and has become the model bacterium for most LAB research in biotechnology. The two subspecies of *L. lactis* were initially designated as *Streptococcus lactis* and *Streptococcus*

cremoris but were later reclassified as *L. lactis* ssp. *lactis* and *L. lactis* spp. *cremoris* [4]. *L. lactis* has been exploited for a vast variety of biotechnological applications during the past few decades especially in the production of cheese. Its homofermentative nature and long history of safe usage in food preparation combined with its generally regarded as safe (GRAS) status have made *L. lactis* an economically important microorganism in the dairy food industry. In addition, a large amount of research on the metabolic engineering of *L. lactis* has been performed for the production of compounds such as diacetyl, alanine, and exopolysaccharides [5]. *L. lactis* has also emerged as an efficient cell factory for the production of food ingredients [6], nutraceuticals [7], heterologous proteins [8, 9], and vaccine delivery platforms [10–12].

The genomes of several *L. lactis* strains have been completely sequenced, such as IL1403 [1], KF147 [13, 14], and CV56 [15] from *L. lactis* subsp. *lactis*, as well as strains such as MG1363 [16, 17] and SK11 [3] from *L. lactis* subsp. *Cremoris*, and the recently sequenced *L. lactis* IO-1 [18]. The availability of the complete genome sequence of multiple *L. lactis* strains enables the powerful application of proteomics to investigate the global cellular protein expression profiles of different *L. lactis* strains. Proteomics can be defined as the analysis of the whole-protein complement of the genome that is expressed by a cell or any biological sample at a given time under specific conditions [19]. Proteomics refers to large-scale protein studies with a particular emphasis on protein expression, structure, and function [20]. Although genomics and proteomics are complementary techniques, proteomics has the advantage of directly accessing biological processes at the protein level. Proteins are responsible for the structure, energy production, communication, movement, and division of all cells; thus, gaining a comprehensive understanding of proteins using systems biology is of utmost importance [21].

During the past decade, a number of proteomics studies on *L. lactis* have been performed. Initially, proteomics analyses were performed to set up reference maps [22, 23] and to better understand the cellular pathways of *L. lactis* related to important physiological processes and technological properties at the protein level [2]. Later, additional proteomics research on *L. lactis* focused on comparative expression studies [24–33] and stress-response studies [34–38] due to *L. lactis* economic importance. Novel applications for LAB as living vehicles for the targeting of antigens or therapeutics to the digestive mucosa are being developed and proteomics methods are being used to investigate and identify new markers of *L. lactis* adaptation to the mouse digestive tract [39, 40]. To date, only one proteomics study of *L. lactis* on differences between multiple strains [2] and posttranslational modifications [41] has been reported.

Conventional two-dimensional polyacrylamide gel electrophoresis (2-DE) coupled with mass spectrometry is commonly used for proteomics analysis. However, with the advancement of technology, high throughput instruments such as liquid chromatography tandem mass spectrometry (LC-MS/MS) have been invented to facilitate the research and development of protein studies. This gel-free, label-free, highly sensitive, and specific LC-MS/MS approach enables the effective separation of complex protein mixtures by eluting peptides and their corresponding fragment ions.

In the present work, the target microorganism was the Malaysian isolated *L. lactis* M4 dairy strain obtained from the milk of locally bred cows. The *L. lactis* strain M4 was identified as *L. lactis* spp. *lactis* based on the molecular analysis of the16S rRNA gene. It was chosen as our target microorganism because it has been proven to be a potential host for the expression of heterologous proteins [42]. The proteomes of the *L. lactis* M4 dairy strain at mid-exponential and early stationary growth phases were determined and the physiological changes between the growth phases were elucidated using highly sensitive and specific ultra-performance liquid

chromatography nanoflow electrospray ionization tandem mass spectrometry (UPLC-nanoESI-MS/MS). The resultant proteomes are essential in providing an improved dynamic and global view of the *L. lactis* cellular machinery during growth phase transition. These proteomes may also have predictive value for the modeling of biological processes. They could also be utilized as a gold standard and make it possible to understand the differences in the protein expression patterns of local *L. lactis* strains between the growth phases, which in turn could provide insight into various applications such as the identification of target proteins that respond to alterations in diet or treatments, the development of effective cell factories for biorefinery through metabolic engineering, and the development of lactococcal expression systems in the future.

2. Materials and Methods

2.1. Bacterial Strains and Growth Conditions. The *Lactococcus lactis* M4 dairy strain (Malaysian isolated strain obtained from the milk of locally bred cows) was obtained courtesy of the Universiti Putra Malaysia (Serdang, Selangor, Malaysia). Frozen cell stocks of *L. lactis* M4 were streaked onto M17 plate supplemented with 0.5% glucose and grown at 30°C for overnight. Seed cultures were generated by transferring a single colony to a flask with M17 broth supplemented with 0.5% glucose. Seed cultures were incubated at 30°C without agitation for 16 h. The seed cultures were then diluted (2 mL into 200 mL) in M17 broth supplemented with 0.5% glucose. The cultures were incubated at 30°C without agitation until mid-exponential growth phase ($OD_{600} \sim 0.8$) or early stationary growth phase ($OD_{600} \sim 1.8$) and then harvested by centrifugation (7000 ×g for 15 minutes) using an Eppendorf 5810R centrifuge (Hamburg, Germany). The culture media were discarded, and the cells were frozen at −20°C until needed for protein extract preparation.

2.2. Protein Extraction and Trypsin Digestion. Protein extract was prepared using previously reported protocols [43, 44]. The cells were collected by centrifugation at 7000 ×g for 15 min, and the pellet was washed in phosphate buffer (1.24 g/L K_2HPO_4; 0.39 g/L KH_2PO_4; 8.8 g/L NaCl, pH 7.2). The pellet was then suspended in 1 mL of extraction buffer (0.7 M sucrose; 0.5 M Tris-HCl, pH7; 30 mM HCl; 50 mM EDTA; 0.1 M KCl; 40 mM DTT). The cell suspension was incubated for 15 min at room temperature and sonicated on ice (10 times for 10 s at 30 s intervals). The cell debris was removed by centrifugation at 10000 ×g for 30 min at 4°C. Protein concentration was estimated using the Quick Start Bradford Protein Assay (Bio-Rad, Hercules, CA, USA). The resulting protein extract was dispensed and kept at −20°C for subsequent analysis. To prepare the protein digest, approximately 250 μg of total protein was denatured using 12.5 μL of denaturing buffer (8 M urea and 25 mM NH_4HCO_3, pH 8.0). The mixture was then reduced with 10 mM DTT at 37°C for 1 h and alkylated with 50 mM iodoacetamide in the dark for 30 min. The urea concentration was reduced to 1 M by dilution with 50 mM NH_4HCO_3, pH 7.8. Proteolytic

Growth Phase-Dependent Proteomes of the Malaysian Isolated Lactococcus lactis Dairy Strain M4 Using
Label-Free Qualitative Shotgun Proteomics Analysis

175

digestion was initiated by adding mass spectrometry grade Trypsin Gold (Promega, Madison, WI, USA) at a ratio of 1 : 50 (trypsin : protein) followed by incubation at 37°C overnight. Tryptic digestion was terminated by acidification with 0.1% trifluoroacetic acid. The tryptic peptide solution was kept at −80°C until further analysis.

2.3. Ultra-Performance Liquid Chromatography Tandem Mass Spectrometry (UPLC-MS/MS). Tryptic digests of total protein obtained during both growth phases (mid-exponential and early stationary) were generated, separated, and analyzed using LC separation and MS analysis. Injections of 500 ng of total peptide were used in triplicate analyses (each biological replicate was subjected to two technical runs).

LC-nanoESI-MS^E (alternating low-energy MS and elevated energy MS^E) analysis was performed using a SYNAPT G2 HDMS mass spectrometer equipped with a nanoAC-QUITY UPLC separation system. The aqueous mobile phase (mobile phase A) was water with 0.1% formic acid and the organic mobile phase (mobile phase B) was acetonitrile with 0.1% formic acid. Peptide separations were performed with an acetonitrile gradient in 0.1% formic acid (1–50% in 120 min). A total of 500 ng of peptides (partial loop, 1 μL injection) was loaded onto a nanoACQUITY UPLC Symmetry C_{18}, 180 × 20 mm trap column at 15 μL/min for subsequent separation by LC-nanoESI using a nanoACQUITY UPLC BEH 130 C_{18}, 75 μm × 200 mm reverse phase column (Waters Corporation, Milford, MA, USA) with 1% mobile phase B at 0.25 μL/min. Peptides were eluted from the column with a gradient of 1–50% mobile phase B over 90 min at 0.25 μL/min followed by a 10 min rinse of 85% mobile phase B. The column was immediately reequilibrated under the initial conditions (1% mobile phase B) for 20 min. The lock mass, [Glu^1]-fibrinopeptide at 100 fmol/μL, was delivered from the fluidic system at 5 μL/min to the reference sprayer of the NanoLockSpray source.

Mass spectrometry analysis of the separated tryptic digested peptides was performed using a SYNAPT G2 HDMS mass spectrometer. The instrument control and data acquisition were conducted using the MassLynx data system, version 4.1 (Waters Corporation, Milford, MA, USA). The electrospray voltage was set at 3 kV, the source temperature was 80°C, and the cone voltage was 40 V. For all experiments, the mass spectrometer was operated in resolution mode with a typical resolving power of at least 20000, and all analyses were performed using positive mode ESI using a NanoLockSpray source. The lock mass channel was sampled every 30 s. The spectral acquisition time in each mode was 1.0 s, and the mass spectra were acquired from m/z 50 to 2000. In low energy MS mode, data were collected at constant collision energy of 15 eV. In MS^E mode, the collision energy was ramped from 15 to 40 eV during each 1.0 s data collection cycle. One cycle of MS and MS^E data was acquired every 10.0 s.

2.4. Data Analysis. The continuum LCMS^E data were processed and searched using ProteinLynx Global Server (PLGS) version 2.4. Protein identifications were assigned by searching against the *Lactococcus* protein database available at UniProt. The search parameter values for each precursor and associated fragment ion were set by the software using the measured mass error and intensity error obtained from the processing of the raw continuum data. The mass error tolerance values were typically less than 5 ppm. Peptide identifications were restricted to tryptic peptides with no more than one missed cleavage. Cysteine carbamidomethylation was considered as a fixed peptide modification, whereas methionine oxidation, asparagines deamination, and glutamine deamination were considered as variable peptide modifications. Peptides with PLGS scores greater than 200 were considered for charting the proteome of the strain M4 at both growth phases (late-exponential and early stationary). PLGS score is calculated by the ProteinLynx Global Server (PLGS 2.4) software using a Monte Carlo algorithm to analyze all available mass spectrometry data and is a statistical measure of accuracy of identification. A high score implies greater confidence of protein identity [45]. Protein identification was manually validated for proteins with PLGS scores less than 200.

The identified proteins obtained from both growth phases (late-exponential and early stationary) were categorized based on their molecular functions. The functional classification of proteins was performed according to a previously published list of categories [1, 46]. The consensus lists of proteins were then analyzed to determine their cellular localization based on gene ontology using the STRAP software [47]. The proteomes were then compared and further analyzed for their association with different processes, networks, and pathway maps.

3. Results and Discussion

3.1. Identification of the Growth Phase-Dependent Proteomes. Our study presents the first comparative investigation of the Malaysian isolated *L. lactis* strain M4 proteome at two time points, the mid-exponential and early stationary growth phases, using a label-free shotgun proteomics approach. *L. lactis* M4 was grown at 30°C in M17 media. The optical density of cells at 600 nm was recorded hourly to construct the growth curve of *L. lactis* M4 and to identify the time points for the mid-exponential growth phase (OD_{600} ~ 0.8) and early stationary growth phase (OD_{600} ~ 1.8). To assure the comparable numbers of living cells for the subsequent extraction steps and comparative proteomics analysis, the OD_{600} reading was verified before the cells were harvested. These two growth phases were chosen because the production of organic acids (mainly lactic acid), ethanol, aroma compounds, bacteriocins, exopolysaccharides, and several enzymes is of importance to dairy and fermented food industry, and they are reported to be produced during the transition of these phases [48]. M17 medium was chosen as it is the commonly used growth medium for *L. lactis*. The protein extracts were proteolyzed and the tryptic digests were separated by ultra-performance liquid chromatography nanoflow electrospray ionization tandem mass spectrometry (UPLC-nanoESI-MS/MS).

TABLE 1: List of the proteins identified at the midexponential (ME) and early stationary (ES) growth phases of *L. lactis* strain M4.

Gene name	Protein name	ME	ES
	Amino acid biosynthesis		
glnA	Glutamine synthetase	•	•
luxS	S-Ribosylhomocysteine lyase	•	•
	Biosynthesis of cofactors, prosthetic groups, and carriers		
panE	2-Dehydropantoate-2-reductase	•	
trxH	Thioredoxin H-type		•
	Cellular processes		
ahpC	Alkyl hydroperoxide reductase subunit C		•
ftsZ	Cell division protein GTPase FtsZ	•	•
divIVA	Cell division initiation protein DivIVA		•
dnaK	Chaperone protein DnaK	•	•
sodA	Superoxide dismutase [Mn]		•
tig	Trigger factor	•	•
	Cell envelope		
murD	UDP-N-acetylmuramoylalanine-D-glutamate ligase		•
hasC/galU	UDP-glucose-1-phosphate uridylyltransferase	•	•
	Energy metabolism		
pmg	2,3-Bisphosphoglycerate-dependent phosphoglycerate mutase	•	•
pfk	6-Phosphofructokinase	•	•
gnd	6-Phosphogluconate dehydrogenase decarboxylating	•	•
adhE	Alcohol-acetaldehyde dehydrogenase	•	•
enoA	Enolase	•	•
pfl	Formate acetyltransferase	•	•
fbaA	Fructose-bisphosphate aldolase	•	•
pgiA	Glucose-6-phosphate isomerase	•	•
gapB	Glyceraldehyde-3-phosphate dehydrogenase	•	•
ldhA	L-Lactate dehydrogenase 1	•	•
arcB	Ornithine carbamoyltransferase 2	•	
pgk	Phosphoglycerate kinase	•	•
enoB	Phosphopyruvate hydratase	•	•
pdhC	Pyruvate dehydrogenase complex E2 component		•
LACR_0691	Pyruvate-formate lyase		•
pyk	Pyruvate kinase	•	•
tpiA	Triosephosphate isomerase	•	•
	Central intermediary metabolism		
metK	S-Adenosylmethionine synthase		•
	Fatty acid and phospholipid metabolism		
fabF	3-Oxoacyl-acyl carrier protein synthase 2	•	•
fabG1	3-Oxoacyl-acyl carrier protein reductase		•
accB	Acetyl-CoA carboxylase biotin carboxylase subunit		•
acpA	Acyl carrier protein	•	•
fabI	NADH-dependent enoyl-acyl carrier protein reductase		•
	Purines, pyrimidines, nucleosides, and nucleotides		
adk	Adenylate kinase	•	•
purA	Adenylosuccinate synthase		•
carB	Carbamoyl phosphate synthase large chain		•
rmlB	dTDP-glucose-4,6-dehydratase		•
rmlA	Glucose-1-phosphate thymidylyltransferase		•
guaA	GMP synthase glutamine hydrolyzing	•	•
guaB	Inosine-5-monophosphate dehydrogenase	•	•
pyrE	Orotate phosphoribosyltransferase	•	•

Growth Phase-Dependent Proteomes of the Malaysian Isolated Lactococcus lactis Dairy Strain M4 Using
Label-Free Qualitative Shotgun Proteomics Analysis

177

TABLE 1: Continued.

Gene name	Protein name	ME	ES
deoB	Phosphopentomutase	•	•
prsA	Ribose-phosphate pyrophosphokinase		•
prsB	Ribose-phosphate pyrophosphokinase		•
upp	Uracil phosphoribosyltransferase	•	•
pyrH	Uridylate kinase		•
	Regulatory functions		
pyrR	Bifunctional protein pyrR	•	•
typA	GTP-binding protein TypA/BipA	•	•
codY	Transcriptional regulator	•	•
codZ	Transcriptional regulator		•
llrC	Two-component system regulator	•	•
	Replication		
hslA	Hu-like DNA-binding protein	•	•
	Transcription		
rheA	ATP-dependent RNA helicase	•	
rpoA	DNA-dependent RNA polymerase alpha subunit	•	•
rpoB	DNA-dependent RNA polymerase beta subunit		•
nusG	Transcription antitermination protein	•	
greA	Transcription elongation factor GreA		•
	Translation		
rpsA	30S ribosomal protein S1	•	•
rpsB	30S ribosomal protein S2	•	•
rpsC	30S ribosomal protein S3	•	•
rpsD	30S ribosomal protein S4	•	•
rpsE	30S ribosomal protein S5	•	•
rpsF	30S ribosomal protein S6	•	•
rpsG	30S ribosomal protein S7	•	•
rpsH	30S ribosomal protein S8	•	•
rpsI	30S ribosomal protein S9	•	•
rpsJ	30S ribosomal protein S10	•	•
rpsL	30S ribosomal protein S12	•	•
rpsM	30S ribosomal protein S13	•	•
rpsN	30S ribosomal protein S14	•	•
rpsP	30S ribosomal protein S16	•	•
rpsR	30S ribosomal protein S18	•	•
rpsS	30S ribosomal protein S19	•	
rpsT	30S ribosomal protein S20	•	•
rpsU	30S ribosomal protein S21	•	
rplA	50S ribosomal protein L1	•	•
rplB	50S ribosomal protein L2	•	•
rplC	50S ribosomal protein L3	•	•
rplD	50S ribosomal protein L4	•	•
rplE	50S ribosomal protein L5	•	•
rplF	50S ribosomal protein L6	•	•
rplL	50S ribosomal protein L7/L12	•	•
rplJ	50S ribosomal protein L10	•	•
rplK	50S ribosomal protein L11	•	•
rplM	50S ribosomal protein L13	•	•
rplO	50S ribosomal protein L15	•	•
rplP	50S ribosomal protein L16	•	•
rplQ	50S ribosomal protein L17	•	•

TABLE 1: Continued.

Gene name	Protein name	ME	ES
rplR	50S ribosomal protein L18	•	•
rplS	50S ribosomal protein L19	•	•
rplT	50S ribosomal protein L20		•
rplU	50S ribosomal protein L21	•	•
rplV	50S ribosomal protein L22	•	•
rplW	50S ribosomal protein L23	•	•
rpmA	50S ribosomal protein L27	•	
rpmB	50S ribosomal protein L28	•	•
rpmC	50S ribosomal protein L29	•	•
rpmD	50S ribosomal protein L30	•	•
rpmE2	50S ribosomal protein L31 type B	•	•
rpmF	50S ribosomal protein L32	•	•
rpmG1	50S ribosomal protein L33 1	•	
rpmJ	50S ribosomal protein L36	•	•
pepC	Aminopeptidase C		•
pepN	Aminopeptidase N	•	•
argS	Arginyl-tRNA synthetase	•	•
pepV	Dipeptidase		•
fusA	Elongation factor G	•	•
efp	Elongation factor P		•
tsf	Elongation factor Ts	•	•
tuf	Elongation factor Tu	•	•
ileS	Isoleucyl-tRNA synthetase	•	
pepT	Peptidase T		•
ppiB	Peptidyl-prolyl *cis-trans* isomerase	•	•
pheT	Phenylalanyl-tRNA synthetase beta chain		•
pepQ	Proline dipeptidase	•	•
pepO	Prolidase		•
frr	Ribosome recycling factor	•	•
serS	Seryl-tRNA synthetase	•	•
thrS	Threonyl-tRNA synthetase	•	
infA	Translation initiation factor IF-1		•
tyrS	Tyrosyl-tRNA synthetase	•	
LACR_1813	Xaa-Pro aminopeptidase		•
	Transport and binding proteins		
ptsl	Phosphoenolpyruvate protein phosphotransferase	•	•
malE	Maltose ABC transporter substrate binding protein		•
	Other categories		
clpB	ClpB protein		•
cspE	Cold shock protein E	•	•
grpE	Stress response protein E	•	
	Hypothetical proteins		
yhjA	General stress protein, CsbD superfamily		•
ytjD	Nitroreductase family protein		•
llmg_1773	Putative uncharacterized protein	•	•
llmg_2049	Putative uncharacterized protein	•	
SA8A11-2	SA8A11-2 protein		•
LACR_1462	UDP-glucose pyrophosphorylase	•	

Growth Phase-Dependent Proteomes of the Malaysian Isolated Lactococcus lactis Dairy Strain M4 Using
Label-Free Qualitative Shotgun Proteomics Analysis

179

Approximately 73158 and 75494 MS/MS peptide spectra were collected and analyzed for the mid-exponential and early stationary growth phases, respectively. The UPLC-nanoESI-MS/MS analysis yielded a total of 100 and 121 proteins from the mid-exponential and early stationary growth phases of *L. lactis* M4, respectively. The proteome dataset identified in the present study was small, corresponding to only 4.3–5.3% of all proteins predicted from the genome of *L. lactis* spp. *lactis* IL1403. However, this level of coverage is comparable to the proteome dataset of *Bifidobacterium longum* strain NCC2705, which was reported to correspond to 7.57% of the probable coding regions of *B. longum* NCC2705 [49, 50]. The search parameter values fixed in PLGS were stringent to increase the confidence level in protein identification in our qualitative proteomics study. This could be the main reason that low protein coverage was identified in our study. The low protein coverage may also be due to the lack of a published genome sequence specific to *L. lactis* strain M4. Protein coverage could be greatly improved if the genome of the Malaysian isolated *L. lactis* strain was completely sequenced and a multidimensional liquid chromatography tandem mass spectrometry approach was used. From the 100 and 121 proteins identified for the mid-exponential and early stationary growth phases, >50% of proteins were detected in three biological replicates (Figure 1(a)). When the proteomes from both growth phases were compared, 85 proteins were found to be present in both datasets (Figure 1(b)).

3.2. Functional Categorization of the Identified *L. lactis* Proteins.

The identified proteins from both growth phases were categorized according to their annotated functions (Figure 2, Table 1) and cellular localization based on gene ontology (Figure 3). The identified proteome dataset of *L. lactis* strain M4 was mainly associated with translation and energy metabolism, which comprised 56% and 15% of the identified proteins, respectively (Figure 2(a)), for the mid-exponential growth phase and 48% and 13% of the identified proteins, respectively, (Figure 2(b)) for the early stationary growth phase. The remaining identified proteins were categorized as purines, pyrimidines, nucleosides, and nucleotides or as being associated with regulatory functions, cellular processes, replication, transcription, amino acid biosynthesis, the biosynthesis of cofactors, prosthetic groups and carriers, cell envelope, fatty acid and phospholipid metabolism, transport and binding, hypothetical proteins, or as others (Figure 2).

The majority of the *L. lactis* strain M4 proteome for the mid-exponential growth phase was localized in the ribosome (59%), cytoplasm (31%), macromolecule complex (7%), cell surface (7%), extracellular space (7%), and other areas (7%) (Figure 3(a)). Similarly, the majority of the identified *L. lactis* strain M4 proteins from the early stationary growth phase were found to be localized in the ribosome (51%) followed by the cytoplasm (39%), macromolecule complex (6%), extracellular space (2%), cell surface (1%), and other areas (1%) (Figure 3(b)). Low percentage of extracellular proteins and cell surface proteins

were identified in our study because the sample preparation was designed for the extraction of cytoplasmic proteins.

3.2.1. Translation.

Both the mid-exponential and early stationary growth phases *L. lactis* M4 proteome datasets were dominated by translation machinery proteins. A transcriptomic study of *L. lactis* MG1363 reported recently had also shown that genes involved in "translation, ribosomal structure, and biogenesis" were upregulated at the transition point from exponential to stationary growth [51]. Ribosomal proteins, which are some of the most highly expressed proteins in the cell, were identified in large numbers in the *L. lactis* M4 proteome for both growth phases. Based on the genome sequence of *L. lactis* subsp. *lactis* IL1403, 58 proteins were predicted to be ribosomal proteins [1]. In this study, there were 44 and 41 ribosomal proteins identified from the mid-exponential and early stationary growth phases, respectively.

In addition to ribosomal proteins, other proteins involved in translation, including peptidases (aminopeptidase C, aminopeptidase N, dipeptidase, peptidase T, proline dipeptidase, prolidase, and Xaa-Pro aminopeptidase), aminoacyl tRNA synthetases (arginyl-tRNA synthetase, isoleucyl-tRNA synthetase, phenylalanyl-tRNA synthetase beta chain, seryl-tRNA synthetase, threonyl-tRNA synthetase, and tyrosyl-tRNA synthetase), translation factors (elongation factors G, P, Ts, and Tu, translation initiation factor IF-1, and ribosome recycling factor), and proteins involved in modification (peptidyl-prolyl *cis-trans* isomerase), were identified in the proteomes from both growth phases. Previous studies have shown that a high expression of aminoacyl tRNA synthetases and ribosomal proteins is one of the characteristics of the early-exponential growth phase [52]; however, the expression level of these proteins could not be determined through our qualitative proteomics analysis. Nevertheless, it is important to note that a greater number of peptidases (aminopeptidase C, dipeptidase, peptidase T, prolidase, and Xaa-Pro aminopeptidase) were identified during the early stationary phase than during the mid-exponential phase. The expression of these peptidases only in the early stationary phase suggested that the selective degradation of proteins, peptides, or glycopeptides at this time is likely related to the cessation of growth.

3.2.2. Energy Metabolism.

Overall, 15% and 13% of the proteins from the mid-exponential and early stationary growth phases of the *L. lactis* M4 proteome datasets, respectively, were involved in energy metabolism. The majority of these expressed proteins were involved in glycolysis (enolase, fructose-bisphosphate aldolase, glyceraldehyde-3-phosphate dehydrogenase, L-lactate dehydrogenase, 6-phosphofructokinase, glucose-6-phosphate isomerase, phosphoglycerate kinase, phosphopyruvate hydratase, pyruvate kinase, triosephosphate isomerase, and 2,3-bisphosphoglycerate-dependent phosphoglycerate mutase) and the remaining proteins were involved in fer-mentation (alcohol-acetaldehyde dehydrogenase, formate ace-tyltransferase, and pyruvate-formate

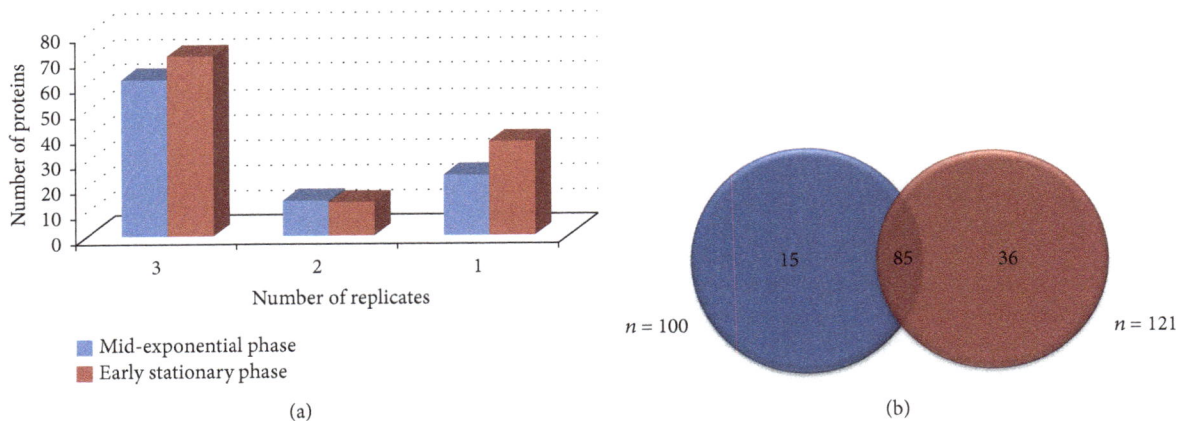

FIGURE 1: Comparison of the growth phase-dependent proteomes of *L. lactis* strain M4. Panel (a), the bar charts show the comparison of the number of proteins found in the biological replicates. Panel (b), the Venn diagrams show the comparison of the proteomes during the mid-exponential and early stationary growth phases.

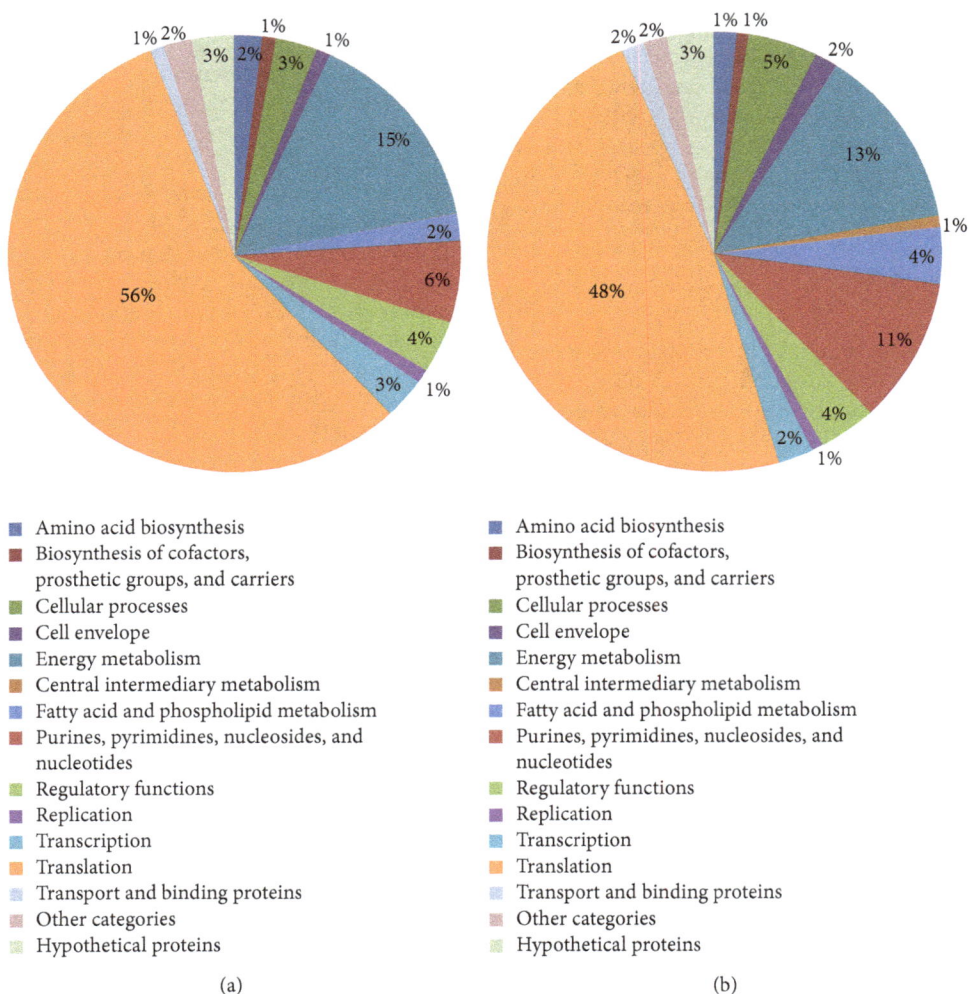

FIGURE 2: Functional attributions of the growth phase-dependent proteomes of *L. lactis* strain M4. Panel (a), the pie chart shows the functional attributions of *L. lactis* M4 proteome during the mid-exponential growth phase ($OD_{600} \sim 0.8$); Panel (b), the early stationary growth phase ($OD_{600} \sim 1.8$).

Growth Phase-Dependent Proteomes of the Malaysian Isolated Lactococcus lactis Dairy Strain M4 Using
Label-Free Qualitative Shotgun Proteomics Analysis

181

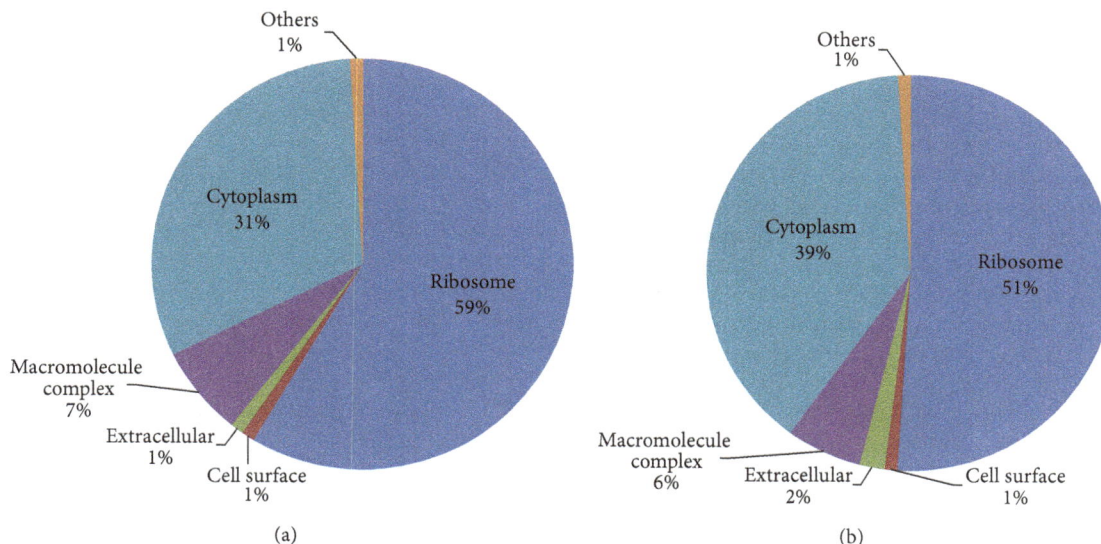

FIGURE 3: Cellular localization of the growth phase-dependent proteomes of *L. lactis* strain M4 based on gene ontology using STRAP software. Panel (a), the pie chart shows the cellular localization of *L. lactis* M4 proteome during the mid-exponential growth phase ($OD_{600} \sim 0.8$); Panel (b), the early stationary growth phase ($OD_{600} \sim 1.8$) based on gene ontology analysis.

lyase), the pentose phosphate pathway (6-phosphogluconate dehydrogenase [decarboxylating]), pyruvate decarboxylation (pyruvate dehydrogenase complex E2 component), and the urea cycle (ornithine carbamoyltransferase). A greater number of proteins were involved in energy metabolism during mid-exponential growth than early stationary growth, which provides sufficient energy for growth [53]. Previous transcriptomic study also showed that genes involved in "energy production and conversion" respond during the exponential growth phase, but not so in the stationary phase [51]. The presence of L-lactate dehydrogenase indicates that lactic acid fermentation occurred in *L. lactis* M4 during both of the growth phases, and the presence of formate acetyltransferase indicates that mixed acid fermentation rather than homolactic fermentation occurred in *L. lactis* M4 during both of the growth phases [54].

3.2.3. Purines, Pyrimidines, Nucleosides, and Nucleotides.
In this class, 6 and 13 proteins from the mid-exponential and early stationary growth phases, respectively, were found to be involved in purine ribonucleotide *de novo* biosynthesis (GMP synthase, IMP dehydrogenase, and adenylosuccinate synthase), pyrimidine ribonucleotide *de novo* biosynthesis (carbamoyl phosphate synthase, orotate phosphoribosyltransferase, and uridylate kinase), the salvage of nucleosides and nucleotides (adenylate kinase, phosphopentomutase, ribose-phosphate pyrophosphokinases, and uracil phosphoribosyltransferase), and sugar-nucleotide biosynthesis and interconversion (dTDP-glucose-4,6-dehydratase, glucose-1-phosphate thymidylyltransferase). A greater number of proteins involved in purines, pyrimidines, nucleosides, and nucleotides biosynthesis were expressed during the early stationary phase, indicating that the exhaustion of metabolites

during this time necessitates the resynthesis of these proteins [52].

3.2.4. Cellular Processes.
In this study, 3% of proteins from the mid-exponential growth phase and 5% of proteins from the early stationary growth phase of the *L. lactis* M4 proteome were identified as chaperones or are involved in cellular processes such as cell division, detoxification, and protein and peptide secretion. Cell division protein GTPase FtsZ, chaperone protein DnaK, and trigger factor were detected for both growth phases. Previous studies have shown that FtsZ protein is more highly expressed during the beginning of the exponential phase for several bacteria species [55]. Additionally, DnaK protein has been reported to be expressed at different levels during different phases of growth which could reflect the adaptation of the cultures to changes in pH and the nutrient compositions [29]. Furthermore, trigger factor and DnaK protein cooperate in the folding of newly synthesized proteins [56]. Thus, our findings are relevant and well substantiated by these studies.

In addition to these findings, two proteins involved in detoxification, alkyl hydroperoxide reductase subunit C and superoxide dismutase [Mn], were identified only in the early stationary growth phase proteome for *L. lactis* M4. Superoxide dismutase [Mn] is only expressed under aerobic conditions [57] and is believed to be involved in the prevention of DNA damage [58]. Superoxide dismutase [Mn] converts oxygen radicals to hydrogen peroxide, which is important for the oxidative stress response. Studies have also shown that alkyl hydroperoxide reductase plays a pivotal role in oxidative stress resistance and acts on hydrogen peroxide [59, 60]. Reactive oxygen species (ROS) and reactive nitrogen intermediates (RNI) accumulate during the early stationary phase [61], and the expression of these detoxification proteins protects the cells from the oxidative stress caused by ROS and RNI.

3.2.5. Regulatory Functions. Proteins involved in regulatory functions were identified during both growth phases of *L. lactis* M4 (4% of the proteome dataset for both phases). The function of transcriptional regulators such as transcriptional regulator CodY, transcriptional regulator CodZ, and bifunctional protein pyrR is to control the expression level of genes by modifying the rate of transcription during different growth phases. At different growth phases, two-component system regulators and transcriptional regulators are believed to act as a simple stimulus-response coupling mechanism to allow the cells to detect, act on, and adapt to various environmental changes, stresses, and growth conditions [62]. In addition, transcriptional regulator CodY was identified as the first global regulator with a low G+C content conserved in Gram-positive bacteria that regulates amino acid biosynthesis [63].

3.2.6. Fatty Acid and Phospholipid Metabolism. A greater number of proteins involved in fatty acid and phospholipid metabolism were identified in the *L. lactis* M4 proteome during the early stationary growth phase (~4%) than during the mid-exponential growth phase (2%). These proteins include 3-oxoacyl-acyl carrier protein synthase 2,3-oxoacyl-acyl carrier protein reductase, the acetyl-CoA carboxylase biotin carboxylase subunit, the acyl carrier protein, and NADH-dependent enoyl-acyl carrier protein reductase. Enzymes involved in fatty acid and phospholipid metabolism are most likely expressed at a higher level during the early stationary growth phase due to the reinforcement of the cell membrane that occurs during the transition of growth phases where large numbers of phospholipids are needed [53].

3.2.7. Replication and Transcription. A small number of proteins from the proteome datasets for both growth phases were involved in replication and transcription. Only one protein involved in the replication process, Hu-like DNA-binding protein, was identified in both growth phases. The transcriptional machinery proteins ATP-dependent RNA helicase, DNA-dependent-RNA polymerase alpha subunit, RNA polymerase beta subunit transcription antitermination protein, and transcription elongation factor GreA were identified in the proteome dataset from either the mid-exponential phase or early stationary growth phase.

3.2.8. Amino Acid Biosynthesis. Two proteins involved in amino acid biosynthesis, glutamine synthetase and S-ribosylhomocysteine lyase, were identified in the *L. lactis* M4 proteome dataset for both the mid-exponential and early stationary growth phases. It was reported that glutamine synthetase (GS) coded by the *glnA* gene was highly expressed and increased in parallel to high consumption of glutamine. Glutamine is the most consumed amino acid in which it is used for synthesis of biomass proteins and it is the donor of amino groups in purine, pyrimidine, and amino

sugar production pathways [28]. In addition, another study using 2-DE gel analysis reported an overexpression of GS when *L. lactis* NCDO763 was grown in milk medium. The authors suggested that this enzyme plays a vital role in the growth of *L. lactis* in milk medium and that glutamine production through GS activity is an important component of the metabolic adaptation of *L. lactis* tomilk medium. Overexpression was not observed in *L. lactis* NCDO763 grown in a rich medium such as M17 [27]. Nevertheless, from our proteome analysis, GM is believed to be synthesized by *L. lactis* M4 strain throughout its growth period as a part of its adaptation to varied culture conditions during different growth phases.

S-Ribosylhomocysteine lyase is encoded by *luxS* gene and is involved in the synthesis of autoinducer 2 (AI-2), which is secreted by bacteria during quorum sensing [64, 65]. Changes in the cell density during the transition of growth phases are likely to stimulate the expression of S-ribosylhomocysteine lyase and the subsequent synthesis of AI-2 in *L. lactis* M4.

3.2.9. Biosynthesis of Cofactors, Prosthetic Groups, and Carriers. Only 1% of the proteome dataset of *L. lactis* M4 at both the mid-exponential growth phase and early stationary growth phase was associated with the biosynthesis of cofactors, prosthetic groups, and carriers. Enzyme 2-dehydropantoate-2-reductase was identified only during the mid-exponential growth phase. 2-Dehydropantoate-2-reductase is the second enzyme in the pantothenate biosynthesis pathway (vitamin B_5) [66] and catalyzes the reduction of NADPH-dependent ketopantoate to pentanoate [67]. Cell division rates are increased when the bacterial growth enters the mid-exponential phase, and, thus, bacterial cells are required to synthesize large quantities of metabolites such as vitamins to support growth.

The thioredoxin system plays a crucial role in maintaining the redox conditions of the cells and protecting them against oxidative stress by reducing reactive oxygen species (ROS) and reactive nitrogen intermediates (RNI) through various mechanisms. ROS and RNI accumulate at the final stage of growth [61] and, therefore, thioredoxin was expressed by *L. lactis* M4 strain during the early stationary growth phase to maintain the homeostasis of the cell cytoplasm.

3.2.10. Cell Envelope. Only 1% and 2% of the mid-exponential and early stationary growth phases of the *L. lactis* M4 proteome datasets, respectively, were categorized as being involved in cell envelope biosynthesis. UDP-N-acetylmuramoylalanine-D-glutamate ligase was found only in the early stationary growth phase proteome dataset; this protein belongs to the murCDEF family and is involved in peptidoglycan synthesis. The induction of enzymes involved in the synthesis of cell wall structures during the stationary phase is thought to strengthen the cell wall and maintain the bacterial morphology [53]. UDP-glucose-1-phosphate uridylyltransferase (GalU) was identified in the proteome datasets for both phases. GalU is needed for capsular polysaccharide biosynthesis [68], in which it catalyzes the reversible formation of uridine-diphosphate

Growth Phase-Dependent Proteomes of the Malaysian Isolated Lactococcus lactis Dairy Strain M4 Using
Label-Free Qualitative Shotgun Proteomics Analysis

183

glucose (UDP-Glc) and inorganic pyrophosphate (PPi) from uridine-3-phosphate (UTP) and glucose-1-phosphate (Glc-1P). UDP-Glc is a substrate for the synthesis of UDP-glucuronic acid and it is also important for the interconversion of galactose and glucose via the Leloir pathway.

3.2.11. Transport and Binding Proteins. Only 1% and 2% of the mid-exponential and early stationary growth phases proteome datasets of *L. lactis* M4, respectively, were found to be involved in the transport and binding of proteins. Phosphoenolpyruvate protein phosphotransferase PtsI was identified in the proteome datasets for both growth phases. PTS systems contain phase-dependent proteins with the highest expression occurring during the exponential growth phase and decreased expression occurring progressively thereafter. However, continuous PtsI expression has been shown to occur throughout both growth phases [52], which explains the detection of PtsI in the *L. lactis* M4 proteome dataset for both growth phases. Maltose ABC transporter substrate-binding protein, on the other hand, was found only in the early stationary growth phase proteome dataset. The appearance of maltose ABC transporter substrate-binding protein during the early stationary phase suggests that *L. lactis* shifted from utilizing of glucose as a carbon source to alternative carbon sources such as maltose due to the decreasing glucose content in the medium during the stationary growth phase.

3.2.12. Central Intermediary Metabolism. S-Adenosylmethiosynthase was the only protein identified in the *L. lactis* M4 proteome during the early stationary growth phase that was classified under the category of central intermediary metabolism. S-Adenosylmethionine synthase is involved in the biosynthesis of S-adenosyl-L-methionine and converts L-methionine and ATP to S-adenosyl-L-methionine which is required for methyltransferase reactions in the cell and for polyamine biosynthesis [69].

3.2.13. Other Categories. A small number of proteins implicated in the stress response were identified in the *L. lactis* M4 proteome for both growth phases (~2%), including ClpB protein, cold shock protein E, and stress response protein E. Usually, stress proteins are induced and expressed more highly during stationary phase due to unfavorable conditions such as diminishing nutrient levels and high concentrations of lactic acid in the medium [53]. However, the expression pattern could not be observed in our qualitative proteomics study.

3.2.14. Hypothetical Proteins. Approximately 3% of the *L. lactis* M4 proteome datasets for the mid-exponential and early stationary growth phases were identified as hypothetical proteins. These are putative proteins that have not been purified and experimentally identified.

4. Conclusions

The present highly sensitive label-free qualitative shotgun comparative proteome analysis is the first study of the phase-dependent proteomes of *L. lactis* strain M4. A qualitative difference between the phase-dependent proteomes of *L. lactis* M4 in terms of the absolute numbers of proteins identified was studied, the identified proteins were categorized according to annotated functions, and the proteomes were compared and analyzed. In summary, the most important reaction occurring during the exponential phase was the generation of sufficient energy, whereas the metabolic energy pathways decreased and the biosynthesis of proteins became more important during the early stationary phase. Thus, the metabolism of the cells shifted from energy production in the exponential phase to the synthesis of macromolecules in the stationary phase. Additionally, the accumulation of ROS and RNI in the late stage of growth induced the expression of proteins that can protect the cells against oxidative stress. Many stress proteins were also strongly induced in the stationary phase due to unfavorable conditions such as high acid and diminished nutrient levels.

In the future, more advanced proteomics platforms, such as multidimensional liquid chromatography tandem mass spectrometry, can be used to increase the protein coverage and to reveal the quantitative differences between the *L. lactis* M4 proteome at different growth phases.

Conflict of Interests

The authors declare that they have no conflict of interests.

Acknowledgments

The authors express gratitude to the Universiti Putra Malaysia (Serdang, Selangor, Malaysia) which graciously supplied the *L. lactis* strain M4. The authors also acknowledge research officer Asmahani Azira Abdu Sani of the Proteomics Laboratory at the Malaysia Genome Institute (Bangi, Selangor, Malaysia) for technical support regarding the nano-ultra-performance liquid chromatography and mass spectrometry and the Malaysia Genome Institute (Bangi, Selangor, Malaysia) for the advanced facilities. This work was supported by the Ministry of Science, Technology and Innovation (MOSTI), Malaysia, under the Grant 09-05-MGI-GMB003, and Universiti Kebangsaan Malaysia under the Grant DPP-2013-022.

References

[1] A. Bolotin, P. Wincker, S. Mauger et al., "The complete genome sequence of the lactic acid bacterium *Lactococcus lactis* ssp. *lactis* IL1403," *Genome Research*, vol. 11, no. 5, pp. 731–753, 2001.

[2] A. Guillot, C. Gitton, P. Anglade, and M.-Y. Mistou, "Proteomic analysis of *Lactococcus lactis*, a lactic acid bacterium," *Proteomics*, vol. 3, no. 3, pp. 337–354, 2003.

[3] K. Makarova, A. Slesarev, Y. Wolf et al., "Comparative genomics of the lactic acid bacteria," *Proceedings of the National Academy*

of Sciences of the United States of America, vol. 103, no. 42, pp. 15611–15616, 2006.

[4] K. H. Schleifer, J. Kraus, and C. Dvorak, "Transfer of *Streptococcus lactis* and related Streptococci to the genus *Lactococcus* gen. nov.," *Systematic and Applied Microbiology*, vol. 6, no. 2, pp. 183–195, 1985.

[5] M. Kleerebezemab, P. Hols, and J. Hugenholtz, "Lactic acid bacteria as a cell factory: rerouting of carbon metabolism in *Lactococcus lactis* by metabolic engineering," *Enzyme and Microbial Technology*, vol. 26, no. 9-10, pp. 840–848, 2000.

[6] H. Jeroen, "The lactic acid bacterium as a cell factory for food ingredient production," *International Dairy Journal*, vol. 18, no. 5, pp. 466–475, 2008.

[7] J. Hugenholtz, W. Sybesma, M. N. Groot et al., "Metabolic engineering of lactic acid bacteria for the production of nutraceuticals," *Antonie van Leeuwenhoek*, vol. 82, no. 1-4, pp. 217–235, 2002.

[8] Y. le Loir, V. Azevedo, S. C. Oliveira et al., "Protein secretion in *Lactococcus lactis*: an efficient way to increase the overall heterologous protein production," *Microbial Cell Factories*, vol. 4, no. 1, article 2, 2005.

[9] E. Morello, L. G. Bermúdez-Humarán, D. Llull et al., "*Lactococcus lactis*, an efficient cell factory for recombinant protein production and secretion," *Journal of Molecular Microbiology and Biotechnology*, vol. 14, no. 1–3, pp. 48–58, 2008.

[10] M. Bahey-el-Din, B. T. Griffin, and C. G. M. Gahan, "Nisin inducible production of listeriolysin O in *Lactococcus lactis* NZ9000," *Microbial Cell Factories*, vol. 7, article 24, 2008.

[11] A. R. Raha, N. R. S. Varma, K. Yusoff, E. Ross, and H. L. Foo, "Cell surface display system for *Lactococcus lactis*: a novel development for oral vaccine," *Applied Microbiology and Biotechnology*, vol. 68, no. 1, pp. 75–81, 2005.

[12] L. Steidler, "*Lactococcus lactis*, a tool for the delivery of therapeutic proteins treatment of IBD," *TheScientificWorldJournal*, vol. 1, pp. 216–217, 2001.

[13] R. J. Siezen, J. Bayjanov, B. Renckens et al., "Complete genome sequence of *Lactococcus lactis* subsp. *lactis* KF147, a plant-associated lactic acid bacterium," *Journal of Bacteriology*, vol. 192, no. 10, pp. 2649–2650, 2010.

[14] R. J. Siezen, M. J. C. Starrenburg, J. Boekhorst, B. Renckens, D. Molenaar, and J. E. T. van Hylckama Vlieg, "Genome-scale genotype-phenotype matching of two *Lactococcus lactis* isolates from plants identifies mechanisms of adaptation to the plant niche," *Applied and Environmental Microbiology*, vol. 74, no. 2, pp. 424–436, 2008.

[15] Y. Gao, Y. Lu, K.-L. Teng et al., "Complete genome sequence of *Lactococcus lactis* subsp. *lactis* CV56, a probiotic strain isolated from the vaginas of healthy women," *Journal of Bacteriology*, vol. 193, no. 11, pp. 2886–2887, 2011.

[16] U. Wegmann, M. O'Connell-Motherway, A. Zomer et al., "Complete genome sequence of the prototype lactic acid bacterium *Lactococcus lactis* subsp. *cremoris* MG1363," *Journal of Bacteriology*, vol. 189, no. 8, pp. 3256–3270, 2007.

[17] D. M. Linares, J. Kok, and B. Poolman, "Genome sequences of *Lactococcus lactis* MG1363 (revised) and NZ9000 and comparative physiological studies," *Journal of Bacteriology*, vol. 192, no. 21, pp. 5806–5812, 2010.

[18] H. Kato, Y. Shiwa, K. Oshima et al., "Complete genome sequence of *Lactococcus lactis* IO-1, a lactic acid bacterium that utilizes xylose and produces high levels of l-lactic acid," *Journal of Bacteriology*, vol. 194, no. 8, pp. 2102–2103, 2012.

[19] R. L. J. Graham, C. Graham, and G. McMullan, "Microbial proteomics: a mass spectrometry primer for biologists," *Microbial Cell Factories*, vol. 6, article 26, 2007.

[20] X. Zhang, A. Fang, C. P. Riley, M. Wang, F. E. Regnier, and C. Buck, "Multi-dimensional liquid chromatography in proteomics—a review," *Analytica Chimica Acta*, vol. 664, no. 2, pp. 101–113, 2010.

[21] Y. V. Karpievitch, A. D. Polpitiya, G. A. Anderson, R. D. Smith, and A. R. Dabney, "Liquid chromatography mass spectrometry-based proteomics: biological and technological aspects," *The Annals of Applied Statistics*, vol. 4, no. 4, pp. 1621–2205, 2010.

[22] P. Anglade, E. Demey, V. Labas, J. P. le Caer, and J. F. Chich, "Towards a proteomic map of *Lactococcus lactis* NCDO 763," *Electrophoresis*, vol. 21, no. 12, pp. 2546–2549, 2000.

[23] O. Drews, G. Reil, H. Parlar, and A. Görg, "Setting up standards and a reference map for the alkaline proteome of the Gram-positive bacterium *Lactococcus lactis*," *Proteomics*, vol. 4, no. 5, pp. 1293–1304, 2004.

[24] R. Mazzoli, E. Pessione, M. Dufour et al., "Glutamate-induced metabolic changes in *Lactococcus lactis* NCDO 2118 during GABA production: combined transcriptomic and proteomic analysis," *Amino Acids*, vol. 39, no. 3, pp. 727–737, 2010.

[25] J. Palmfeldt, F. Levander, B. Hahn-Hägerdal, and P. James, "Acidic proteome of growing and resting *Lactococcus lactis* metabolizing maltose," *Proteomics*, vol. 4, no. 12, pp. 3881–3898, 2004.

[26] M. Willemoës, M. Kilstrup, P. Roepstorff, and K. Hammer, "Proteome analysis of a *Lactococcus lactis* strain overexpressing gapA suggests that the gene product is an auxiliary glyceraldehyde 3-phosphate dehydrogenase," *Proteomics*, vol. 2, no. 8, pp. 1041–1046, 2002.

[27] C. Gitton, M. Meyrand, J. Wang et al., "Proteomic signature of *Lactococcus lactis* NCDO763 cultivated in milk," *Applied and Environmental Microbiology*, vol. 71, no. 11, pp. 7152–7163, 2005.

[28] P.-J. Lahtvee, K. Adamberg, L. Arike, R. Nahku, K. Aller, and R. Vilu, "Multi-omics approach to study the growth efficiency and amino acid metabolism in *Lactococcus lactis* at various specific growth rates," *Microbial Cell Factories*, vol. 10, no. 1, article 12, 2011.

[29] N. Larsen, M. Boye, H. Siegumfeldt, and M. Jakobsen, "Differential expression of proteins and genes in the lag phase of *Lactococcus lactis* subsp. *lactis* grown in synthetic medium and reconstituted skim milk," *Applied and Environmental Microbiology*, vol. 72, no. 2, pp. 1173–1179, 2006.

[30] D. Magnani, O. Barré, S. D. Gerber, and M. Solioz, "Characterization of the CopR regulon of *Lactococcus lactis* IL1403," *Journal of Bacteriology*, vol. 190, no. 2, pp. 536–545, 2008.

[31] R. K. R. Marreddy, E. R. Geertsma, H. P. Permentier, J. P. C. Pinto, J. Kok, and B. Poolman, "Amino acid accumulation limits the overexpression of proteins in *Lactococcus lactis*," *PLoS ONE*, vol. 5, no. 4, Article ID e10317, 2010.

[32] K. Vido, D. le Bars, M.-Y. Mistou, P. Anglade, A. Gruss, and P. Gaudu, "Proteome analyses of heme-dependent respiration in *Lactococcus lactis*: involvement of the proteolytic system," *Journal of Bacteriology*, vol. 186, no. 6, pp. 1648–1657, 2004.

[33] N. H. Beyer, P. Roepstorff, K. Hammer, and M. Kilstrup, "Proteome analysis of the purine stimulon from *Lactococcus lactis*," *Proteomics*, vol. 3, no. 5, pp. 786–797, 2003.

[34] A. Budin-Verneuil, V. Pichereau, Y. Auffray, D. Ehrlich, and E. Maguin, "Proteome phenotyping of acid stress-resistant mutants of *Lactococcus lactis* MG1363," *Proteomics*, vol. 7, no. 12, pp. 2038–2046, 2007.

Growth Phase-Dependent Proteomes of the Malaysian Isolated Lactococcus lactis Dairy Strain M4 Using
Label-Free Qualitative Shotgun Proteomics Analysis

185

[35] A. Budin-Verneuil, V. Pichereau, Y. Auffray, D. S. Ehrlich, and E. Maguin, "Proteomic characterization of the acid tolerance response in *Lactococcus lactis* MG1363," *Proteomics*, vol. 5, no. 18, pp. 4794–4807, 2005.

[36] Y. Zhang, Y. Zhang, Y. Zhu, S. Mao, and Y. Li, "Proteomic analyses to reveal the protective role of glutathione in resistance of *Lactococcus lactis* to osmotic stress," *Applied and Environmental Microbiology*, vol. 76, no. 10, pp. 3177–3186, 2010.

[37] B. Cesselin, D. Ali, J.-J. Gratadoux et al., "Inactivation of the *Lactococcus lactis* high-affinity phosphate transporter confers oxygen and thiol resistance and alters metal homeostasis," *Microbiology*, vol. 155, no. 7, pp. 2274–2281, 2009.

[38] N. García-Quintáns, G. Repizo, M. Martín, C. Magni, and P. López, "Activation of the diacetyl/acetoin pathway in *Lactococcus lactis* subsp. *lactis* bv. diacetylactis CRL264 by acidic growth," *Applied and Environmental Microbiology*, vol. 74, no. 7, pp. 1988–1996, 2008.

[39] J. Beganović, A. Guillot, M. van de Guchte et al., "Characterization of the insoluble proteome of *Lactococcus lactis* by SDS-PAGE LC-MS/MS leads to the identification of new markers of adaptation of the bacteria to the mouse digestive tract," *Journal of Proteome Research*, vol. 9, no. 2, pp. 677–688, 2010.

[40] K. Roy, M. Meyrand, G. Corthier, V. Monnet, and M.-Y. Mistou, "Proteomic investigation of the adaptation of *Lactococcus lactis* to the mouse digestive tract," *Proteomics*, vol. 8, no. 8, pp. 1661–1676, 2008.

[41] B. Soufi, F. Gnad, P. R. Jensen et al., "The Ser/Thr/Tyr phosphoproteome of *Lactococcus lactis* IL1403 reveals multiply phosphorylated proteins," *Proteomics*, vol. 8, no. 17, pp. 3486–3493, 2008.

[42] N. Noreen, W. Y. Hooi, A. Baradaran et al., "*Lactococcus lactis* M4, a potential host for the expression of heterologous proteins," *Microbial Cell Factories*, vol. 10, no. 1, article 28, 2011.

[43] J. C. Silva, R. Denny, C. Dorschel et al., "Simultaneous qualitative and quantitative analysis of the *Escherichia coli* proteome: a sweet tale," *Molecular & Cellular Proteomics*, vol. 5, no. 4, pp. 589–607, 2006.

[44] M. R. Soares, A. P. Facincani, R. M. Ferreira et al., "Proteome of the phytopathogen *Xanthomonas citri* subsp. *citri*: a global expression profile," *Proteome Science*, vol. 8, no. 1, article 55, 2010.

[45] D. Rosenegger, C. Wright, and K. Lukowiak, "A quantitative proteomic analysis of long-term memory," *Molecular Brain*, vol. 3, no. 1, article 9, 2010.

[46] A. Bolotin, S. Mauger, K. Malarme, S. D. Ehrlich, and A. Sorokin, "Low-redundancy sequencing of the entire *Lactococcus lactis* IL1403 genome," *Antonie van Leeuwenhoek*, vol. 76, no. 1-4, pp. 27–76, 1999.

[47] V. N. Bhatia, D. H. Perlman, C. E. Costello, and M. E. McComb, "Software tool for researching annotations of proteins: opensource protein annotation software with data visualization," *Analytical Chemistry*, vol. 81, no. 23, pp. 9819–9823, 2009.

[48] L. de Vuyst and F. Leroy, "Bacteriocins from lactic acid bacteria: production, purification, and food applications," *Journal of Molecular Microbiology and Biotechnology*, vol. 13, no. 4, pp. 194–199, 2007.

[49] E. Guillaume, B. Berger, M. Affolter, and M. Kussmann, "Label-free quantitative proteomics of two *Bifidobacterium longum* strains," *Journal of Proteomics*, vol. 72, no. 5, pp. 771–784, 2009.

[50] M. A. Schell, M. Karmirantzou, B. Snel et al., "The genome sequence of *Bifidobacterium longum* reflects its adaptation to

the human gastrointestinal tract," *Proceedings of the National Academy of Sciences of the United States of America*, vol. 99, no. 22, pp. 14422–14427, 2002.

[51] A. de Jong, M. E. Hansen, O. P. Kuipers, M. Kilstrup, and J. Kok, "The transcriptional and gene regulatory network of *Lactococcus lactis* MG1363 during growth in milk," *PLoS ONE*, vol. 8, no. 1, Article ID e53085, 2013.

[52] K. M. Koistinen, C. Plumed-Ferrer, S. J. Lehesranta, S. O. Kärenlampi, and A. von Wright, "Comparison of growth-phase-dependent cytosolic proteomes of two *Lactobacillus plantarum* strains used in food and feed fermentations," *FEMS Microbiology Letters*, vol. 273, no. 1, pp. 12–21, 2007.

[53] D. P. A. Cohen, J. Renes, F. G. Bouwman et al., "Proteomic analysis of log to stationary growth phase *Lactobacillus plantarum* cells and a 2-DE database," *Proteomics*, vol. 6, no. 24, pp. 6485–6493, 2006.

[54] C. Garrigues, P. Loubiere, N. D. Lindley, and M. Cocaign-Bousquet, "Control of the shift from homolactic acid to mixed-acid fermentation in *Lactococcus lactis*: predominant role of the NADH/NAD+ ratio," *Journal of Bacteriology*, vol. 179, no. 17, pp. 5282–5287, 1997.

[55] R. B. Weart and P. A. Levin, "Growth rate-dependent regulation of medial FtsZ ring formation," *Journal of Bacteriology*, vol. 185, no. 9, pp. 2826–2834, 2003.

[56] E. Deuerling, A. Schulze-Specking, T. Tomoyasu, A. Mogk, and B. Bukau, "Trigger factor and DnaK cooperate in folding of newly synthesized proteins," *Nature*, vol. 400, no. 6745, pp. 693–696, 1999.

[57] D. Touati, "Transcriptional and posttranscriptional regulation of manganese superoxide dismutase biosynthesis in *Escherichia coli*, studied with operon and protein fusions," *Journal of Bacteriology*, vol. 170, no. 6, pp. 2511–2520, 1988.

[58] K. A. Hopkin, M. A. Papazian, and H. M. Steinman, "Functional differences between manganese and iron superoxide dismutases in *Escherichia coli* K-12," *The Journal of Biological Chemistry*, vol. 267, no. 34, pp. 24253–24258, 1992.

[59] J. W. Sanders, K. J. Leenhouts, A. J. Haandrikman, G. Venema, and J. Kok, "Stress response in *Lactococcus lactis*: cloning, expression analysis, and mutation of the lactococcal superoxide dismutase gene," *Journal of Bacteriology*, vol. 177, no. 18, pp. 5254–5260, 1995.

[60] M. Wasim, A. N. Bible, Z. Xie, and G. Alexandre, "Alkyl hydroperoxide reductase has a role in oxidative stress resistance and in modulating changes in cell-surface properties in *Azospirillum brasilense* Sp245," *Microbiology*, vol. 155, no. 4, pp. 1192–1202, 2009.

[61] N. C. Soares, M. P. Cabral, C. Gayoso et al., "Associating growth-phase-related changes in the proteome of *Acinetobacter baumannii* with increased resistance to oxidative stress," *Journal of Proteome Research*, vol. 9, no. 4, pp. 1951–1964, 2010.

[62] J. M. Skerker, M. S. Prasol, B. S. Perchuk, E. G. Biondi, and M. T. Laub, "Two-component signal transduction pathways regulating growth and cell cycle progression in a bacterium: a system-level analysis," *PLoS Biology*, vol. 3, no. 10, Article ID e334, 2005.

[63] E. Guédon, B. Sperandio, N. Pons, S. D. Ehrlich, and P. Renault, "Overall control of nitrogen metabolism in *Lactococcus lactis* by CodY, and possible models for CodY regulation in Firmicutes," *Microbiology*, vol. 151, no. 12, pp. 3895–3909, 2005.

[64] J.-G. Cao and E. A. Meighen, "Purification and structural identification of an autoinducer for the luminescence system of

Vibrio harveyi," *The Journal of Biological Chemistry*, vol. 264, no. 36, pp. 21670–21676, 1989.

[65] M. E. Taga, J. L. Semmelhack, and B. L. Bassler, "The LuxS-dependent autoinducer Al-2 controls the expression of an ABC transporter that functions in Al-2 uptake in *Salmonella typhimurium,*" *Molecular Microbiology*, vol. 42, no. 3, pp. 777–793, 2001.

[66] M. E. Webb, A. G. Smith, and C. Abell, "Biosynthesis of pantothenate," *Natural Product Reports*, vol. 21, no. 6, pp. 695–721, 2004.

[67] R. Zheng and J. S. Blanchard, "Identification of active site residues in *E. coli* ketopantoate reductase by mutagenesis and chemical rescue," *Biochemistry*, vol. 39, no. 51, pp. 16244–16251, 2000.

[68] L. Bonofiglio, E. García, and M. Mollerach, "Biochemical characterization of the pneumococcal glucose 1-phosphate uridylyltransferase (GalU) essential for capsule biosynthesis," *Current Microbiology*, vol. 51, no. 4, pp. 217–221, 2005.

[69] S. Roje, "S-Adenosyl-l-methionine: beyond the universal methyl group donor," *Phytochemistry*, vol. 67, no. 15, pp. 1686–1698, 2006.

Permissions

The contributors of this book come from diverse backgrounds, making this book a truly international effort. This book will bring forth new frontiers with its revolutionizing research information and detailed analysis of the nascent developments around the world.

We would like to thank all the contributing authors for lending their expertise to make the book truly unique. They have played a crucial role in the development of this book. Without their invaluable contributions this book wouldn't have been possible. They have made vital efforts to compile up to date information on the varied aspects of this subject to make this book a valuable addition to the collection of many professionals and students.

This book was conceptualized with the vision of imparting up-to-date information and advanced data in this field. To ensure the same, a matchless editorial board was set up. Every individual on the board went through rigorous rounds of assessment to prove their worth. After which they invested a large part of their time researching and compiling the most relevant data for our readers. Conferences and sessions were held from time to time between the editorial board and the contributing authors to present the data in the most comprehensible form. The editorial team has worked tirelessly to provide valuable and valid information to help people across the globe.

Every chapter published in this book has been scrutinized by our experts. Their significance has been extensively debated. The topics covered herein carry significant findings which will fuel the growth of the discipline. They may even be implemented as practical applications or may be referred to as a beginning point for another development. Chapters in this book were first published by Hindawi Publishing Corporation; hereby published with permission under the Creative Commons Attribution License or equivalent.

The editorial board has been involved in producing this book since its inception. They have spent rigorous hours researching and exploring the diverse topics which have resulted in the successful publishing of this book. They have passed on their knowledge of decades through this book. To expedite this challenging task, the publisher supported the team at every step. A small team of assistant editors was also appointed to further simplify the editing procedure and attain best results for the readers.

Our editorial team has been hand-picked from every corner of the world. Their multi-ethnicity adds dynamic inputs to the discussions which result in innovative outcomes. These outcomes are then further discussed with the researchers and contributors who give their valuable feedback and opinion regarding the same. The feedback is then collaborated with the researches and they are edited in a comprehensive manner to aid the understanding of the subject.

Apart from the editorial board, the designing team has also invested a significant amount of their time in understanding the subject and creating the most relevant covers. They scrutinized every image to scout for the most suitable representation of the subject and create an appropriate cover for the book.

The publishing team has been involved in this book since its early stages. They were actively engaged in every process, be it collecting the data, connecting with the contributors or procuring relevant information. The team has been an ardent support to the editorial, designing and production team. Their endless efforts to recruit the best for this project, has resulted in the accomplishment of this book. They are a veteran in the field of academics and their pool of knowledge is as vast as their experience in printing. Their expertise and guidance has proved useful at every step. Their uncompromising quality standards have made this book an exceptional effort. Their encouragement from time to time has been an inspiration for everyone.

The publisher and the editorial board hope that this book will prove to be a valuable piece of knowledge for researchers, students, practitioners and scholars across the globe.

List of Contributors

Lerato Mogotsi, Olga De Smidt and Willem Groenewald
Unit for Applied Food Science and Biotechnology, Central University of Technology, Free State, P.O. Box 20539, Bloemfontein 9300, South Africa

Pierre Venter
Fonterra Co-Operative Group Limited, Private Bag 11029, Palmerston North 4442,Dairy FarmRoad 1, Palmerston North, New Zealand

Elena S. Medvedeva, Natalia B. Baranova, Alexey A. Mouzykantov, Tatiana Yu. Grigorieva, Marina N. Davydova, Maxim V. Trushin, Olga A. Chernova and Vladislav M. Chernov
Kazan Institute of Biochemistry and Biophysics, Kazan Scientific Centre of Russian Academy of Sciences, Kazan 420111, Russia

Maxim V. Trushin, Olga A. Chernova and Vladislav M. Chernov
Kazan Federal University, Kazan 420008, Russia

Shuwahida Shuib, Wan Nazatul Naziah Wan Nawi, Ekhlass M. Taha, Othman Omar and Aidil Abdul Hamid
School of Biosciences and Biotechnology, Faculty of Science and Technology, Universiti Kebangsaan Malaysia (UKM), 43600 Bangi, Selangor, Malaysia

Abdul Jalil Abdul Kader
Faculty of Science and Technology, Universiti Sains Islam Malaysia, Bandar Baru Nilai, 71800 Nilai, Negeri Sembilan, Malaysia

Mohd Sahaid Kalil
Department of Chemical and Process Engineering, Faculty of Engineering, Universiti Kebangsaan Malaysia (UKM), 43600 Bangi, Selangor, Malaysia

Pierangeli G. Vital, Kris Genelyn B. Dimasuay and Windell L. Rivera
Institute of Biology, College of Science, University of the Philippines, Diliman, 1101 Quezon City, Philippines
Natural Sciences Research Institute, University of the Philippines, Diliman, 1101 Quezon City, Philippines

Kenneth W. Widmer
International Environmental Analysis and Education Center, Gwangju Institute of Science and Technology, 261 Cheomdan-Gwagiro, Buk-gu, Gwangju 500-712, Republic of Korea

Irais Sánchez-Ortega, Blanca E. García-Almendárez, Aldo Amaro-Reyes and Carlos Regalado
DIPA, PROPAC, Facultad de Química, Universidad Autónoma de Querétaro, 76010 Querétaro, QRO, Mexico

Irais Sánchez-Ortega and Eva María Santos-López
Area Académica de Química, Instituto de Ciencias Básicas e Ingeniería, Universidad Autónoma del Estado de Hidalgo, Ciudad del Conocimiento, Carr. Pachuca-Tulancingo Km 4.5 Col Carboneras, 42184 Mineral de la Reforma, HGO, Mexico

J. Eleazar Barboza-Corona
División Ciencias de la Vida, Universidad de Guanajuato, Campus Irapuato-Salamanca, 36500 Irapuato, GTO, Mexico

Dhanya N. Nair and S. Padmavathy
Research Department of Botany, Nirmala College forWomen, Coimbatore, Tamil Nadu 641018, India

Patil Chandrashekhar Devidas, Borase Hemant Pandit and Patil Satish Vitthalrao
School of Life Sciences, North Maharashtra University, P.O. Box 80, Jalgaon, Maharashtra 425001, India

Patil Satish Vitthalrao
North Maharashtra Microbial Culture Collection Centre (NMCC), North Maharashtra University, P.O. Box 80, Jalgaon, Maharashtra 425001, India

Ponnuswamy Vijayaraghavan and Samuel Gnana Prakash Vincent
International Centre for Nanobiotechnology, Centre for Marine Science and Technology, Manonmaniam Sundaranar University, Rajakkamangalam, Kanyakumari District, Tamil Nadu 629 502, India

Liping Xu, Yunhai Lu, Qian You, Xiaolan Liu and Youxiong Que
Key Laboratory of Sugarcane Biology and Genetic Breeding, Ministry of Agriculture/Fujian Agriculture and Forestry University, Fuzhou 350002, China

Michael Paul Grisham and Yongbao Pan
USDA-ARS, Sugarcane ResearchUnit, Houma, LA 70360, USA

Min Wang, Ping Feng, Xiaofang Xie and Hong Du
Clinical Laboratory,The Second Affiliated Hospital of Soochow University, Suzhou 215004, China

Xun Chen
Clinical Laboratory Center, Xiyuan Hospital, China Academy of Chinese Medical Sciences, Beijing 100091, China

Haifang Zhang and Bin Ni
Department of Biochemistry and Molecular Biology, School of Medical Technology, Jiangsu University, Zhenjiang 212013, China

Gunashree B. Shivanna and Govindarajulu Venkateswaran
Department of Food Microbiology, Central Food Technological Research Institute, Mysore, Karnataka-570 020, India

Albert Mas and María Jesús Torija
Facultad de Enolog´ıa, Universitat Rovira i Virgili, Marcel·l´ı Domingo s/n, 43003 Tarragona, Spain

María del Carmen García-Parrilla and Ana María Troncoso
Facultad de Farmacia, Universidad de Sevilla, Profesor Garc´ıa Gonz´alez 2, 41012 Sevilla, Spain

Civan Gojkovic, Carlos Vílchez, Javier Vigara and Ines Garbayo
Algal Biotechnology Group, Department of Chemistry and Material Sciences, Faculty of Experimental Sciences, University of Huelva, Campus el Carmen, Avenida de las Fuerzas Armadas s/n, 21007 Huelva, Spain

Civan Gojkovic and Ivana Márová
Department of Food Technology and Biotechnology, Faculty of Chemistry, Brno University of Technology, Purkynova 118, 61200 Brno, Czech Republic

Carlos Vílchez
Algal Biotechnology Group, International Centre for Environmental Research (CIECEM), Parque Dunar s/n, Matalascanas, Almonte, 21760 Huelva, Spain

Rafael Torronteras
Department of Environmental Biology and Public Health, University of Huelva, Campus el Carmen, 21007 Huelva, Spain

Veronica Gómez-Jacinto and José-Luis Gómez-Ariza
Department of Chemistry and Material Sciences, University of Huelva, Campus el Carmen, 21007 Huelva, Spain

Nora Janzer
University of Applied Science, Giessen, Wiesenstrasse 14, 35390 Giessen, Germany

Mallikarjun C. Bheemaraddi, Santosh Patil, Channappa T. Shivannavar and Subhashchandra M. Gaddad
Department of Post Graduate Studies and Research in Microbiology, Gulbarga University, Gulbarga, Karnataka 585106, India

H. N. H. Veras, F. F. G. Rodrigues and J. G. M. da Costa
Laboratory of Natural Products Research (LPPN), Regional University of Cariri, Crato, CE, Brazil

M. A. Botelho
Northeast Biotechnology Network, Brazil

I. R. A. Menezes
Laboratory of Pharmacology and Molecular Chemistry (LFQM), Regional University of Cariri, Crato, CE, Brazil

H. D. M. Coutinho
Laboratory of Microbiology and Molecular Biology (LMBM), Department of Biological Chemistry, Regional University of Cariri, Cel. Antonio Luis Street, 1161 Pimenta, 63105-000 Crato, CE, Brazil

F. A. G. Gonçalves and G. Colen
Faculdade de Farm´acia, Universidade Federal de Minas Gerais, 31270-901 Belo Horizonte, MG, Brazil

J. A. Takahashi
Departamento de Qu´ımica, Instituto de Ci^encias Exatas, Universidade Federal de Minas Gerais, 31270-901 Belo Horizonte, MG, Brazil

Osman Erganis, Zafer Sayin, Hasan Huseyin Hadimli and Asli Sakmanoglu
Department of Microbiology, Faculty of Veterinary Medicine, University of Selcuk, 42075 Konya, Turkey

Yasemin Pinarkara
Program of Food Technology, Sarayonu Vocational School, University of Selcuk, 42075 Konya, Turkey

Ozgur Ozdemir
Department of Pathology, Faculty of Veterinary Medicine, University of Selcuk, 42075 Konya, Turkey

Mehmet Maden
Department of Internal Medicine, Faculty of Veterinary Medicine, University of Selcuk, 42075 Konya, Turkey

Chao Kang, Ting-Chi Wen, Ji-Chuan Kang, Ze-Bing Meng and Guang-Rong Li
The Engineering and Research Center of Southwest Bio-Pharmaceutical Resources, Ministry of Education, Guizhou University, Guiyang, Guizhou 550025, China

Chao Kang
Institute of Biology, Guizhou Academy of Sciences, Guiyang, Guizhou 550009, China

Kevin D. Hyde
Institute of Excellence in Fungal Research, School of Science, Mae Fah Luang University, Chiang Rai 57100,Thailand
Botany and Microbiology Department, College of Science, King Saud University, Riyadh 11442, Saudi Arabia

Theresa Wan Chen Yap, Amir Rabu, Farah Diba Abu Bakar and Abdul Munir Abdul Murad
School of Biosciences and Biotechnology, Faculty of Science and Technology, Universiti Kebangsaan Malaysia, 43600 Bangi, Selangor, Malaysia

Raha Abdul Rahim
Faculty of Biotechnology and Biomolecular Sciences, Universiti Putra Malaysia (UPM), 43400 Serdang, Selangor, Malaysia

Nor Muhammad Mahadi
Malaysia Genome Institute, Jalan Bangi, 43000 Kajang, Selangor, Malaysia

Rosli Md. Illias
Department of Bioprocess Engineering, Faculty of Chemical Engineering, Universiti Teknologi Malaysia, 81310 Skudai, Johor, Malaysia

www.ingramcontent.com/pod-product-compliance
Lightning Source LLC
Chambersburg PA
CBHW050454200326

41458CB00014B/5174